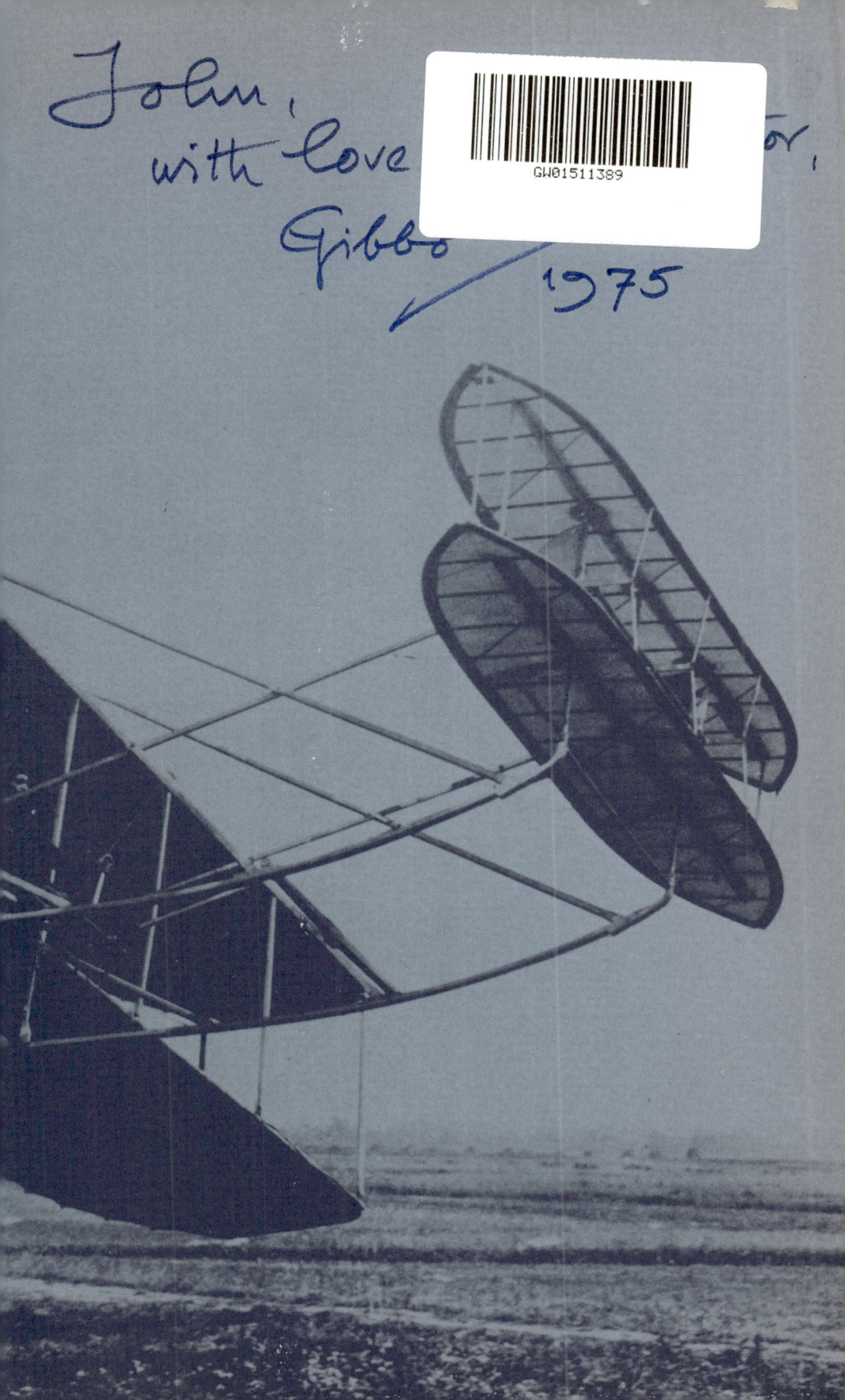

The Wright modified No 3 Glider of 1902 at the Kill Devil Hills

Science Museum

THE REBIRTH OF EUROPEAN AVIATION 1902-1908

A Study of the Wright Brothers' Influence

By
CHARLES HARVARD GIBBS-SMITH
Honorary Companion
of the Royal Aeronautical Society

London
Her Majesty's Stationery Office
1974

© Crown copyright 1974

Typographic design by Her Majesty's Stationery Office

ISBN 0 11 290180 8*

TO
IVONETTE
MILLER
OF DAYTON, OHIO
*First Lady of the
Wright
Descendants*

Must we one day read in history that aviation, born in France, only became successful thanks to the Americans, and that the French only obtained results by slavishly copying them?

VICTOR TATIN (1904)

Without this man (Wilbur Wright) I would be nothing, for I should not have dared, in 1902, to trust myself on a flimsy fabric if I had not known from his accounts and his photographs that it would carry me. Think, that without him my experiments would not have taken place, and I should not have had Voisin as a pupil. Capitalists like Archdeacon and Deutsch de la Meurthe would not in 1904 have established the prizes you know of.

CAPTAIN F. FERBER (1907)

The man (Wilbur Wright), one must honestly recognise, is indeed the father of aviation.

LEON DELAGRANGE (1908)

For a long time, for too long a time, the Wright brothers have been accused in Europe of bluff—even perhaps in the land of their birth. They are today hallowed by France, and I feel an intense pleasure in counting myself among the first to make amends for that flagrant injustice.

FRANÇOIS PEYREY (1908)

The Wrights keep the incomparable glory of an undeniable priority in the creation of a true flying-machine . . .

COMTE DE LA VAULX

The moment has come to reveal, at least briefly, the characteristics of this machine which had just revolutionised the aviators' world.

COMTE DE LA VAULX, writing of the Wright machine in 1908

Wilbur Wright came to France, and gave at the Hunaudières race-course, and then at the Camp d'Auvours—in the neighbourhood of Le Mans—a series of flights which were sensational, and constitute a decisive epoch of history. The first flight took place on August 8th 1908. Numerous flights followed, revolutionising aviation by the excellence of the pilotage, the manageableness, and the versatility of the machine.

CHARLES DOLLFUS (1965)

Contents

Foreword		*page* xvii
Introduction		xix
1	The Wright Brothers and the 'anti-Wright Faction'	1
2	European Aviation after the Death of Lilienthal	7
3	The Contrasting Attitudes of the 'Chauffeurs' and the 'Airmen'	12
4	The Chanute Gliders: 1896-1902	14
5	The Wright Gliders: 1900-1902	17
6	Chanute and the Wright Brothers	23
7	The question of Wing-Trussing and the Biplane in History	29

The year 1902

8	The Wrights' Influence starts with Ferber: 1902	35
9	Ferber and Control in Roll: 1902	41
10	The Wrights' No 3 Glider of 1902	44
11	The Non-Use of the Hargrave Box-Kite: 1902	49

The year 1903

12	Ferber's Article in *L'Aérophile*: February 1903	53
13	Chanute's Lecture in Paris: April 2nd 1903	56
14	Ferber's Letter to Archdeacon, with its Exhortations, and the Suggestion for a Gliding Competition: April 1903	64
15	Archdeacon's Call to Action: April 1903	65
16	Archdeacon forms the Aviation Committee of the Aéro-Club: May 1903	67
17	Ferber's Gliding during the Year 1903	69
18	Chanute's first French Article: August 1903	71
19	Further Reminders of the Wrights: August 1903	74
20	Chanute's second French Article: November 1903	76
21	The Wrights build their first Powered *Flyer*: 1903	82

22 The Wrights' first Powered Flights: December 17th 1903   *page* 84
23 The Wrights and Chanute: 1903   86
24 The first Impact on Europe of the Wrights' first Powered Flights: 1903   91

The year 1904

25 The Wrights' Powered Flights of 1903: the second (and major) Impact on Europe: January 1904   97
26 The first published Picture of the Wright *Flyer I*: January 1904   102
27 Archdeacon's Wright-type Glider is announced: January 1904   103
28 Archdeacon's continued Efforts to launch Gliding Competitions: January 1904   105
29 Tatin's Protest and Exhortation: February 4th 1904   106
30 Archdeacon's Lecture on the Wrights and on the Position on Aviation in France: February 4th 1904   108
31 Ferber writes to Archdeacon about the Wrights: February 1904   111
32 The Aéro-Club's Tribute to Ferber and Archdeacon: March 3rd 1904   112
33 *L'Aérophile* publishes an Article on How to fly a Chanute Glider: March 1904   114
34 Further Efforts on Behalf of Gliding: also Proposals for a Grand Prix: March 1904   116
35 Ferber's first Article in the *Revue d'Artillerie*: March 1904   117
36 Other French Experimenters: March 1904   120
37 The Archdeacon Wright-type Glider and its Tests: April 1904   122
38 Gabriel Voisin enters professional Aviation: 1904   126
39 Esnault-Pelterie tests his Wright-type Glider, and first fits Ailerons: May and October 1904   128
40 The Wrights' Work in 1904, and the Arrival of the News of it in Europe: 1904   129

41 The Wrights' Influence on Flight-Control Principles and Practice: 1904 and on	page 133
42 Ferber's tailed Glider: October 1904	136
43 Prizes are offered to stimulate European Aviation: October 1904	137
44 Aeronautical Comment in Britain: 1904	139
45 The Wrights' first Impact on Britain: 1904-05	141

The year 1905

46 The first Illustration to be published of Ferber's tailed Glider: February 1905	145
47 The Aéro-Club's Model Competition: February 1905	146
48 The second Archdeacon Wright-type Glider: March 1905	147
49 Archdeacon challenges the Wrights: March 1905	149
50 Another Picture of a Wright Aeroplane is published: March 1905	151
51 Esnault-Pelterie's Article on his Wright-type Glider: June 1905	152
52 The Voisin-Archdeacon and the Voisin-Blériot Float-Gliders: June, July 1905	157
53 Hargrave's Box-Kite enters full-size Aviation: June, July 1905	159
54 Ferber's Work in 1905, and his second Article in the *Revue d'Artillerie* (August 1905)	161
55 The Wrights' Work during the Year: the first practical Aeroplane in History: 1905	166
56 The Wrights send News of their Success with *Flyer III* to Ferber: October 1905	170
57 Archdeacon attacks the Wrights, and again issues Exhortations: October 1905	173
58 The Wrights' new Impact on Europe: November, December 1905	177
59 The Wrights; the Aéro-Club; and Archdeacon's Enmity: December 1905	180

60 A pirated Sketch of the Wright *Flyer III*: December 1905 *page* 182
61 A Summary of Ideas and Achievements in 1905   184

The year 1906

62 Discussion on the Wrights at the Aéro-Club:
     January 4th 1906   189
63 The first *Aérophile* Article on the Wrights: the December
     1905 Number, published in January 1906   190
64 The second *Aérophile* Article on the Wrights: January 1906   193
65 French Scepticism and British Belief: January 1906   195
66 Re-publication in France of the Wright Patent: January 1906 199
67 Tatin's Admission, Censure of the French, and Exhortation:
     January 1906   201
68 Santos-Dumont turns to Aviation: January 1906   204
69 Miscellanea: 1906   207
70 Santos-Dumont and his No. 14-*bis*: July 1906   212
71 The Great French Manifesto: August 1906   213
72 Santos-Dumont makes his first Tests, and first free
     Take-Off: August, September 1906   218
73 Santos-Dumont wins his first Prize: October 23rd 1906   220
74 The Banquet in honour of Santos-Dumont: November
     10th 1906   222
75 Santos-Dumont wins his second Prize: November 12th 1906   225
76 The European Attitude to Flight-Control: 1906   227
77 Summing-Up for 1906   231

The year 1907

78 General Survey of 1907   235
79 Ferber's Tribute to Wilbur Wright: June 1907   241
80 Henri Farman's first Flights: October, November 1907   244
81 European Stagnation and Procrastination: 1907   247

The year 1908

82	European Biplanes: January to July 1908	page 253
83	European Monoplanes: January to July 1908	258
84	Farman flies the first European Circle: January 13th 1908	261
85	Ferber lectures in London: February 1908	265
86	The Wright Brothers in 1908 (General Survey)	271
87	Wilbur Wright's Flights in France in 1908	273
88	Orville Wright's Flights at Fort Myer: September 1908	275
89	The Wright Brothers and aerial Records	277
90	The Press Coverage of Wilbur's Flights	279
91	French Recognition and Repentance: 1908	284
92	First and second Thoughts of the French Pioneers: 1908	286
93	Gabriel Voisin on Wilbur Wright in France: 1908	293
94	The Question of assisted Take-off in History	295
95	Could the Wright Machines only be flown by Acrobats?	301
96	Flying in the Wind	304
97	Passenger-Carrying	306
98	Gabriel Voisin and the Wright Engines: 1908	307
99	Gabriel Voisin and the Wright Propellers: 1908	309
100	Wilbur Wright's Influence on Europe in 1908 (Summary)	311
101	The Influence of the Wright Machine	312
102	Conclusions	316
103	Synopsis of Quotations on Aircraft, Influence, and Procrastination, 1902-1908	319
Appendix A	A Brief Synopsis of Aviation History prior to the Death of Pilcher in 1899	347
Appendix B	Mr Chanute in Paris (1903). By Ernest Archdeacon	354

Appendix C Aerial Navigation in the United States. By Octave
                Chanute   *page* 359

Bibliography   369

Conversion Tables   375

Index   379

Illustrations

Frontispiece: Wright's No 3 glider (modified): 1902

1a	Lilienthal monoplane hang-glider: 1894	*between pages 28 and 29*
b	Lilienthal monoplane hang-glider: 1894	
c	Lilienthal biplane hang-glider: 1895	
2	Chanute multiple-wing hang-glider: 1896	
3	Chanute oscillating wing hang-glider: 1902	
4a	Chanute biplane hang-glider: 1896	
b	Chanute biplane hang-glider: 1896	
c	Chanute biplane hang-glider: 1896	
d	Chanute biplane hang-glider: 1896	
e	Chanute biplane hang-glider: 1896	
5	Wright warping kite: 1899	
6	Wright No 1 glider: 1900	*between pages 44 and 45*
7	Wright No 2 glider: 1901	
8	Wright No 3 glider: 1902	
9a	Wright No 3 glider (modified): 1902	
b	Wright No 3 glider (modified): 1902	
c	Wright No 3 glider (modified): 1902	
d	Wright No 3 glider (modified): 1902	
e	Wright No 3 glider (modified): 1902	
10	Ferber Lilienthal-type hang-glider: 1901	
11a	Ferber Wright-type glider: 1902	
b	Ferber Wright-type glider: 1902	
12	Hargrave multi-wing float-glider design: 1902	*between pages 112 and 113*
13	Ferber Wright-type glider with wing-tip rudders: 1903	
14	Ferber powered Wright-type glider on a whirling arm: 1903	
15	First flight by the Wright *Flyer I*: December 17th 1903	
16a	Wright *Flyer I* (general arrangement drawings): 1903	
b	Wright *Flyer I* (general arrangement drawings): 1903	
17	First pirated picture of the Wright *Flyer I*: 1904	

18 Page from *L'Aérophile* showing drawings of Chanute's 1896 hang-glider: 1904

19a Archdeacon first Wright-type glider: 1904
  b Archdeacon first Wright-type glider: 1904

between pages 112 and 113

20a Esnault-Pelterie aileroned Wright-type glider: 1904
  b Esnault-Pelterie aileroned Wright-type glider: 1904

21 Wright *Flyer II*: 1904

between pages 128 and 129

22 Ferber tailed Wright-type glider: 1904

23 Archdeacon second Wright-type glider: 1905

24 Pirated picture of the Wright *Flyer II*: 1905

25 Voisin-Archdeacon float-glider: 1905

26 Voisin-Blériot float-glider: 1905

27 Ferber powered Wright-type tailed glider: 1905

28a Wright *Flyer III* (general arrangement drawings): 1905
  b Wright *Flyer III*: 1905
  c Wright *Flyer III*: 1905

29 Pirated drawing of the Wright *Flyer III*: 1905

30 Cody aileroned glider: 1905

31 Page from *L'Aérophile* showing the article on the Wrights: 1905-06

32 Santos-Dumont monoplane design: 1906

33 Two pirated drawings of the Wright *Flyer III*: 1906

between pages 224 and 225

34 Vuia I monoplane: 1906

35 Blériot IV biplane: 1906

36 Ellehammer tethered biplane: 1906

37 Santos-Dumont 14-*bis*: 1906

38 Santos-Dumont 14-*bis* (modified, with ailerons): 1906

39 Blériot VI *Libellule*: 1907

40 Blériot VII monoplane: 1907

41 REP No 1: 1907

Illustrations

42	Santos-Dumont No 19: 1907	*between pages*
43	Voisin Chanute-type glider: 1907	224
44	Voisin-Delagrange I: 1907	*and* 225
45	De Pischoff biplane: 1907	
46	Voisin-Farman I (modified): 1908	
47	Roe biplane: 1908	
48	Cody "British Army Aeroplane No 1": 1908	
49	Blériot VIII-*bis*: 1908	*between pages*
50	Gastambide-Mengin I: 1908	272
51a	Wright Type A: 1908	*and* 273
b	Wright Type A: 1908	
c	Wright Type A: 1908	
d	Wright Type A: 1908	
52	Wright Type A flying at Fort Myer: 1908	
53	Voisin-Farman *I-bis* (2nd modification): 1908	
54	Henry Farman III: 1909	
55	Voisin biplane: 1909	
56	Antoinette IV (ailerons): 1909	
57	Antoinette VI (warping): 1909	
58	Blériot XI: 1909	
59	Wright Type A at Reims: 1909	

Foreword

The Science Museum is already greatly indebted to Charles Gibbs-Smith for the series of publications he has written on aeronautical history. In this new volume he is concerned with the question of the extent to which the work of the Wright brothers in America influenced the work of the early pioneers in Europe. Only someone with the author's immense and detailed knowledge of the source material available could have brought together the evidence presented here. He shows that in fact the European workers were greatly influenced by the Wrights and that although this influence was extended by the Wrights' well-reported demonstration flights in France in mid-1908, it had originated some years earlier when news and reports of their work had first reached Europe.

This volume which can but further enhance Mr Gibbs-Smith's reputation as an aeronautical historian, increases our knowledge of a crucial period in the development of aviation and we are pleased to add it to the range of Science Museum publications.

London, 1973   MARGARET WESTON
                                Director
                                Science Museum

Author's Introduction

It might be supposed that the entire history of the aeroplane had, by this time, been worked over in great detail; but, curiously enough, this is not true of one of the most vital periods in the development of aviation, the years 1902 to 1908, when practical full-scale flying was born, and the aeroplane came to take its place as the latest of the world's self-propelled vehicles.

This pivotal period saw the Wright brothers achieve the first powered, sustained, and controlled flights in 1903, and the first fully practical aeroplane in 1905: it also saw the Wright's influence spark into new life the near-moribund school of European aviators. This impact was then followed by a period of inexplicable torpor and procrastination among the British and Continental pioneers, which lasted until they were finally galvanised into effective action by Wilbur Wright's spectacular flying in France in 1908.

This critical period of aeronautical history has never before been properly scrutinised and assessed; and what little coverage there has been in the past has so often been bedevilled by displays of national prejudice and antipathy which have confused the true issues and distorted the historical realities beyond all recognition.

In this book, I have attempted a long overdue analysis of the forces which went to mould the modern powered aeroplane, and, with it, much of our modern world. It has been my task to set out the facts, and to establish what happened—and why—in this veritable rebirth of practical European aviation.

I would thank my friend and scientific adviser, Mr John Bagley, AFRAeS, of the Aerodynamics Department of the Royal Aircraft Establishment at Farnborough, for his most generous interest in the project, for reading the book in typescript, and for his many suggestions and improvements to the text. I also wish to thank another old Farnborough friend, Dr P. B. Walker, CBE, formerly Head of the Structures Department at the RAE, for help and advice in his special fields.

As always, I owe much to the constant beneficence and encouragement of my "near colleagues" in the Science Museum, particularly the former and present Directors, Sir David Follett and Miss Margaret Weston, and to the Keeper and Deputy Keeper of Aeronautics, Mr Brian Lacey and Lieutenant-Commander Walter Tuck. Also to Mr Marvin McFarland, Chief of Science and Technology at the Library of Congress in Washington (USA), the editor of one of our great aeronautical bibles, *The Papers of Wilbur and Orville Wright*.

The translation of passages from the French is often supposed to be a simple task; but this is far from being the case, especially where early technical phrases and comments are concerned: so I owe an especial debt of gratitude to two more friends, Max and Henriette Samné, for their careful checking and correcting of my translations from their precise and elegant language.

I wish it were possible to thank individually by name the numerous members of Her Majesty's Stationery Office, who have for many years borne so stoically, and with great good humour, the quite dreadful demands I have made on them, and who have produced my books so admirably; but official etiquette denies me this pleasure, and I can only pay them collectively the high tribute which is their due.

Finally I would like to express much gratitude to the lady who has allowed me to dedicate this book to her, the First Lady of the Wright Descendants, Ivonette Miller—Mrs Harold Miller—of Dayton, Ohio. Mrs Miller is the daughter of Lorin Wright, brother of Wilbur and Orville, and, with her husband, is one of the principal Trustees of the Wright Estate; she maintains a deep, active and expert interest in everything concerning the famous Wright Brothers, and I am most fortunate in being able to count on her friendship, encouragement and advice.

Autumn, 1973   CHARLES H. GIBBS-SMITH
                                United Service and Royal Aero Club
                                London SW1

I

The Wright Brothers and the 'anti-Wright Faction'

This monograph has grown out of the belief that as the aeroplane has come to play such an important role in civilisation, the process of its gestation, birth, and coming of age deserves to be the subject of detailed consideration in all its aspects. In the present work I attempt, so to say, to look through a magnifying glass at what was undoubtedly the most crucial period of the aeroplane's development, the years 1902 to 1908.

It is strange how time and events can become telescoped in the mind; how difficult it is to build up an image of the teeming events which can occupy a single month—or even a week—when they happened long ago. If the reader will refer to the literature of flight, he will find it difficult to discover more than a handful of books or articles which devote even a dozen lines to the overall picture of what happened in the sphere of the conception and construction of aeroplanes during these early and critical years of aviation history; yet it was the most vital of periods. So my aim will be to transport the reader back into the very swim of events during those years, and to show him how the early pioneers thought, spoke and acted, from month to month, on these subjects which were so near their hearts and aspirations.

In the present work these three main themes will be dealt with:

(a) The decisive influence of the Wright Brothers' work in precipitating the re-birth of practical man-carrying aviation in Europe; and in conditioning the forms which that re-birth was to assume;

(b) The well-nigh inexplicable procrastination and 'torpor'—their own word for it—on the part of the European pioneers in the development of practical aviation;

(c) The major influence of Wilbur Wright on European aviation in 1908, when his display of mastery in flight-control finally set the Continent on the road to success.

Where the direct influence of the Wrights is concerned—which is so often the subject of controversy in certain quarters—I would summarise this influence, or impact, as follows:

(1) In general, their influence was directly responsible for precipitating the re-birth of practical man-carrying aviation in Europe between 1902 and 1908; the first impact was brought about in France through Captain Ferber—the only European pioneer actively productive at the time—being persuaded to change from a primitive Lilienthal-type hang-glider to an admitted copy of the Wright glider No 2 of 1901. Ferber was, indeed, to pay a handsome tribute to the Wrights for their influence on him, and on his followers.

(2) The second and more powerful influence of the Wrights on Europe was made by Octave Chanute's illustrated lecture to the Aéro-Club de France on April 2nd 1903, which was followed by the content of the lecture being reported in a number of important illustrated magazine-articles during 1903. This influence was exerted by descriptions and photographs of the Wrights' modified No 3 glider of 1902, which led directly to both Ernest Archdeacon and Robert Esnault-Pelterie building and testing admitted copies of the Wright machine.

(3) The third direct impact of the Wrights on Europe was in the sphere of flight-control, particularly control in roll, and led directly to the first use of ailerons in 1904; but this influence was to play a decisive role later (see paragraph 6).

(4) The fourth sphere of influence of the Wrights was caused by news of the brothers' first powered flights on December 17th 1903, which proved profoundly disturbing to the Europeans.

(5) The fifth Wright impact was the news—first received in France by letters from the Wrights to Ferber and Besançon—of the brothers' triumphant 1905 season with their powered *Flyer III*: this news, after it had been checked by the French, resulted in an *Aérophile* editorial of January 1906 which freely admitted that the Wrights had now first achieved the conquest of the air in heavier-than-air machines. The effect of this news, and of the long *Aérophile* report confirming it, was climactic in its effect on Europe: it produced the most violent reactions ever felt—before or since—over an aeronautical matter. There immediately appeared in France a dichotomy of attitude towards the Wrights, with those who sadly accepted the news ranged on one side, and those who simply refused to believe what had happened, drawn up on the other. But both of these opposing factions came together in their determination to goad themselves and their friends into effective action; the French Aéro-Club's President came out with the biting "self-accusation" that his fellow-countrymen were suffering from an "inexcusable torpor"; and he delivered an impassioned exhortation to the French pioneers to drive ahead, and at least achieve the first public experiments in practical aviation.

(6) The sixth, and final, sphere of the Wright brothers' work to decisively influence European aviation, was Wilbur Wrights' superb demonstration of flight-control in action—particularly roll-control—when he flew at Hunaudières and Auvours (near Le Mans) from August to the end of December 1908, and broke every record. It was a great Frenchman, Count Henry de La Vaulx—founder in 1905 of the FAI (Fédération Aéronautique Internationale)—who described Wilbur's classic 1908 aircraft as the machine which "revolutionised the aviators' world."

The Wrights' work of invention, insemination, and influence then came to an end. From 1909 on, the Europeans steadily overhauled the Wrights.

But there was one extraordinary "non-event" so to speak—or "non-impact"—which must be recorded here, following the positive impacts. For in the January 1906 issue of *L'Aérophile*, there was published the Wrights' basic patent, which described in detail not only the operational details of the Wrights' three-axis flight-control system, but its aerodynamic *raison d'être*. This was the first time in history that Europe had been presented "free of charge", so to speak, in a key periodical, with the solution to the problem of the flight-control of aeroplanes.[1] But the Europeans, stung to action and struggling for achievement, took absolutely no notice of these vital revelations. There had been the first, but weak, impact of Chanute's references to the Wrights' roll-control in 1903; but now that the complete explanation of the how and why of roll-control was published, it was not acted upon. Europe had to wait until it saw Wilbur Wright fly in France in 1908, to get the message loud and clear. It is, of course, from the Wrights that the basic modern flight-control principles derive. Apart from this fact, their specific method of control in roll—by means of a helical twisting of the wings (warping)—was adopted by a number of the early Europeans, in particular by Esnault-Pelterie, Blériot and Levavasseur; the standard Blériot type XI machines even retained wing-warping right into World War I. And from the Wrights' wing-warping there have developed directly the ailerons of today.

After being the first in the world to make powered flights (1903); the first to build and fly an all-round practical aeroplane (1905); and having directly precipitated man-carrying powered aviation in Europe, the "men of Dayton"—as Gabriel Voisin was to call them—could safely rest on their laurels; for, as Charles Dollfus has said,

"Ils ont changé la face du monde."

---

[1] It had been initially published by the French patent office in 1904, but evidently the pioneers in such a new sphere as flying took no interest in scrutinising patents.

But despite the fact that all the "impacts" mentioned above are abundantly demonstrated by the contemporary documents and events—which will be set out in the course of this work—there are those who form what can only be called the "anti-Wright faction", on both sides of the Atlantic. This "anti-Wright faction" may be found taking a fairly well-defined form from 1906 onwards, based mostly on a patriotic attempt by the French to keep the aeroplane to themselves, so to speak, as indeed they virtually admitted.

The arguments which this "faction" have adopted may be summarised as follows:

(a) A denial that the first Wright flights on December 17th 1903 *were* proper flights; and/or

(b) A denial that the Wrights flew properly in 1904 and 1905, based on the fact that they used an assisted take-off method from 1904 onwards; and/or

(c) An acceptance of their flights in 1903, 1904 and 1905, but with the assertion that—although they might have been the first to fly—they were totally isolated from Europe; therefore, that they influenced nobody at any time, and stood quite aside from the main flow of history; also, that when Wilbur came to Europe in 1908, the Europeans had by then advanced so far on their own that the Wrights had nothing of value to impart to them.

One of the most influential members of the anti-Wright faction has always been the eminent French pioneer Gabriel Voisin; but his undoubted importance in history clothes some of his statements with an authority which they do not deserve. His autobiography has been translated into English, and published both in the USA and the United Kingdom[1]. As he is one of the most venerable and respected of the French pioneer fraternity, he is, of course, listened to with respect by many lay readers. For this reason, I shall include in the following pages a considerable number of quotations from this book to show what the historian has to face in such cases. Here is the first batch, the italics being Gabriel's:

> "America, with unbelievable insolence, claims to have been the birthplace of aviation. It is inconceivable that France should bow before so naive a claim. Aviation was born in France, and not one of our great men, true pioneers of the air, borrowed anything at all from the men of Dayton. The Super-Fortresses, the Constellations and the jet-driven Boeings, owe their origins to Ader's *Avion* of 1897 and to that of Blérot in 1909, while the most up-to-date *Lightning* leaves in its slipstream remembrances of the old Voisin of 1908, just

---

[1] Voisin (G). *Men, Women and 10,000 Kites.* Translated by Oliver Stewart. London (Putman), 1963. (From *Mes dix mille cerfs volants.* Paris, 1961.)

as the latest steam locomotives, in their pride of six thousand horsepower, had been foreshadowed by Stephenson's Rocket." (page 160)

"In order to understand fully the Wright 'swindle' it is necessary to read *The Wright Papers* attentively, and to analyse the statements which are to be found there in the spirit of a judge impartially seeking the truth." (page 240)

"We never had at any time during our work communications relating to their (the Wrights') arrangements, and when my brother finally succeeded, on 30 March 1907, at Bagatelle, in making the first French powered flight recorded on film[1], the totality of French constructors *had no knowledge of the Wrights or of their work*. In a word, the existence of the two Americans never influenced our researchers in any way." (page 237)

"No technician of real standing can admit that the Wrights inspired anything at all, and that for two reasons. The first is important: the Wrights kept their secret so well that it remained impenetrable from 1903 to 8 August 1908, a date by which French aviation was definitely under way. The second reason also has its significance: *the Wright aircraft had no future*. When the Wrights came to France, financed by influential people, and with resources we could never have hoped for, praised to the skies by a (largely paid) press, it taught us nothing." (page 237)

"At the moment when his boat left the shores of France—which had welcomed him with an almost unbelievable enthusiasm—Wilbur Wright, had he been able to peer through the Channel fog, would have distinguished the shadow of the Blériot XII [XI], the aircraft which was to be carried to England on the wings of victory. Then he might have realised the uselessness of his efforts, the poverty of his devices, and the futility of his secrets." (page 238)

The next quotation is from a letter published in the British journal *Flight International* in 1964[2]:

"These facts emerge inexorably as a result of a dispassionate approach to the subject and are as follows:

(a) the aeroplane as we know it today is wholly European (and primarily French) both in concept and development;

(b) that if the Wright brothers had never lived, the aeroplane would still have been conceived in Europe by the same people at the same times and would have proceeded through the same stages of development that, in fact, it did.

Any attempt to deny the foregoing is not only a distortion of history, but inflicts a grave injustice on the true creators of the modern aeroplanes—people like the Voisin brothers, Blériot, Levavasseur, Breguet, Goupy, A. V. Roe, and a host of others."

[1] Brother Charles was airborne for 6 seconds, and covered 60 metres.
[2] This letter appeared in the issue of August 13th 1964.

One is continually conscious of these attitudes when encountering members of the "anti-Wright faction"; *ie* the chauvinistic bias at one end of the scale, and the emotional antipathy to the Wrights themselves —or perhaps to America in general—at the other: more often, these attitudes emerge as blended, one with the other.

❧ 2 ❧

European Aviation after the Death of Lilienthal

After Lilienthal was killed in 1896, it might have been supposed that a host of disciples would immediately have arisen to carry on and complete his work: but this did not happen. Of his few followers, the only outstanding disciple was Percy Pilcher, who might well have conquered the air before the Wrights, had he, too, not been killed gliding in 1899. European aviation thereupon became virtually moribund. In 1897 Clément Ader had made two tests with his powered *Avion III*; and it had failed to fly.[1] In 1901 the Austrian Wilhelm Kress had completed his ambitious tandem-wing float-plane, derived from Langley, which was wrecked during that same year before its first take-off attempt. In Germany, a name comparatively new to aeronautical history has recently emerged—Karl Jatho—to take a small place in the early history of the aeroplane. Jatho was a civil servant in Hanover, and in 1903 completed what was little more than a large powered kite: it had a 9 hp petrol engine and a primitive pusher propeller, but there were neither tail-unit nor controls forward of the 'planes', and only rudimentary rudder and elevator devices. On August 18th he made a "running jump" claimed to be 18 metres; then, in November, with the structure modified to biplane form, the machine made another hop of 60 metres. These tests took place, probably downhill, on the Vahrenwalder Heide, north of Hanover: they are not claimed as true flights, even in Germany, where the word 'Flugsprung' (leap into the air) has been used for them: he therefore may make the minor claim to be the first German to leave the ground in a powered aeroplane.

The experiments of Ader, Kress and Jatho led nowhere and influenced nobody. The field of aviation during these years was a desert, with only a few model aeroplane competitions, a number of visionary schemes—including some for helicopters and even ornithopters—and

---

[1] See my Science Museum monograph, *Clément Ader* (1967).

many pious hopes and exhortations, to break the monotonous landscape. As far as flying in general was concerned, it was the heyday of the free balloon—our own Aero Club (later Royal) was founded in London in 1901—and of the airship. The first, very tentative, Zeppelin was tested in 1900; Santos-Dumont caused a sensation in 1901 by flying round the Eiffel Tower in his airship No 6; and the French Lebaudy airship was showing great promise in 1902. But most attention in the world of adventure was being paid to the automobile, which in 1896 had received the freedom of the roads in Britain; and automobilism had become by far the leading mechanical sport in Europe at the turn of the century.

There was but one man in the whole of Europe, the French Captain of Artillery Ferdinand Ferber, who was endeavouring to keep alive the main tradition of aviation during 1901 and 1902. But only a few people knew of him at first, and it appeared to the majority of air-minded men in 1902 that nothing of note was happening in the heavier-than-air sphere. We have the significant evidence of the Secretary-General of the Aéro-Club de France on this subject. But before quoting him, it is important to note the unique position and status of that Club in Europe.

Far from being a mere social society, the Aéro-Club de France was the central focus, breeding ground, and exchange-and-mart for both practical and theoretical aeronautics in Europe, with the practical side naturally predominating: it was the equivalent of the Aeronautical Society and the Aero Club in London combined. The Club was dedicated to "la Grand Cause de la Conquête de l'Air par le Génie humain". All the leading French balloonists, air-minded patrons, and many foreigners, were members; and naturally enough, it was passionately patriotic. Its speakers, aware of the multifarious aeronautical activities espoused by France in the previous century, were later apt to remind each other—and the world—that "aviation is indeed a French science" (aviation est bien une science française), although at this time it was hard to find more than a handful of members with any real interest in aviation.

The Aéro-Club de France was doubly fortunate in that it possessed not only a large and talented membership, but the best aeronautical periodical of the time as its official organ; this was the monthly *L'Aérophile*, edited by Georges Besançon, who was also Secretary-General of the Aéro-Club. This journal was read throughout the world; and its articles, illustrations, and reports of discussions—which covered world aeronautics—were of vital importance in the development of flying.

Throughout the year 1902, the Aéro-Club, however, was preoccupied as ever with lighter-than-air aeronautics, with only an occasional article or reference to helicopters—still a minor French preoccupation—or other heavier-than-air experiments. Aviation was all but a dead letter. It is significant that when the Secretary-General presented his report on the Club's activities in 1902—at the General Assembly on March 5th 1903—*he only devoted 3 printed lines (out of over 250) to heavier-than-air activities.* He simply mentioned the Vicomte Descazes, who nursed a visionary helicopter scheme, and Ferber, whose work he did not as yet think it even worthwhile to describe. But, in order to put it as colourfully as he could, in his three lines, the Secretary-General described the two men as "militant aviators" (aviateurs militants), not "militaires"! He closed his brief sentence with the elegant statement that their new activities merited the "highest hopes" (les plus belles espérances).

One of the great Frenchmen of this period was the Comte Henry de La Vaulx; at this time he was Vice-President of the Aéro-Club de France, and he was at the centre of European aeronautical events during the whole of this vital period. In 1905, he founded the FAI (Fédération Aéronautique Internationale) which flourishes to this day. The Count published, in 1911, one of the most authoritative of the early histories of flying, entitled *Le Triomphe de la Navigation Aérienne* (1911), from which a number of passages will be quoted. The first of these, which now follows, is an interesting description of the state of French aeronautics at the turn of the century:

"While the American aviators, following the way opened by Lilienthal, arrived in a few years—by the 'detour' of gliding—at establishing the full-size powered aeroplane, what was happening elsewhere, and particularly in France? With us, Lilienthal did not create a school. Our aviators preferred to approach the problem direct, without passing through the school of gliding; they had to struggle against the double difficulty of the flying-machine itself, and of a motor light and powerful enough to propel it. They had the idea that the progress of the automobile was about to be able to resolve the second problem; and the small model aeroplane, created many years ago by Pénaud, indicated at least one method of satisfying the first. Moreover, many inventors neglected the aeroplane altogether, orientating their researches in the direction of other types of flying machines, such as the ornithopter and helicopter. In all this, there was no collective effort, no real feeling that there was a possibility of impending achievement. What went on was mostly theoretical studies, projects, and highly erudite discussions, but which did not appear to be conducive to a speedy solution. In a word, French aviation—despite the efforts of the early pioneers like Victor Tatin—seemed to be dozing. The present belonged to the balloonists, proud of the sporting successes of their free balloons, and the first results obtained by our dirigibles. Into this conspicuous apathy cut the activity of a man who, perhaps alone in France, had properly understood the full range and interest of Lilienthal's experiments: this man was the Captain of Artillery, Ferdinand Ferber."

The state of aviation in all other European countries was even more parlous. Britain may be taken as a typical example, and to watch what was going on, and to test the climate of opinion, we may go through two periodicals; first, the *Aeronautical Journal*, the official publication (then issued quarterly) of the Aeronautical Society of Great Britain—now the Royal Aeronautical Society; and second, *The Automotor and Horseless Vehicle Journal*, which appeared monthly under this title until March 1902; then, from the issue of April 19th 1902 it became a weekly with the title abbreviated to *The Automotor Journal*, with the same sub-title as before, i.e. "A Record and Review of applied automatic Locomotion". Not only was the *Automotor Journal* an admirably edited and produced publication, getting steadily larger with every issue; but, from the first, it showed a close interest in aeronautics: in its enterprise and coverage of all aeronautical subjects, it put to ever greater shame, I regret to say, the *Aeronautical Journal*.

Starting with the *Aeronautical Journal* of January 1900, we have to go all the way through to the issue of April 1901 before there is any item dealing with aviation, except for kites. In this April 1901 issue there is a notice just 27 lines long, headed "Soaring Machines"; it is a report of Professor Bryan's lecture at the Royal Institution, another report of which was to "spark" Ferber in France (see Section 8). It refers to Chanute as one who

> "considers there is a future for the soaring machine in spite of the tragedies which have been connected with its manipulation. . . ."

The notice goes on to say that in view of these "peculiarly disastrous" machines, investigators should be protected from danger by performing above safety nets, as acrobats do in circuses! In the July issue was a brief article by Wilbur Wright, entitled "Angle of Incidence", but with no explanation of who he was.

Then came "aviation silence" again, until the October 1901 issue of the *Journal*, where we find a review of an article by Chanute published in the current number of *Cassier's Magazine*, which dealt with balloons, airships, and gliders. The review in the *Aeronautical Journal* briefly notes Chanute's opinions on gliders, especially the "maintenance of equilibrium." Next, in the issue of January 1902, is a long paper by Sir Hiram Maxim entitled "Aerial Navigation by Bodies Heavier than the Air". This deals chiefly with his own experiments in 1894, and makes no mention of gliding; it ends with prophecies that the practical aeroplane will certainly be achieved, and will be chiefly useful for "military purposes".

Then we turn to the *Automotor Journal* for February 1902, and find with astonishment that there is not only a laudatory editorial on the Wright brothers, but that Wilbur's entire first Chicago paper—together with its illustrations—is reprinted in this and the next (March) issue. It had appeared in print in America for the first time only in the previous December (1901); here now was the Editor of the *Automotor Journal*, in February 1902, not only praising the Wrights, but asserting their importance and publishing their first paper in full. Incidentally, the Editor did not make the blunder of calling them Chanute's pupils. This was almost certainly the first time in Europe that a mechanically-minded audience—indeed an audience of any kind—was made fully conscious of the name of Wright. Wilbur's brief article on "Angle of Incidence" had appeared in the *Aeronautical Journal* for July 1901; but there was no further mention of the Wright name in the *Journal* until the issue of July 1902, when it appeared in a short article on Hargrave's powered aeroplane being built near Sydney, which unfortunately came to nothing. One paragraph started as follows:

"The presumption today is that all engineers have read about the experiments made by Maxim, Langley, Chanute, Walker, Wright, Pilcher, Lilienthal, Kress, Phillips, and others. . . ."

But there was no further mention of the Wrights; indeed there was to be no description of the brothers' gliding in the *Aeronautical Journal* until April 1904, and then only a brief one, accompanied by a statement from Orville on the powered flights of December 17th 1903. There was no more to be read on aviation in general in the *Aeronautical Journal* until January 1903, which will be noted later.

It is rather a bitter pill for us to swallow, even today, to realise that the historic Wright paper, with its description of wing-warping, should first be launched in a magazine almost wholly devoted to the automobile, and that our own technical society, devoted exclusively to the study of aeronautics, never even mentioned it, let alone quoting from it or reproducing it at length. Admittedly the *Journal* was a quarterly, but one would have hoped the significance of the paper might have struck the editor, and caused him to print it, even though it might be considered by some to be "stale news".

3

The Contrasting Attitudes of the 'Chauffeurs' and the 'Airmen'

As man-carrying aviation came within sight of success during the last decade of the nineteenth century, there were emerging two vitally different streams of pioneer, whose concepts of flying differed radically one from the other.

First there were what we may call the 'chauffeurs'; and second came the stream of true 'airmen'.

The chauffeur attitude to aviation regarded the flying-machine as a winged automobile, to be driven off the ground and into the air by brute force of engine and propeller, so to say, and sedately steered about the sky as if it were a land—or even a marine—vehicle which had simply been transferred from a layer of earth to a similarly flat layer of air. They sought inherent stability in the air at all costs, and neglected proper flight-control, which they thought was unnecessary except for the elevator to take them up or down, and the rudder to steer them to left or right. They had no conception of proper flight-control about three axes, and roll-control meant nothing to them: stability in roll *via* the dihedral angle they understood, but not control in roll. These men had little or no idea of the vagaries of the wind, nor of what the pilot could do if he was at their mercy. The chauffeur attitude was a static attitude, and the chauffeurs were often enthusiastic model-makers.

The true airman's attitude was one of identifying himself with his machine—"je veux faire corps avec la machine", as Françoise Sagan put it—and he wished to partake in the real experience of flying. He looked toward the bird, not the flying automobile, for his inspiration, because he wanted to control his machine as perfectly as the birds fly. Sometimes the airmen seemed to look upon the aeroplane as an aerial steed, to be ridden and controlled in the air as a living animal.

The 'chauffeurs' came to devote themselves mainly to the pursuit of thrust and lift, and thereby proved singularly unfruitful: they invariably tried to take off in powered machines before they had any true idea of flight-control. Whereas the 'airmen' thought primarily in terms of control in the air, and quickly realised that the unpowered glider was the vehicle of choice, in which a man might emulate the technique of gliding birds, and learn to ride the air successfully before having himself precipitated into the incorporeal atmosphere in a powered flying-machine without knowing what would happen, or what he should do, when once airborne. This distinction between chauffeurs and airmen was to prove pivotal in the final conquest of the air.

4

The Chanute Gliders: 1896-1902

As the great American civil engineer and aviation pioneer Octave Chanute was intimately involved with the early career of the Wright brothers, and with the birth of practical aviation in Europe, a word should be said here about this remarkable man. Born in France of French parents in 1832, he was taken when a baby to the USA in 1838, and became by training and environment a typical pioneering American. After a very busy and successful life as a civil engineer, and having become interested in aeronautics along the way, he collected every available piece of information on aviation history; he wrote and published a series of articles, which he revised and re-published in book-form in New York, entitled *Progress in Flying Machines* (1894). This work—a classic of aviation history—presented to the would-be pioneers, for the first time, the whole picture of aviation up to date. Chanute's vital importance to history was threefold: (1) he was the world's chief disseminator of aeronautical information; (2) he finalised the Lilienthal-type hang-glider in biplane form; and (3) he was the most valued encourager of the Wright brothers.

It was in 1896-97—directly inspired by Lilienthal (Fig 1)—that Chanute himself designed and built his gliders, and publicly exhorted inventors to build full-size gliders, and get into the air to experiment. This exhortation did not, by the way, affect the Wright brothers, as they had already decided to take up full-scale aviation before they had heard of Chanute.

Chanute's whole and declared intention in his practical work was the attainment of an inherently stable glider: he said he wished "to study equilibrium, and that alone", and to achieve "an apparatus with automatic stability in the wind"; he had no interest at all in pilot-operated controls, and none of his machines incorporated such controls, despite an old tradition to the contrary. He designed and built four types of glider, three in 1896-97, and one in 1902, and had them flown—

since he was too old at 64 to fly himself—by his assistants A. M. Herring and W. Avery. The testing ground selected by Chanute for his gliders was an area of sandhills near Miller (Indiana), on the south shore of Lake Michigan, 30 miles east of Chicago.

MODIFIED LILIENTHAL HANG-GLIDER: 1896
He first had tested a Lilienthal monoplane glider there at the end of June 1896; this machine is often believed to have been designed by Chanute; but it was in fact a glider bought from Lilienthal himself by Chanute's assistant A. M. Herring, and altered by the latter, we know not in what respects. In the *Aeronautical Annual* for 1897, Chanute writes: "He (Herring) rebuilt for me his Lilienthal apparatus, with which he had made some gliding flights in 1894". This machine was quickly abandoned after being found "unstable and dangerous", evidently as a result of Herring's modifications, since they were safe when built and flown by Lilienthal himself.

CHANUTE MULTIPLE-WING HANG-GLIDER: 1896 (Fig 2)
Called his "multi-wing machine". It was a machine whose wings could swing backwards in the horizontal plane "if the relative wind increases". and which were restrained by rubber springs in front. It was first tested in July 1896: it made some 300 glides, but was not a practicable type. A replica was built in 1902.

CHANUTE BIPLANE HANG-GLIDER: 1896
Called his "two-surface" machine. This is described in detail later.

CHANUTE OSCILLATING-WING HANG-GLIDER: 1902 (Fig 3)
Called his "oscillating-wing machine". A triplane, of which Chanute wrote: "In 1902 I caused to be built an apparatus to obtain automatic stability by a third method. This consists in pivoting the sustaining surfaces about 4/10 of their width from the front, and restraining them by springs. If the relative wind increases, the center of pressure moves backward and tends to give the surfaces a smaller angle of attack." This machine was tested by Avery in October 1902 at the Wrights' camp at the Kill Devil Hills, and was a failure.

* * * * *

CHANUTE BIPLANE HANG-GLIDER: 1896 (Fig 4)
This glider—called by Chanute his "two surface" (or "double-deck") machine—started, in fact, as a triplane; but the wing area was found to be excessive, and the lowest wing was too near the ground; so it was removed, leaving the pilot hanging well below the biplane "residue". The wings were braced according to the Pratt-truss system—which Chanute introduced into aviation from his own field of civil engineering —and a cruciform tail-unit, comprising tailplane-cum-fin, was set at a slight negative angle of incidence to form a dihedral angle in relation to

the main wings, and thus make for longitudinal stability. There were no pilot-operated control surfaces, despite an old and erroneous belief to the contrary. Control was exercised entirely by body-movements, similar to the Lilienthal gliders. But there was a gust-damping device attached to the tail (see below). The ancestry of this machine goes straight back to Stringfellow's triplane model of 1868, which owed its configuration to Cayley, who specifically suggested a triplane to Stringfellow, and to the paper read by Wenham to the first meeting of the Aeronautical Society in 1866, in which he illustrated superposed structures of high aspect-ratio. Chanute was a friend, and great admirer, of Wenham, and the two men had corresponded regularly since 1892.

This classic 1896 biplane of Chanute's was given its main series of tests, with success—after its first brief tests as a triplane at the end of August—in the first half of September 1896; it was tested again later in the same year; then in 1897; and finally in 1904. There seem to have been at least two of these machines built in 1896-97, one of which was re-built for Chanute's mechanical launching tests at the St Louis Exposition of 1904.

The significant improvements on Lilienthal's machines which were made by Chanute in his 1896 hang-glider, were as follows:

(a) The Pratt-truss system of wing-bracing;

(b) the rectangular wings of higher aspect-ratio, with a more efficient section;

(c) the position of the pilot, who not only hung below the wings (thus improving the pendulum stability of the machine), but hung on horizontal bars, so that greater pre-flight, and in-flight, adjustments to the centre of gravity could be made.

(d) the more efficient tail-unit.

Herring made one modification of his own, which seems to have been a doubtful improvement, with his application of "elastic fastenings" on the tail-unit, as a would-be gust-damper, which incorporated rubber 'inserts' in the bracing wires, to operate either vertically or laterally.[1] This was instead of Lilienthal's tail-unit, which consisted of a fixed vertical fin, 'surrounded' by a tailplane hinged to the leading edge of the fin, half-way down, so that it could move up freely but was held by wires from descending lower than the horizontal. Lilienthal, after experiencing a painful nose-dive, decided on this device in the hope that if the machine was gusted to a standstill and started mushing, the freely up-moving tailplane would move up, and not—as it might if it were fixed—tip the machine into a nose-down attitude.

But Chanute was aiming at too much "automatic stability", and too little controllability by the pilot.

[1] Chanute said that, on Avery's showing, this device would not work properly until he (Avery) had introduced improvements.

5

The Wright Gliders: 1900-1902

The history of the Wright gliders is available in detail elsewhere, and the following notes will, I hope, suffice here.

From the very first the "flight philosophy" of the Wrights was entirely different from Chanute's; and, as the published papers show, the brothers entered the active sphere of aviation in 1899, quite independently of Chanute. Although they became close friends, and the brothers benefited from the older man's many kindnesses and continual encouragement, they owed only two debts to him, and fairly minor ones, *ie* the Pratt-truss method of rigging a biplane, and the idea of assisted take-off pioneered by Chanute with an electric winch, and modified by the Wrights in their weight-and-derrick method. It is not known what caused the Wrights to choose a biplane configuration (see Section 7), but when they finally saw the Chanute form and its rigging, they chose to adopt it as the most practical basis for their own machines. But, in any case, the biplane configuration was not, of course, original to Chanute, as he was following Lilienthal; and even the Pratt-truss rigging had to be modified by the Wrights to allow for their wing-warping. But it was undoubtedly the Chanute wing-configuration that appealed to the Wrights, as opposed to the fragile Lilienthal biplanes.

The Wright brothers were the sons of a United Brethren Church Bishop, and lived at Dayton, Ohio. They had progressed from the selling of bicycles to their manufacture, and thereby made a comfortable living; this business alone provided the funds for their aviation. It was the work and death (in 1896) of Lilienthal that focused their mature attention on aviation, the initial inspiration probably arising from an excellently illustrated article on Lilienthal in *McClure's Magazine* in 1894, a magazine to which they are known to have subscribed.

In their subsequent study of Lilienthal's gliding, the Wrights became fully alive to the inefficiency of the latter's method of seeking balance and control solely by body-movements. They were, of course, equally alive to the effortless mastery of these problems exhibited by the birds. Then, about 1899, Wilbur Wright made his first two decisive moves: (a) he observed that gliding and soaring birds—evidenced especially by their local 'expert' the buzzard—"regain their lateral balance ... by a torsion of the tips of the wings";[1] and (b) he decided to apply this bird-practice to aeroplane wings, in order to effect control in roll. At first he conceived the idea—but abandoned it for structural reasons—of pivoting the wings about full-length span-wise shafts, so that the angle of incidence of the whole of each wing could be altered to produce more lift on one side and less on the other. Then he—or possibly Orville—hit on the idea of helically *twisting* the wings, after toying with a long narrow cardboard box. The Wrights' "basic idea was the adjustment of the wings to the right and left sides to different angles so as to secure different lifts on the opposite wings" (Orville).

The Wrights came to use four words to describe their wing-movements, ie 'torsion', 'bending', 'warping' and 'twisting', the last being their favourite. Unfortunately 'warping' became the common usage, which, when translated into French, could mean something quite different. Chanute's espousal of the word in 1903 seems to have hastened the standardisation of this misleading word (see Postscript on page 22).

Another source of confusion has arisen in aviation history owing to the suggestion by a number of inventors that parts of the wing-ends (or tips) should be bent or hinged upwards or downwards in order to act as steering air-brakes, which were intended to slew the aircraft round and change its heading. These devices have constantly been claimed—particularly in the case of Mouillard—as anticipations of ailerons or of the Wrights' wing-warping; but they were, of course, nothing of the kind, as they were never intended to exercise any control in roll.

The French verb that should have been used for helical twisting was. of course, "tordre".[2] The result was that Clément Ader, who in his 1890 aeroplane had provided that the wing-tips of his machine should be capable of simple *flexing*—ie up-and-down bending, with no intended change in their angles of incidence—proceeded later, and mendaciously, to claim that he had invented the Wrights' wing-warping, since the word he had used for his flexing ('gauchir') had now been adopted by the French to describe the totally dissimilar 'wing-warping' of the Wrights.[3] It would have been much better if everyone had stuck to the

---

[1] Ohio possesses one of the finest 'flocks' of buzzards in the world.

[2] The words used in the Wrights' French patent were 'tordre' and 'torsion'.

[3] This point was decisively settled by the French Courts when they fully upheld (in 1911) the Wrights' contention that no "prior art" existed: this was re-affirmed in 1913 by the Paris Court of Appeal, after three French experts had failed to find any previously similar invention. It is discussed in detail in my Science Museum monograph *Clément Ader: his Flight-Claims and his Place in History* (1968).

better word "twist"; for the words "warp", "warping" and "wing-warping"—which the Wrights themselves were forced to use after they had become common currency—led to some curious misunderstandings, and one particularly unpleasant and unnecessary controversy. Owing to Chanute's use of the word 'warp', it was translated into French literally by the verb 'gauchir', and the noun 'gauchissement', which, before then, did not mean a helical twisting, but an up-and-down flexing, as already mentioned.

The bare idea of altering the angle of incidence of the wings, or of small auxiliary surfaces, had been previously envisaged; but it had never been applied successfully in practice, and never even been imagined in combined action with the rudder, as the Wrights were soon both to envisage it, and to apply it successfully in man-carrying aircraft.

Like Lilienthal, the Wrights were true airmen, who were determined to get up into the air and fly; but they were also determined methodically to master control in the air with gliders before attempting any experiments with powered machines.[1]

The Wrights built their first aircraft in August 1899 (Fig 5); it was a biplane kite of 5 feet span, designed to test the efficiency of wing-warping, with a horizontal tailplane which was braced to the wings: the latter could also—apart from being warped—be moved fore and aft in relation to one another, the tailplane moving accordingly.

The Wright gliders appeared as follows:

WRIGHT NO 1 GLIDER: 1900 (Fig 6)

The effect of the warping on the kite was a success, and the brothers decided to build their first full-size glider as soon as their business affairs permitted; this machine was completed by September 1900. It was a biplane of 17 feet span and 165 square feet area, but with the horizontal surface placed out front, and now transformed into an elevator; and with wing-warping. They believed that a front elevator would provide a safer fore-and-aft control than a rear one. The pilot lay prone in a gap in the lower wing, to reduce head-resistance. Owing to the small wing area, this No 1 glider was flown (in October 1900) chiefly as an unmanned kite, with its controls operated from the ground. Only a few manned "pilot-controlled" kite-flights were made, as well as a few free piloted glides. Most of these tests took place at Kitty Hawk (North Carolina), a desolate location chosen not for its secrecy, but as a result of their enquiries about suitable locations; it had

---

[1] They were in wholehearted agreement with what Lilienthal had written: "One can get a proper insight into the practice of flying only by actual flying experiments. . . . The manner in which we have to meet the irregularities of the wind, when soaring in the air, can only be learnt by being in the air itself. . . . The only way which leads us to a quick development in human flight is a systematic and energetic practice in actual flying experiments." (1896).

steady strong winds, and there was also a wealth of soft sand into which crashing would be comparatively painless.[1]

WRIGHT NO 2 GLIDER: 1901 (Fig 7)
Encouraged by their experience with their No 1, the Wrights were more determined than ever to fly; but they were equally determined to proceed carefully and logically. Next year (1901) they built their No 2 glider, following the same basic configuration as No 1, but with a wing area increased to 290 square feet, a span of 22 feet, a wing camber of 1 in 12, and—for the first time—an anhedral 'droop' of 4 inches: also for the first time, a hip-cradle was introduced—swung to right or left—to operate the warping cables. They took this new glider to the Kill Devil (sand) Hills, 4 miles south of Kitty Hawk, and tested it in piloted flight during July and August 1901. Finding the wing camber too pronounced—with the centre of pressure moving too rapidly backwards at small angles of incidence—they reduced it to 1 in 19, and so improved the machine's performance; glides of up to 389 feet were made, and control was well maintained in winds of up to some 27 mph. But the machine was far from satisfactory. The Wrights began to suspect the accuracy of Lilienthal's calculations, upon which they had relied until now. The aircraft also showed an alarming tendency—when being warped—for the down-warped wings to swing *back*, and for the whole machine to slew round in their direction, and crash. "Having set out with absolute faith in the existing scientific data", wrote Wilbur, "we were driven to doubt one thing after another, till finally, after two years of experiment, we cast it all aside, and decided to rely entirely upon our own investigations".

WRIGHT NO 3 GLIDER: 1902
Between September 1901 and August 1902, the Wrights carried out an intensive programme of research, including the testing of aerofoil sections in a wind-tunnel, and on a bicycle, as well as thoroughly re-working all their aerodynamic problems. The result was their No 3 glider of 1902, built during August and September 1902, and tested at the Kill Devil Hills during September and October (Fig 8). It was a robust biplane with a wing area of 305 square feet, a span of 32 feet 1 inch, and a shallow camber of 1 in 24 to 1 in 30; it had the same warping system, the same anhedral droop, and the same type of forward elevator, as the No 2; but there was added at the rear a double fixed fin to counteract—by its weather-vane action—the swing-back of the wings on the down-warped side, in the case either of regaining lateral balance after being gusted out of the horizontal, or of the pilot initiating a bank.

[1] The free piloted glides were made on a brief visit to the Kill Devil Hills.

The machine, which was man-launched like its predecessor, behaved well in some of its glides; but severe trouble arose with deliberately warped banks, or gust-produced (non-warped) banks, when the pilot—in trying to hold off the bank, or in trying to return to the horizontal—applied down-warp on the lowered wings in order to raise them: instead of coming up, these lowered wings would sink even lower, or swing back, and the machine would spin and crash.

The Wrights diagnosed the basic trouble as warp-drag (aileron-drag today) where the positive warp increased the resistance of the wings on one side, and the negative warp decreased it on the other, causing them not only to bank, but to turn about their vertical axis in the opposite direction to that originally anticipated. The addition of the fixed rear fin caused them much more trouble than it cured. For, in the acts of banking noted above, the resulting side-slip caused the fin to act as a lever, and rotate the wings about their vertical axis, thus increasing the speed (and thus the lift and height) of the raised wings, whilst retarding and lowering the dropped wings. When the pilot applied positive warp to the dropped wings, it simply aggravated the situation, and produced a spin, by swinging back the dropped wings through warp-drag, thus increasing their incidence beyond the stalling angle.

This problem was solved by converting the fixed double fin into a single movable rudder, which—with its cables fastened to the warp-cradle—was always turned towards the direction of bank, thus counteracting the warp-drag (Fig 9). As the rudder was also (but unintentionally) adjusted to more than compensate for the warp-drag, the machine could also be made to perform a smooth banked turn. Experience with this glider also convinced them that all their machines should continue to be made inherently unstable to allow of sensitive and immediate response to the controls.

After this vital step, the Wrights had a practical and highly successful glider, with which they made over 800 perfectly controlled glides often in winds of up to 35 mph; they set a distance record of $622\frac{1}{2}$ feet, and a duration record of 26 seconds. "The flights of 1902", wrote Orville, "demonstrated the efficiency of our system of control for both longitudinal and lateral stability. They also demonstrated that our tables of air pressure which we made in our wind-tunnel would enable us to calculate in advance the performance of a machine."

Justly elated by the success of this last glider, the brothers—in March 1903—applied for a patent based on it:[1] this was granted in 1906. They had already determined to build a powered aeroplane, but they did not, as often said, put an engine into one of their gliders; this machine was constructed during the Summer of 1903. But they had had

---

[1] It was Chanute who urged them to apply for a patent.

to surmount two formidable obstacles before their first powered *Flyer* was ready for testing; (a) the lack of any available light, yet powerful, engine; (b) the lack of data on airscrews. They thereupon designed and built their own 12 hp motor; and—an outstanding achievement—carried out basic and original research to produce highly efficient airscrews, geared down to rotate at about 350 rpm.

As the time factor in these early days of flying is of such interest and importance, the following points about the Wright brothers should perhaps be emphasised:

(a) Throughout the whole of their early years of work in aviation (1899-1905), they were earning their living at their cycle business, and at first could only afford their spare time and holidays for aviation.

(b) Apart from the local flying—near Dayton—of their warping kite in 1899, they spent $3\frac{1}{2}$ weeks in their camp at Kitty Hawk during 1900; $5\frac{1}{2}$ weeks at the Kill Devil Hills in 1901; and 2 months—also at the Kill Devil Hills—in the crucial year 1902. This is an aggregate of some 18 weeks "in the field", in some of the most intensive flight-testing ever undertaken; during these weeks they mastered gliding flight, and evolved the basic three-axis flight-control system used in aviation ever since. What is more, this 18 weeks included the assembling of all the three gliders, and making the necessary modifications as they became necessary.

The part played by the Wrights' powered *Flyers*, and the factors involved, will be described later.

Postscript. In their basic patent (No 821,393) which was applied for in 1903 and granted in 1906, the Wrights used three words to describe their wing-movements, 'bend', 'twist', and 'warp', often in association with the word 'helicoidal'; they did not use the word 'torsion' in their patent. It is a great pity they used such a word as 'warp', and it may have been Chanute who influenced them to adopt it as the standard word. This may seem unfair to Chanute, but I cannot find any instance of 'warp' being used by the Wrights prior to 1903, when Chanute seems first to have used it.

6

Chanute
and the Wright Brothers

The Wright brothers and Octave Chanute came to be lifelong friends after Wilbur Wright had written to introduce himself to Chanute in 1900. But it should be made clear from the outset—and this will be dealt with in detail later—that:

(1) the Wrights did not take up aviation at the suggestion, or from the example of, Chanute;

(2) The Wrights were never in any sense pupils of Chanute;

(3) the Wrights were never collaborators with Chanute;

(4) the Wrights were only influenced by Chanute in two aeronautical respects, *ie* the adoption of the Pratt-truss type of biplane rigging (Chanute himself derived his biplane idea from elsewhere) (see Section 4), and the idea of an accelerated take-off;

(5) the Wrights were, from the first, working on completely different lines, and their achievements in aviation were gained by their own original work;

(6) the Wrights owed Chanute a deep debt of gratitude for his constant encouragement, his many kindnesses, and the stimulating comments Chanute brought to their many discussions.

All these points may be easily verified by reference to the voluminous documents which have been published, of which the Chanute-Wright correspondence forms a large and revealing part. They will, of course, be considered in detail in this monograph, as they occur.

One of the most important features of the controversy over the Wrights' influence on European aviation is the confusion of thought which took root in France in 1902-03—and has lasted ever since—concerning the relationship between Chanute and the Wrights. This confusion followed fairly well-defined lines, which were naturally in close proximity, and often overlapped: it may be summarised as follows:

(a) *The Wrights were often believed to have taken up aviation at the suggestion, or instigation, of Chanute*

The Rebirth of European Aviation 1902-1908

As will be seen later, this idea seems—without much doubt—to have been encouraged by Chanute himself. On various occasions it was said in French periodicals that Chanute had, in 1897, invited 'amateurs' to repeat his experiments and progress further, and that in 1900 the Wrights had taken up this invitation. This is entirely untrue, as even the most superficial survey of the documents shows. The Wrights' determination to take up aviation derived from their reading about Lilienthal; and their first important aircraft—the warping kite—was made and tested in 1899, following Wilbur's 'discovery' of the buzzard's "wing-warping".

(b) *The Wrights were often spoken of in Europe as the "pupils" of Chanute, and sometimes as his "collaborators"*

The Wrights were never, in any sense of the words, either the pupils or collaborators of Chanute. This misrepresentation came about primarily, I believe, by the paternalistic attitude which Chanute adopted in speaking of them, especially during his visit to Paris in 1903; despite his later denials to the Wrights, I feel sure that he implied, if not actually said, to the French that the Wrights were his pupils. At any rate, during Chanute's visit, the French pioneers certainly received this impression very clearly, as can be seen from their remarks. Here, for example, is Ernest Archdeacon, in his article in *La Locomotion*, reporting Chanute's lecture:

> "Admitting he was no longer very young, he took pains to train young intelligent, and daring pupils, capable of carrying on his researches by an unlimited number of glider experiments. Principal among them is certainly Mr. Wilbur Wright of Dayton . . ."

(As Archdeacon wrote this report immediately after Chanute's lecture on April 2nd 1903, and published it on April 11th, it is hard to believe that Chanute did not give a clear indication to his audience that Wilbur was his pupil. Whatever he said, no matter how innocent the motive, it certainly was to prove a grave disservice to the Wrights. Chanute was a great man and a great pioneer; but the fact that he was an old and respected figure, of high standing, and speaking of younger and (at that time) obscure men, was no excuse for him to suggest, even indirectly—and the suggestion seems to have been anything but indirect—that the Wrights were his pupils, thereby causing the endless confusion and injustice which followed.

It was even suggested in France that Chanute subsidised the brothers financially: in fact he once offered to do just this, but the offer was politely refused, and the Wrights never accepted a cent from anyone. They used the modest profits from their cycle business to finance all their early aviation.

Another reason for the misapprehension about their being Chanute's pupils was probably due to the simple fact that both Ferber (in 1902) and the other French pioneers (in 1903), first heard of the Wrights through Chanute; it was clear to the Europeans that the Wrights were at least close friends of the "grand old man" of American aviation, if not his pupils. Although the Wrights were always in praise of Chanute for his never failing encouragement, they could claim with perfect justification, that they were—and had been from the first—working on completely different lines to him. The overriding importance of the Wright machines was always to lie in their flight-control by pilot-operated movable surfaces; first by means of wing-warping and elevator, with no rear surfaces (1900-01); then by the addition of fixed rear fins (1902); and finally by the crowning achievement of exchanging the fins for a rear rudder working in concert with the warping (also 1902), which inaugurated the flight-control system of today. At no time did Chanute think of warping, or of any other wing-surface movements for lateral control; nor did he employ any elevator or rudder on his machines. When it came, so to say, to "invisible assets" in aerodynamics, the Wrights had to abandon (in 1901) the aerodynamic tables and formulae which they and Chanute had inherited from others, as they were found to be unreliable, and indeed positively misleading. The Wrights then conducted their own extensive laboratory research, including wind-tunnel tests, as well as flight tests, and arrived at their own tables and formulae, which they used throughout their subsequent work.

(c) *The Wright gliders were generally believed in Europe to be modifications of Chanute's biplane glider of 1896*

Typical of this belief, and the prejudiced wish to perpetuate it, even in our own day, is a statement in Voisin's autobiography. Speaking of the Wrights, he says:

> "They had adopted the Chanute (French) type of machine which consists of two rectangles superposed, with slightly cambered surfaces."

The bracketed word "French" adds a pathetic touch of attempted patriotic annexation not even characteristic of the other contemporary French pioneers, since they all recognised Chanute as an American: he was taken to America when a baby. Voisin's intention here—as in much of his book—is, of course, to denigrate the Wright brothers. As to the type of machine being French, this is one of those strange myths that some people try to propagand and perpetuate for their own partisan purposes. The biplane (and multi-plane) concept was, as it happens, a peculiarly *British* idea in its origins. Starting with Cayley, it evolved through Wenham and Stringfellow, to Maxim and Lilienthal, and then to the Americans.

I have already, I hope, shown that the Wrights' gliders were in no way modifications, technically, of Chanute's glider. What is more, even the visible resemblance between all forms of the Wright gliders and the Chanute machines was limited solely to the generic similarity of their both being rectangular wing biplanes. There was nothing new to the European pioneers about the superposed wing concept as such, since they were all familiar with illustrations of the Wenham, Stringfellow, Maxim, and Lilienthal machines. The visible differences between the Chanute and Wright gliders were striking enough for anyone to take in at a glance, particularly the idiom of the pilot hanging below—and well clear of the wings—in the Chanute machine, and the inconspicuous figure of the Wright pilot—with only his head and shoulders visible in most photographs—lying prone on the lower wing. Just as conspicuous was the presence of the front elevator in all the Wright gliders, and the lack of this feature in the Chanute machine. Even the tail-units of the respective gliders were radically different: in the first Wright glider to influence the Europeans (the No 2 of 1901), there were no tail surfaces at all; and in the modified No 3 of 1902 there was a tall and narrow rectangular rudder as against the bold cruciform tail-unit of the Chanute glider (see the comparative data set out at the end of this section).

To show even further how confused the Europeans could become, there was one occasion, in a 1905 report of the Chanute hang-glider—which was being power-launched at the St Louis Exposition of 1904 by a new method devised by Chanute—when this famous machine was described by one French writer as a "biplane, generally analogous (analgoue dans son ensemble) to the Wright machines ... but without the front elevator".

(d) *The Wright gliders, in photographs published in the European technical press, were incorrectly described as Chanute gliders*

It is not known how this came about, as we do not know exactly what the authors and periodicals had at their disposal; nor do we know what captions were attached to the photographic prints they used. But whatever the answers to these questions, there seems little excuse for miscaptioning the photographs when they accompanied accounts of Chanute's lecture, since he made it perfectly clear which the Wright gliders were, and what they looked like. This mis-captioning was particularly confusing and deplorable when it occurred in the first two authoritative and widely-read reports of the lecture, in *L'Aérophile* and *La Locomotion*. In both of these accounts, the same four photographs were used, all of the Wright No 2 glider of 1901, with the same caption—

"L'appareil Chanute"—on the first two, which was understood to apply also to the other two, where the captions simply read "Le planement", and "La direction". Thus the initial and vital impact on the French pioneers of Chanute's lecture was accompanied visually by photographs of Wright gliders which readers were told were Chanute's. No wonder that for a long time to come there was so much confusion on all sides as to who had produced which aircraft. But the confusion was even worse confounded by the evident non-availability—or misunderstanding—at that moment, of photographs of the modified No 3 Wright glider (with rudder), to which Chanute and the other writers paid most attention, thus resulting in readers being presented with the then relatively unimportant tailless No 2 of 1901.

(e) *The European-built Wright-type gliders were sometimes described as being of the "Chanute-Wright" type*

This misrepresentation grew out of the factors already described, aided by a not-too-careful reading by the French pioneers of the textual accounts they received from Chanute and others. But I feel it is important to point out here, not only that there were no European gliders built combining both Chanute and Wright features, but that no Chanute-type gliders were built in Europe until the Voisin machine of 1907.

(f) *It is sometimes believed today that the Europeans were influenced to adopt purely Chanute-type gliders in 1902-4, alongside the Wright-type machines*

This is, in a way, an extension of the previous point. The first Chanute-type glider to be built in Europe was, as just noted, in 1907. In this year Gabriel Voisin built a modified one for Farman; modified, that is to say, by the cruciform tail-unit being exchanged for a Hargrave cellular unit. This situation has been still further bedevilled by photographs of this 1907 Voisin glider being published in the authorised English translation of Voisin's autobiography, and bearing the date 1899;[1] but this is contradicted by the text, in which Gabriel places the machine in 1903. Whichever date it was that Gabriel intended, both are wrong, and it seems clear enough that he was playing the dangerous game of pre-dating the machine in order to show how advanced he and his brother were in their aviation work. All one can say is that we are lucky to have documentary proof in *L'Aérophile* of the true date as being 1907.

It might seem strange, with Chanute's formidable reputation as a pioneer, and after descriptions, photographs and measured drawings of his classic biplane hang-glider of 1896 had been included in his 1903 lecture—and the follow-up articles in the same year—that some of the leading Europeans did not take up his type of machine in the early days; but the fact is that they did not. To render this situation even more

[1] These photographs were specially provided for the English edition, as the original French edition was unillustrated.

curious and instructive, we find in *L'Aérophile* for March 1904, an article by G. Blanchet devoted to the practice of glider-flying which is illustrated with no less than fifteen ably sketched views, all of the Chanute machine. (Other aspects of this article will be discussed later.) But, despite this, no one was then inspired to build Chanute-type gliders.

(g) *Articles in European periodicals sometimes stated, or implied, that the reported American flying experiments of 1901-05 included experiments by both Chanute and the Wrights*

This is an important point, because when Archdeacon and others were trying in 1902-06 to spur on their fellow French pioneers to emulate and beat the Americans, it was thought that equally important experiments in the USA—of which they were being constantly reminded—were being carried out by both the Wrights and by Chanute.

What had actually happened was that the main test-flying of the Chanute gliders by Herring and Avery took place in 1896 and 1897; then there were brief and abortive tests of Chanute's multi-wing and oscillating-wing machines made by Herring, conducted at the Wright camp at the Kill Devil Hills in October 1902; and finally, in 1904, Chanute had Avery fly one of his re-built biplane hang-gliders at the St Louis Exposition, in order to test a new take-off technique in which the glider was placed on a trolley attached to a cable, the latter being wound in rapidly on a distant drum operated by an electric motor: this resulted in an effective method of launching, which was an improvement on Pilcher's horse-drawn method: it was also an anticipation of modern methods.

When, therefore, we find the constant references in European periodicals to experiments being carried out by the 'Americans', or by "Chanute and the Wrights", these references should be taken to apply only to the Wrights.

THE CHANUTE GLIDER OF 1896 AND THE MODIFIED WRIGHT GLIDER NO 3 OF 1902 COMPARED

	Chanute (1896)	*Wright* (1902)
Configuration	Biplane hang-glider, with cruciform tail-unit comprising fixed surfaces	Biplane glider, with front elevator and rear rudder
Flight-Control	Body movements (no moveable control surfaces). There was an elastic attachment on the tail-unit to act as a gust-damper	Pilot-operated elevator, wing-warping, and rudder
Pilot Position	Hanging by the arms from two bars below the wings	Lying prone on the lower wing

1a Lilienthal Type 11 monoplane Hang-glider: 1894
Otto Lilienthal flying one of his standard Type 11 monoplane hang-gliders in 1894. He is here seen using the artificial hill he had constructed in Berlin, and has reached a point just over half-way down the hill, with his body and legs swung slightly forward to keep the nose down by shifting the centre of gravity of the machine. It was photographs such as this which told the world of Lilienthal's achievements, and helped to inspire the Wright brothers and other inventors.

1b Lilienthal Type 11 monoplane Hang-glider: 1894 Lilienthal flying in 1894. This shot shows admirably how the pilot had to swing his body and legs far towards the side of a raised wing in order to shift the centre of gravity of the machine and bring the wing down.

1c Lilienthal biplane Hang-glider: 1895 Seen from behind, this shot clearly shows the tailplane which was hinged to the leading edge of the fin, and could rise freely upwards (but not downwards) to stop the nose going down in case the machine started to mush.

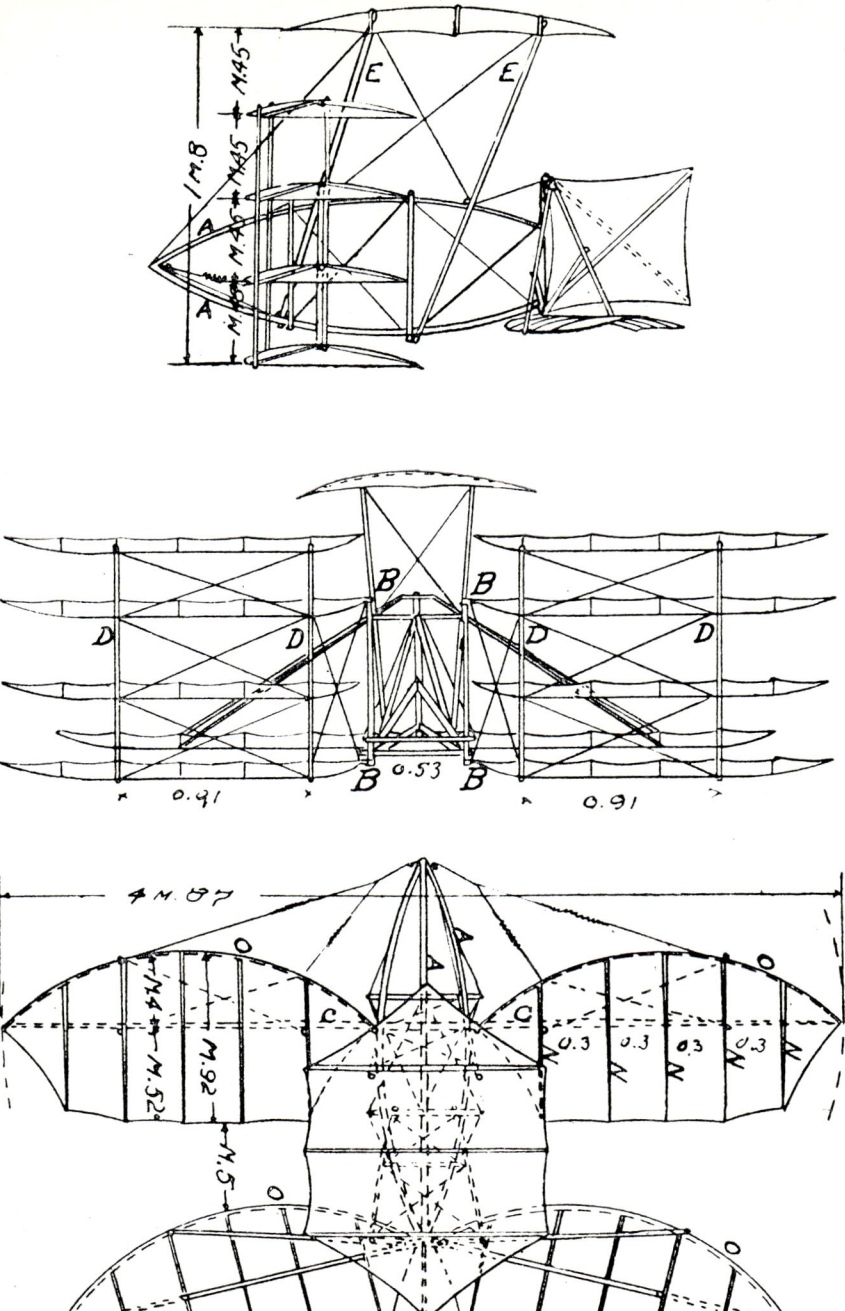

2 Chanute multi-wing Hang-glider: 1896 General arrangement drawings of Octave Chanute's multi-wing hang-glider of 1896, whose wings could swing backwards in the horizontal plane "if the relative wind increases". It was first tested in July 1896, and although it made some 300 glides, it was not a practical machine. From Chanute's article in *L'Aérophile*, August 1903.

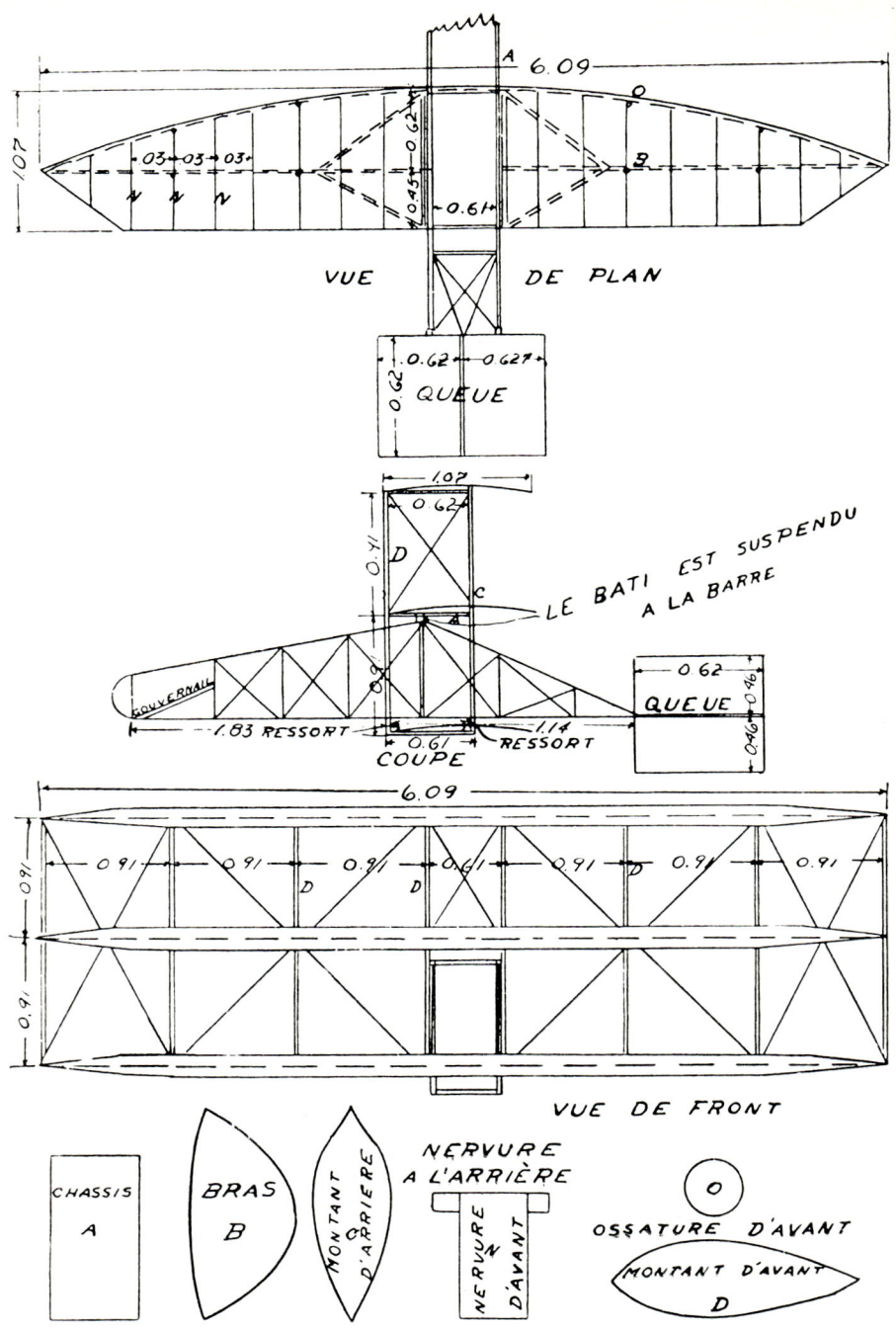

3 Chanute oscillating-wing Hang-glider: 1902 General arrangement drawings of Chanute's oscillating-wing hang-glider of 1902. This was a triplane which Chanute described as "an apparatus to obtain automatic stability", consisting of pivoted wings whose angle of attack was reduced if the relative wind increased. This machine was also a failure. From the same source as Fig 2.

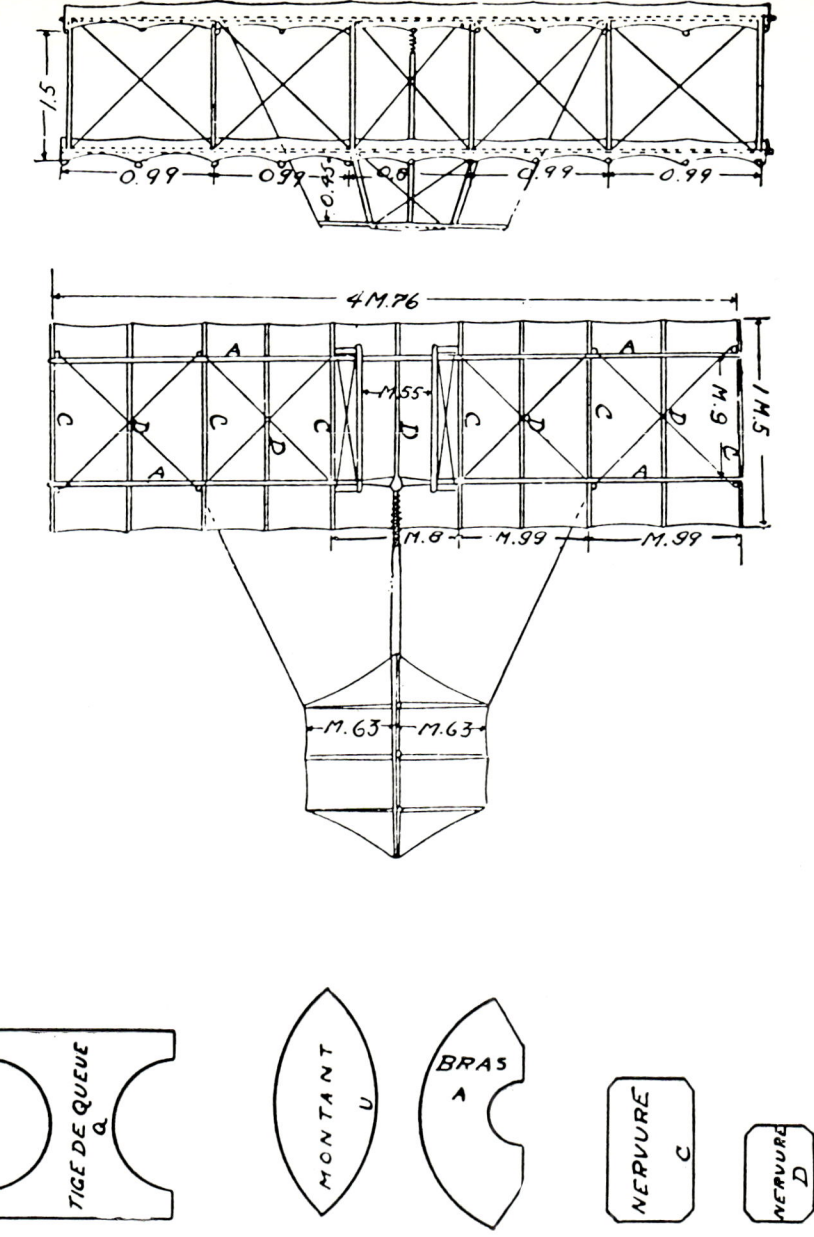

4a Chanute biplane Hang-glider: 1896 General arrangement drawings of Chanute's biplane hang-glider of 1896, from his article in *L'Aérophile,* August 1903. This machine was one of the classic types in aviation history, and represented Chanute's most mature ideas on gliders. It was derived from the Lilienthal biplane gliders, but was rigged according to the Pratt-truss type of structure as often used in bridge building. The tail unit was of the Cayley cruciform type.

Machine a deux surfaces Chanute et Herring, type 1896-1897
Echelle de 1/48

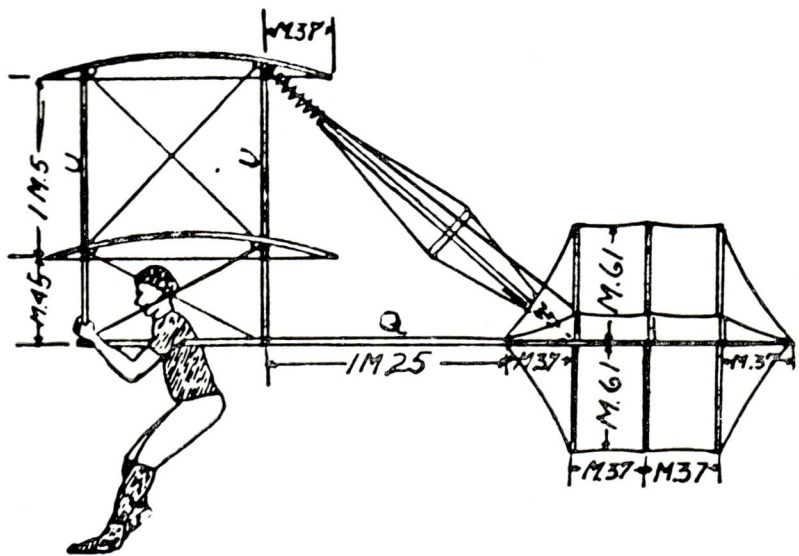

4b Chanute biplane Hang-glider: 1896 The side view general arrangement drawing of Chanute's hang-glider of 1896, which completes the set in Fig 4a, taken from *L'Aérophile*. Clearly shown here is the elastic device which Avery — Chanute's assistant — invented to allow the tail-unit to hinge upward in gusts of wind.

4c Chanute biplane Hang-glider: 1896 Chanute's biplane hang-glider, seen from slightly behind on the port side. Chanute provided parallel bars for the pilot to grip with his arm-pits, so that he could achieve a greater fore and aft movement of his body than was possible with Lilienthal's fixed position. From an illustration in the *Aeronautical Annual* for 1897.

4d Chanute biplane Hang-glider: 1896 Photograph of the Chanute hang-glider of 1896 in flight showing well the configuration and construction of the machine. It was first tested as a triplane, then modified to become the biplane seen in these illustrations. It was flown by Herring and Avery, as Chanute — by this time in his life — was too old for practical experimentation.

4e Chanute biplane Hang-glider: 1896 Avery about to launch himself in Chanute's hang-glider. Note the absence, on this occasion, of the upper fin in the cruciform tail-unit. The glider, in its classic form, was first flown near Chicago in September 1896; then in 1897; and later at the St Louis Exposition of 1904.

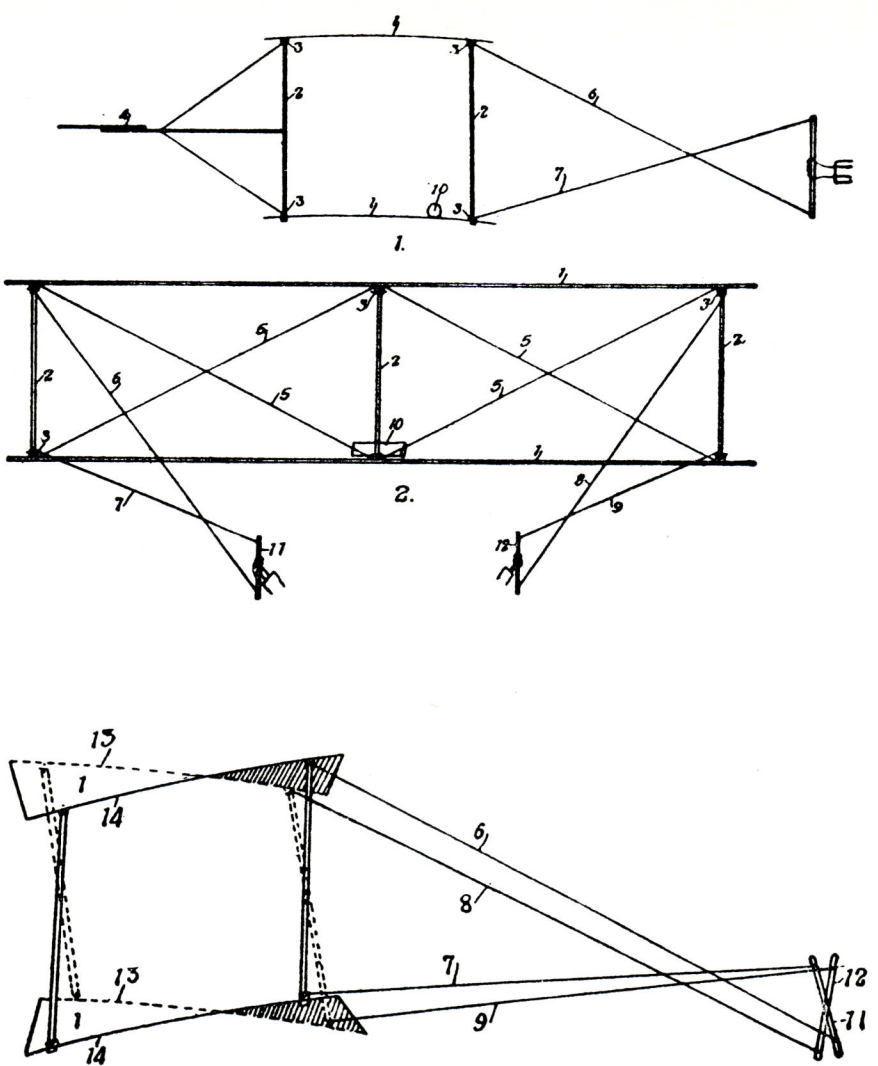

5 Wright warping biplane Kite: 1899 General arrangement drawings of the Wrights' warping biplane kite of 1899. The top view shows the side elevation with the cords (shown diagrammatically) for shifting the wings fore and aft in relation to one another, with the fixed tailplane then acting as an elevator. The centre drawing shows the front elevation, with the cords for warping the wings. The lower drawing shows the same system of cords as in the centre drawing, but now seen from the side, to demonstrate the effect of the wing-warping movements. This was the first machine in which the Wrights' wing-warping was incorporated, and the first time such warping was used anywhere in the world. The machine was tested near Dayton in August of 1899, when the success of the warping led to the brothers deciding to build a full-scale glider the year after.

7

The Question of Wing-trussing and the Biplane in History

A number of commentators place undue importance on biplane wing-trussing and its influence; some of them are inclined to feel that Chanute's use of the so-called Pratt-truss in his 1896 glider was a key influence not only in the Wright Brothers' story, but in early aviation history generally. They even try to trace the main biplane influence from Stringfellow's 1868 triplane onwards, in terms of rigging.

I feel this is a curiously exaggerated point of view. After all, there was nothing new (a) about the desirability of a braced biplane structure, or (b) about the types of bracing which would provide the required rigidity. Unfortunately, it is not known how Stringfellow's triplane model was rigged when it was originally completed in 1868; but he would obviously have incorporated some type of braced frame, as used by engineers.

We are equally ignorant of how Lilienthal rigged his three biplanes of 1895, since it is uncertain whether the single surviving one—in the Deutsches Museum at Munich—is in the state in which Lilienthal left it. But they are known to have been structurally sound.

When Chanute came on the scene, and produced his classic biplane hang-glider in 1896 (Fig 4), he decided to rig it with the Pratt-truss, with which he was familiar as a civil engineer. Rigging adjustments would be aided by having vertical struts incorporated in his triplane.

A braced biplane structure is a form of box girder, and the exact type of the bracing would depend upon the method used for adjusting wing incidence, and the need to ensure adequate rigidity in accordance with the principles of triangulation.

Also important, of course, is the rigidity of the wing-surfaces, which require substantial wing-ribs as well as tensioned fabric. It is obviously no use having a properly braced biplane structure as a frame, if the surfaces it supports are relatively flimsy, as were Ferber's in 1902 and 1903. The need for adequate stiffness of wing-surfaces, and adequate ribs, was first recognised by Henson in the patent drawings for his 'Aerial Steam Carriage' (1842-43).

We come now to the interesting problem of whether or not the Wrights were influenced by Chanute, not only to use his method of rigging—which they always acknowledged—but to adopt the biplane in the first place, which would appear to be so from a passage on pages 8 and 9 of *The Papers of Wilbur and Orville Wright* (see Bibliography); after describing the well-known scene where Wilbur was toying with a cardboard box, and got the idea of the helical twisting of wings from twisting the box, we find the following:

> "From this it was apparent that the wings of a machine of the Chanute double-deck type, with the fore-and-aft trussing removed, could be warped in like manner, . . ."

But this passage was written by Orville in 1919, when he was going back in his mind over those days in 1899 when he and his brother had made their first warping kite.

Wilbur first wrote to the Smithsonian asking for information about aviation on May 30th 1899. The Smithsonian replied on June 2nd, enclosing a list of suggested books, and also enclosing four Smithsonian pamphlets, in none of which appeared any biplanes: incidentally Chanute did not figure in any of them.

Wilbur wrote again to the Smithsonian on June 14th, thanking them for the pamphlets, and enclosing a dollar for *Experiments in Aerodynamics* by Langley. Till now he had received nothing concerning Chanute, except the suggestion of his own book, and the volumes of Means' *Aeronautical Annual*.

Mr McFarland has established (see note to page 8 of *The Papers*) that the 1899 warping kite was partly assembled, if not completed, by July 7th, which means that it would have been designed in June.

Now, of the books recommended by the Smithsonian, only one included illustrations of Chanute's biplane hang-glider of 1896, and that was the *Aeronautical Annual* for 1897. Chanute, in his own book *Progress in Flying Machines*, did not of course deal with it, as he had not then (1894) thought up the idea of his biplane glider, which actually started as a triplane.

The question therefore is, did Wilbur get that volume of the *Aeronautical Annual* from the Dayton public library soon after—say—June 3rd? It is very unlikely that the library would have had the three volumes of the *Annual*, as at that time they represented a highly specialised interest, and very few copies were printed.

It is far more likely that in reading about Lilienthal's death in 1896, and thereafter keeping their eyes open for matters of aeronautical interest, the Wrights saw an illustration of one of the Lilienthal biplanes, although it is also just possible that they might have seen a picture of the Chanute machine; but Lilienthal was news at that time, and Chanute wasn't. So we shall never know for certain what put the idea of a biplane into the Wrights' minds in time to suggest the biplane configuration for their warping kite of 1899.

When Wilbur first wrote to Chanute on May 13th 1900, he had already read the latter's works, and noted his biplane; and he refers to Chanute's machine as the type he proposed to adopt.

When the Wrights adopted Chanute's biplane construction, they did so within the terms of what they themselves had already decided was necessary; that is to say, they wished to produce a biplane wing-structure which could be helically twisted in the manner of the long carboard box which Wilbur found himself toying with in 1899; and which could also be built in such a manner as to preserve the relative positions of the two wings in all positions of straightness or twist. They found that if they braced the inner sections of their biplanes to form a rigid box-like structure, they could operate the twisting (later called 'warping') of the outer portions of the wings with the relative inclinations of upper and lower wings being unchanged. Even if the Wrights had not come to know Chanute, they would inevitably have arrived at a satisfactory system of bracing, since they were both, after all, talented engineers, and the business of bracing was no secret.

THE YEAR 1902

8

The Wrights' Influence starts with Ferber: 1902

Captain Ferdinand Ferber was, during 1902, in command of the 17th Alpine Battery at Nice; but for some years he had been enthusiastic about the possibilities of heavier-than-air flight, and had become—like Chanute in America—a devoted admirer and follower of Lilienthal. As said above, he was the sole European who now preserved the precious thread of continuity in aviation; who fully appreciated the importance of glider experiments before applying power; and who was determined to persist and succeed in his efforts to advance aviation. He was, as the French historian Charles Dollfus says,

> "the great French precursor of practical aviation: he was the first in our country to build, and test methodically, full-size gliders."

It should also be said that it was a great pity his standards of craftsmanship and construction were so poor, for both his success and his influence would have been far greater had his early machines not been made with such extreme crudeness. After examining his own statements, and the many surviving photographs of his machines, one even begins to doubt the quality of Ferber's flair, and his imaginative ability—his "aero-talent"—despite his undoubted enthusiasm and determination: he does not seem to us today to have grasped the full implications of the ideas and configurations he espoused. Nevertheless he was at this time the only significant pioneer in Europe.

In 1901 Ferber had built, and was testing, a Lilienthal-type hang-glider: photographs show it to have been a primitive affair (Fig 10), which is surprising in view of the wealth of fine photographs of Lilienthal gliders already published in France, and the excellent drawings in Lilienthal's patent of 1893, which was readily accessible. Ferber had been experimenting unsuccessfully with this apparatus when, in 1901, he happened to read an article which led to his entire change of direction in aviation, and which was to inaugurate the revival of European aviation and the birth of practical flying on the Continent. For as a result of this article, he finally got in touch with Chanute; and from information supplied by him, Ferber in 1902 changed to a Wright-type glider, despite Chanute informing him about both his own and the Wrights' work (Fig 11). The journal which thus had such a profound influence on Ferber was the *Revue Rose*, which was the generally-used popular name for the Paris scientific weekly called *Revue Scientifique*. This sequence of events was recorded by Ferber himself in his book *L'Aviation* (1909). Unfortunately, Ferber gave the date as early 1902, when he came across this article.

Before detailing the significant train of events which led Ferber to adopt a Wright-type glider—"un planeur biplan en bambou, type Wright" as Dollfus calls it—I quote here Ferber's own unequivocal tribute to the source of his inspiration and success, contained in a letter to Georges Besançon; this was published in the issue of *L'Aérophile* for June 1907. He is here speaking of Wilbur Wright, instead of the two brothers, as at that time he did not realise the part played by Orville in the partnership. Ferber writes:

> "Just think of it, that without this man (Wilbur Wright) I would be nothing (je ne serais rien); for I would not have dared, in 1902, to trust myself to a flimsy fabric if I had not known—from his accounts and from his photographs—that it would bear me. Think of it, that without him, my experiments would not have taken place (mes expériences n'auraient pas eu lieu), I would not have had Voisin as a pupil; the backers such as Archdeacon and Deutsch de la Merthe would not—in 1904—have offered the prizes you know about; the Press would not have sown the good seed everywhere; your journal (*L'Aérophile*) would not have quadrupled its circulation; and other specialised journals would not have been born."

It would be hard indeed to find more categorical admissions of debt from this "grand precurseur français".

Ferber was to enshrine in print another, and just as categorical, tribute when he came to publish his book *L'Aviation* in 1909; for he entitled one of his chapters:

"FERBER IN PURSUIT OF THE WRIGHTS, FROM 1902 TO 1906"

(Ferber à la poursuite des Wright, de 1902 à 1906)

In these two quotations, Ferber had granted the Wrights, in his own words, the title of "precipitator" of the re-birth of European aviation, of which he was the "grand precurseur".

But we now also know in detail how this Wright influence was transmitted. For when carrying out research into this phase of aviation history, I decided to try and identify the magazine article which had thus "sparked" Ferber, and, through him, Europe; for such an item would clearly be of great historical significance in this absorbing story. To my dismay, the four January (1902) issues of the *Revue Rose* revealed nothing remotely resembling the article in question. So, as he might well have been reading an old issue, I went back through those for the winter months of 1901, and still found nothing. Feeling sure that Ferber would not have "mis-remembered" the name of the periodical, but might well have forgotten when he actually read it, I continued with my search backwards through the formidable 1901 volume—formidable in both size and complexity—of the *Revue Rose*. In the issue dated June 1st 1901 (pages 689-692), I at last found the article. The piece was entitled "La Locomotion Aérienne", and was an unsigned resumé of a discourse—"History and Progress of Aerial Locomotion"—given to the Royal Institution in London on February 8th 1901, by Professor G. H. Bryan, FRS, a member of the Aeronautical Society. It dealt with various nineteenth century efforts and accomplishments in gliding, ending with a description of Chanute and Herring; it did not mention the Wrights, as they were as yet unknown to Bryan.

June seemed a long way back in 1901, until I came across the two excellent articles by Ferber in the *Revue d'Artillerie* for March 1904, and August 1905, respectively; the first was entitled "Les progrès de l'aviation par le vol plané depuis 1891"; the second, simply "Les progrès de l'aviation par le vol plané"; both were signed "F. Ferber, Capitaine d'artillerie". In the first of these, he gives a clear account of what happened, similar to that in his book, but saying it was in 1901 that he first heard the name Chanute, and that it took him two months to get in touch with him. By the time he came to write his book *L'Aviation* in 1908-09 (published 1909)—where I first came across the story—Ferber had clearly forgotten the details of this important event. In the *Revue d'Artillerie* (1904) he wrote:

> "The name of M. Chanute was revealed to us in 1901 in an unsigned article in the *Revue Rose;* M. Ch. Richet [the Editor] told us that the article was an extract from a lecture by Professor G. Bryan, of the University of Bangor. The latter very obligingly gave us the address of M. Chanute, who brought to our attention the Wright brothers."

To obtain more details of this sequence of events, I wrote to my friend, the late Miss Pearl Young, in the USA—the biographer of Chanute—asking if the date of the first letter from Ferber was known, and for the dates of subsequent letters. She generously delved into every detail of the Chanute-Ferber correspondence for me, with surprising and important results. Ferber said that it took him two months to get in touch with Chanute after first seeing his name in the *Revue Rose*. Miss Young found that Ferber first wrote to Chanute on November 10th 1901; so Ferber either did not see the number for some time, or else he put off writing to the editor for a while. Ferber's original letter to Chanute does not survive; but the date is known from Chanute's answer, which Miss Young kindly copied for me. It is dated November 24th 1901, and forms one of the most interesting "influence links" in aviation history. Chanute wrote in French (both his parents were French, and they had brought him to the USA in 1838, at the age of six-and-a-half):

> "I have your kind letter of November 10th. As I rarely write in French, I pray you to excuse me if I make mistakes in my reply. I am posting to you four fairly recent brochures which provide a popular account of my experiments. Quite recently, a Mr Wright has done still better than I ("a fait encore mieux que moi") but he has not published anything. I had a book published in 1894 [*Progress in Flying Machines*], of which I send you the title, etc., and I have written several articles in the *Aeronautical Annual* for 1896 and 1897, giving some particulars of my experiments, to which Mr Herring has also contributed. I regret that I have no more copies of these, but I enclose the address of the publisher."

Ferber wrote, thanking Chanute, on September 10th.

So here, in November 1901, was the first significant news of the Wrights to reach the Continent direct. It was actually not the first time the name of Wright had crossed the Atlantic: as already noted, Wilbur had previously had his article "Angle of Incidence" published in the *Aeronautical Journal* for July 1901; and another article, "Die Wagerechte Lage während des Gleitfluges (The horizontal Pilot-Position in Gliding Flight)" published in the *Illustrierte Aeronautische Mitteilungen*, also for July 1901. The first of these articles was a theoretical paper on aerodynamics; it did not mention any practical flying, and was unillustrated. The second did in fact speak of the Wrights' piloted glides in 1900, but only briefly, and in general terms, and was accompanied by one somewhat inappropriate illustration, the photograph of the 1900 glider being flown as an unmanned kite (Fig. 6). No particular notice was taken in Europe of either article; but the photograph in the German periodical has the small historical distinction of being the first illustration to be published anywhere in the world of a Wright aircraft.

Then, on January 8th 1902, Chanute sent to Ferber what was to be the decisive document in the latter's career, and the first true "sparker" of the European revival; it was a copy of the *Journal of the Western Society of Engineers* (Chicago) of December 1901, containing Wilbur

Wright's now famous first paper entitled "Some Aeronautical Experiments", which described and illustrated the work with the 1900 and 1901 Wright gliders; it had been read before the Society on September 18th 1901. In it Wilbur made specific reference to the warping of the wings for control in roll:

> "the lateral equilibrium and the steering to right or left was to be attained by a peculiar torsion of the main surfaces, which was equivalent to presenting one end of the wings at à greater angle than the other. .... Our system of twisting the surfaces to regulate the lateral balance was tried and found to be much more effective than shifting the operator's body."

The presence of this published account of the Wrights' wing-warp control is of particular significance when taken in connection with later events.

On January 22nd 1902, Ferber wrote to Chanute, and included a message for the Wrights. Chanute then wrote to Wilbur Wright on February 13th 1902, saying that Ferber "is in a state of *admiration* of your performances and wishes me to convey his felicitations".

In 1909, Ferber wrote in his *L'Aviation* about this crucial phase in his life, and in the history of aviation:

> "About the same time (February 1902), thanks to Mr O. Chanute, I received brochures and photographs on the work of the Wright brothers in 1900 and 1901; and they were so convincing and so remarkable (si probants et si remarquables) that it was not difficult to gauge that they would arrive easily at a complete flying-machine (qu'ils arriveraient facilement à la machine à voler complète). But at that moment, such was the state of discredit among the public of the idea of aviation, that the contents of these brochures only achieved understanding—following my campaign of popularisation —three years later! In agreement with Mr Chanute over the same idea as myself—that is to say that the Lilienthal method should lead to the flying machine—they [the Wrights] had adopted a practical method of construction, and invented suitable controls, which constituted a great progress. To bring home the invention to France, it was therefore necessary to pursue them on the same lines (sur la même piste), and overtake them. I constructed my aeroplane No 5 on their data, and my flights (mes parcours) improved considerably."

In his London lecture of 1908 (see Section 85) Ferber was to say:

> "I believe, however, that I did not do wrong in trying first to realise that which already worked well elsewhere, for one must always take what exists in order to improve it afterwards."

Ferber thus abandoned his Lilienthal-type machine, and adopted the Wright configuration in his new (1902) glider (Fig 11), based upon the photographs he had received of the No 2 Wright glider of 1901, still with no tail surfaces. It is important to note that *Ferber at no time was to copy the Chanute-type hang-glider*. The Ferber Wright-type glider was a somewhat pathetic contraption, and primitive to a degree, with no rigid surfaces, the latter almost resembling a washing line in their simple 'billowing' attachment to the spars: this is clearly seen in the photographs.[1] It had the familiar Wright forward elevator, no tail-surfaces, and the prone pilot position. But there was no attempt at wing-warping, or any other form of control in roll.

So we have here the first direct impact of the Wright gliders, which was to change the whole face of aviation, and bring about the birth of practical European flying.

As the subject of control in roll is so important in the present context, the next section will be devoted to that, and to other related subjects.

In the year 1908, François Peyrey—the most respected French aviation journalist—was to write in *L'Auto* (August 9th 1908):

". . . the keen and long labour of the two brothers, of which the gliding at Kitty Hawk had sounded the reveille for French aviation."

Interestingly enough, he was to apply the same phrase to the Wrights' powered flying in 1908.

---

[1] It is interesting to find Charles Dollfus (in *L'Homme, l'Air et l'Espace*, 1965) writing as follows: "In 1902, advised by Chanute, Ferber adopted the general form of the Wright gliders . . . but it must be admitted that his machines, ingeniously constructed of bamboo, remained very rudimentary, with wings without ribs, ineffective rudders, and doubtful stability. It is curious to record this state of affairs when one knows the high scientific culture of Captain Ferber."

❧ 9 ❧

Ferber and Control in Roll: 1902

The lack of wing-warping, or other control in roll, on Ferber's first Wright-type glider needs special consideration, as it focuses attention right at the start on one of the most important problems of aviation history: it points to one reason for the extraordinary procrastination which slowed down the whole development of aviation in Europe, *ie* the consistent neglect of control in roll.

As the Wrights saw clearly from the first—gauging it from the birds and from general principles—no aeroplane can fly properly without proper three-axis control; they also saw equally clearly that proper flight-control must first be achieved in gliders before any attempts at powered flight were made: for they well knew that if a man found himself precipitated into the air in a powered machine without knowing how to control it in flight, he would court disaster. Blériot might very well have been killed for precisely this reason when he was testing his forward-looking No VII in 1907.

Considering that Wilbur Wright had drawn special attention to control in roll in his first Chicago paper, and specifically described his wing-warping method of providing it, Ferber's neglect of it seems astonishing. If, by some chance, he had not understood the English words and phrases Wilbur used, he was running a serious risk of injury by not having every word correctly translated before building and testing a glider of completely new configuration: and, being at Nice, he had literally hundreds of bilingual neighbours. As a matter of fact, the relevant passages in Wilbur's paper were in good straight English.

However, we immediately find ourselves in a dilemma here, because Ferber himself gave an explanation of his early neglect of lateral control; but he gave it in 1908, and I feel it must be regarded as suspect, for reasons which will follow. In an article in *L'Aérophile* for July 1st 1908 entitled "Que valent les Brevets Wright?", Ferber wrote:

> "As to warping, I did not wish to employ it in 1902, as I judged it useless to begin with; so my successors, having set off along my track (partis sur ma piste), did not use it either..."

I feel this explanation is not only suspect, but a cover-up either for having not understood the English of Wilbur's paper, or for having inexplicably misunderstood the true nature and purpose of wing-warping. For in his first article for the *Revue d'Artillerie* in 1904 (see Section 35) he included this strange description of the Wright gliders:

> "In 1900 and 1901, they had no vertical rudder; they changed direction by straightening up the elevator (ils changeaient de direction en dégauchissant le gouvernail horizontal).[1] In order to make more important movements (pour pouvoir fair des mouvements plus importants), they added at the back a vertical directional rudder, and commenced to describe quarter-circles."

It is hard to believe this was written by the man who, only a few years before, had received Wilbur's first Chicago paper in which appeared these words:

> "The lateral equilibrium and the steering to right or left was to be attained by a peculiar torsion of the main surfaces, which was equivalent to presenting one end of the wings at a greater angle than the other.... Our system of twisting the surfaces to regulate the lateral balance was tried and found to be much more effective than shifting the operator's body."

With these quotations in mind, the reader is invited to go back to the previous Section, and to the last quotation there, in which Ferber said of the Wrights:

> "they were so convincing and so remarkable ... they had adopted a practical method of construction, and invented suitable controls ..."

I find it hard to believe that Ferber had understood the basic principle of wing-warping, if he could omit to incorporate any roll-control in his first glider; and if he could make no mention of such control in his description of his first Wright-type glider in his *Aérophile* article in February 1903 (see Section 12); and if he could also write his 1904 remarks about the Wrights' method of steering in 1900 and 1901. Nor did Ferber mention roll-control in any other of his writings until 1908.

---

[1] It is hard to understand what Ferber meant by the word "dégauchissant"; "dégauchir" means "to straighten" or possibly "to untwist". But, whatever he meant, it applied clearly enough to the front elevator, and leaves the statement as curious as ever.

We are next faced with the fact that the Wrights' wing-warping was to be mentioned at least twice—and prominently—in the reports of Chanute's vital lecture in Paris on April 2nd 1903 (see Section 13). One of these accounts even mentioned the added fact of the simultaneous use of warping and rudder; it appeared in the April 1903 issue of *L'Aérophile*, which Ferber—as an enthusiastic member of the Aéro-Club—was bound to have read.

Then the position becomes even more confused by his book *L'Aviation*, published in 1909, much of which was written earlier. In this work he repeats his earlier statements about the Wright gliders, but varies the statements about the 1900 and 1901 Wright gliders by saying—with no further comment—that both the elevator and the wings were warped to steer the machine; and he uses the by-now standard verb for Wright-type wing-warping, "gauchir". But to confuse matters even more, Ferber writes of his own conclusions at the end of his own 1904 'season' of gliding, and says that he has used a vertical rudder to control the machine in roll.

Then we find him building, in 1908, his own No IX, a powered biplane (also called the *Antoinette III*, as he had joined the firm), and fitting Wright-type warping to it: but the machine was unsuccessful, even in straight runs, and was never turned or banked.

Finally, in 1909, Ferber abandoned designing aircraft altogether, and purchased the only aeroplane on the market without any roll-control at all, the standard Voisin: it was on this machine, when taxying at speed, that he was killed in 1909 at Boulogne, when the machine ran into a ditch.

This problem of Ferber and control in roll is of great importance in the history of flying, as he was the spearhead of the birth of practical aviation in Europe, and his influence was profound. On the balance of evidence, it is impossible to avoid the conclusion that Ferber simply did not possess enough of what we might call 'aero-talent', and that he had only a partial grasp of the basic realities of aviation. Ferber had one or two good ideas, and he was a man of enthusiasm, resolution, and courage; and—philosophically speaking—of foresight: but this was not enough. Apart from his confusion over lateral control, there were such things as his curious neglect (until 1908) of placing the rudder on the tail-plane which he so wisely added to his Wright-type glider.

The addition of a fixed tailplane in 1904 was, of course, his most productive move, and one which was to exert a profound influence on the development of aviation.

But the significant facts remain that—from the start—Ferber neglected roll-control; that his conception of construction was to remain very crude in all his early work; and that he was never in his life to build either a successful glider or a successful powered aeroplane.

❧ 10 ❧

The Wrights' No 3 Glider of 1902

The following account of the Wright brothers' vitally important No 3 Glider was given by Orville Wright himself in a deposition on January 13th 1920, when he was a witness in a lawsuit brought by the heirs of the pioneer J. J. Montgomery against the US Government, alleging that the latter had a patent claim which had been infringed by aircraft bought by the US Government. This account—it is only a short excerpt of the whole deposition—was first published in *How We Invented the Airplane*, by Orville Wright, edited by Fred C. Kelly (New York, 1953):

"Our experiments of 1901 were rather discouraging to us because we felt that they had demonstrated that some of the most firmly established laws, those regarding the travel of the center of pressure and pressures on airplane surfaces, were mostly, if not entirely, incorrect. At first we had taken up the problem merely as a matter of sport, but now it was apparent that if we were to make much progress it would be necessary to get better tables from which to make our calculations. In September we set up a small wind tunnel in which we made a number of measurements similar to those which we had attempted to make earlier in the year. The earlier measurements had been made in the open air, where it was difficult to determine the exact direction of the wind. The new measurements were made inside of the tunnel, through which a blast of air was forced. The new experiments were conducted with much more care than had been the first, but still they were not entirely satisfactory. We immediately set about designing and constructing another apparatus from which we hoped to secure much more accurate measurements. In this instrument the lift of the surface to be measured was balanced against a pressure created on a screen by the flow of the air through the tunnel. This enabled us to make very accurate comparative measurements of the lift.

We also designed and constructed another instrument for measuring the ratio of the lift to the drift. This utilized an idea which had

6 Wright No 1 Glider: 1900 The Wright No 1 Glider of 1900, here being flown as an unmanned kite at Kitty Hawk, North Carolina. It was tested in October and November 1900 as a manned and as an unmanned kite, and — for one day — at the Kill Devil Hills, as a free-flying piloted glider. There was a forward elevator and wing-warping, but at this time these controls could not be operated simultaneously by the pilot.

7 Wright No 2 Glider: 1901 The Wright No 2 Glider of 1901, seen here after landing, at the Kill Devil Hills, some four miles south of Kitty Hawk. This was a larger and improved version of the No 1 Glider, on which the brothers made many glides, but which they found unsatisfactory in various ways.

8 **Wright No 3 Glider: 1902** The Wright No 3 Glider of 1902 seen in flight, with Wilbur piloting, at the Kill Devil Hills. This much more sophisticated machine was now fitted

9a Wright No 3 Glider (modified): 1902 The Wright No 3 Glider (modified) of 1902, flying at the Kill Devil Hills, with Wilbur piloting. The two fixed fins have now been replaced by a single movable rudder, which counteracted the warp drag, which the fixed fins could not. This photograph well shows the prone pilot position, in which the man's hips fitted into a cradle to operate the warping and rudder cables. The front elevator was worked by the left hand.

9b Wright No 3 Glider (modified) : 1902 This photograph shows the method of launching, in which two men — one at each tip — ran forward into the wind, and then released the machine. Orville is here seen at the controls. Owing to the pilot having to hold on to the machine with one hand, and to operate the elevator with the other, both the rudder and the warping cables had to be fixed to the hip-cradle, and worked together; the rudder was automatically put over to the side of the lowered (up-warped) wings.

9c Wright No 3 Glider (modified) : 1902 Taken from the rear when gliding from the Kill Devil Hills. Hundreds of glides were made with this machine, and it was the first aircraft in history to embody practical three-axis control, *ie* control in pitch, control in yaw, and control in roll. It was basically this machine which formed the subject of the Wrights' patent application of 1903, which was granted in 1906.

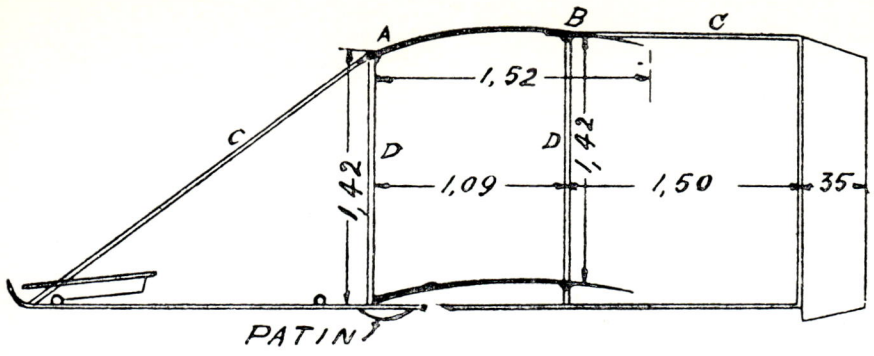

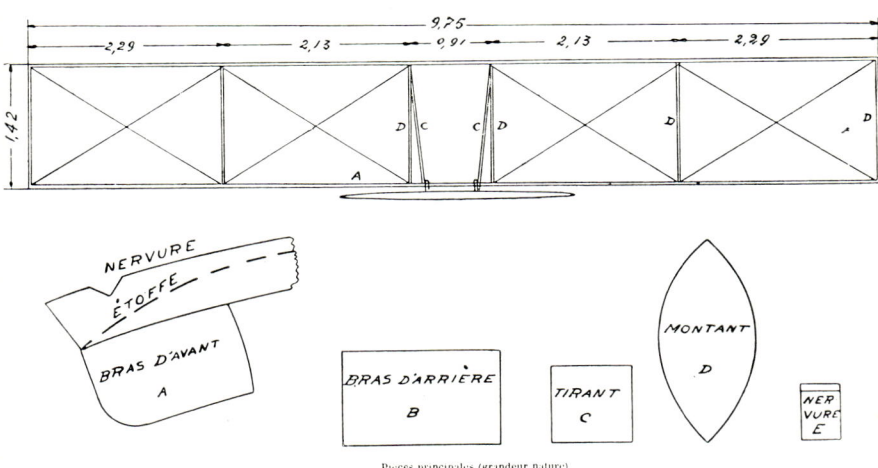

Pièces principales (grandeur nature)

9d Wright No 3 Glider (modified): 1902 General arrangement drawings as shown in Chanute's article in the August 1903 issue of *L'Aérophile*. By this time, the Wrights did not want the secret of their combined rudder and warp controls published; and it will be seen that there is no inclusion of the control cable runs. But Chanute did in fact mention in his lecture, on which this article was based, that the Wrights used their rudder and warping simultaneously; but the Europeans failed to follow up the hint, and therefore did not appreciate the vital importance of what they were being shown. Even without the control runs being shown in these drawings, the reasons for Chanute's statement about the simultaneous operation should have been followed up, and the *raison d'être* sought for. But this did not happen, and the European inventors thus missed the all-important features of roll and yaw control combined.

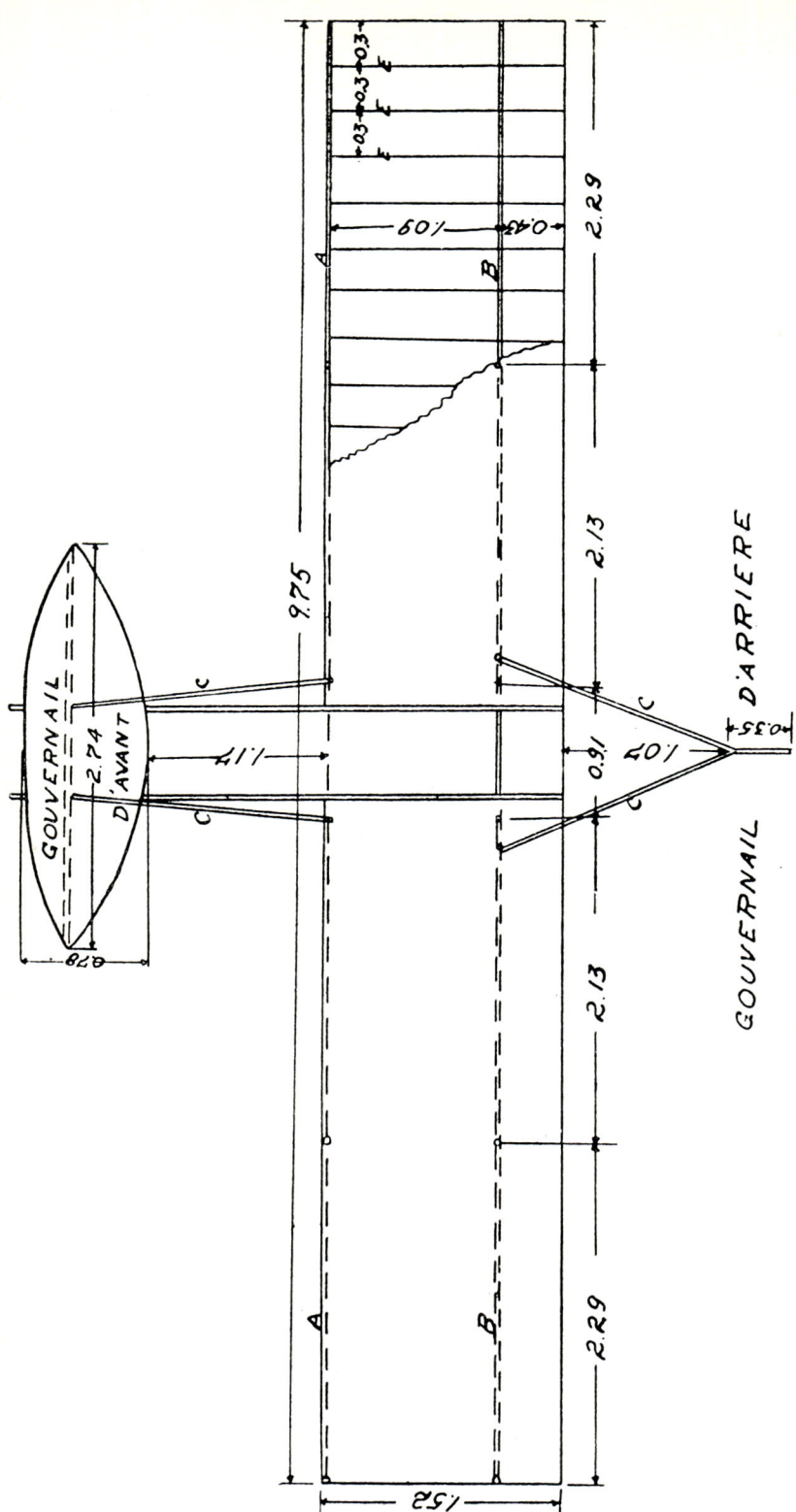

9e Wright No 3 Glider (modified): 1902 The plan view of the Wright No 3 Glider (modified) of 1902, which completed the set of general arrangement drawings published by Chanute in his *Aérophile* article in 1903. It will be noticed that there is no hint either of the control runs, or of the cradle in which the pilot's hips were fitted. It was the news, and photographs, of this glider that made such a profound impact on the Europeans, and helped to re-direct their efforts to achieve aeroplane flight.

10 Ferber Lilienthal-type Hang-glider: 1901 The Lilienthal-type hang-glider built by the French Captain Ferdinand Ferber in 1901. Like all Ferber's machines, the workmanship was deplorable, and what was intended here as a copy of a Lilienthal glider became a mere travesty. The poverty of the construction and workmanship of Ferber's gliders, throughout his career, is a subject for much astonishment and discussion, as he was a highly intelligent and devoted pioneer.

11a Ferber Wright-type Glider: 1902 In 1902, Ferber, after learning through Chanute of the Wright gliders, changed his whole direction of endeavour, and took to building Wright-type gliders. At this time he only knew of the 1900 and 1901 Wright gliders — both without tail-units — but even so, the Ferber glider that resulted was almost pathetic in the crudity of its construction, as shown here.

11b Ferber Wright-type Glider: 1902 Another photograph showing the primitive construction, more reminiscent of clothes on a washing line than of a stiff-winged glider. But it was Ferber's change to Wright-type gliders, and his wide advocacy of them, that changed the whole course of European aviation history. Ferber became the first and most powerful figure to propagate the Wrights' influence.

been suggested by Dr Spratt. During the following three or four months after October 1901 we made thousands of measurements of the lift, and the ratio of the lift to the drift with these two instruments. We measured the lift of square planes and rectangular planes of different aspect ratios, in order to determine the effect of the aspect ratio on the lifting qualities of the plane. We also made measurements of a number of similarly curved surfaces having different cambers to determine the effect of camber on the lift and also on the drift.

We also measured these curved surfaces to determine the effect of aspect ratio on their lifts and drifts. We measured a number of surfaces superposed with gaps ranging from one-fourth of the chord to one and one-eighth times the chord. We measured a series of surfaces having a regular camber like that of a sector of cylinder, [and] having different depths of curvature, as well as a great number of other surfaces having the greatest depth of curvature forward of the center, and some with the greatest depth back of the center. . . .

We decided to build another machine basing it upon calculations to be made from our own tables. We decided to attach a fixed vertical vane in the rear of the main plane, which we thought would maintain an equal velocity of the right and left wings when the wings were warped to different angles. Our tables made it apparent that we would secure a higher dynamic efficiency in the machine by using surfaces of smaller camber and of greater aspect ratio.

We went to Kitty Hawk in the last week of August 1902, and began the assembling of a machine embodying the changes which I have just mentioned. The wings of this machine measured 32 feet from tip to tip, and 5 feet from front to rear. The curvature of the wings measured from one twenty-fourth to one twenty-sixth of the chord of the surfaces. The front rudder, or elevator, contained 15 square feet of surface, and the rear vertical vanes had a total surface of $11\frac{2}{3}$ square feet. The area of the main planes totaled 305 square feet. These planes were spaced one 5 feet above the other. The wiring or trussing of the wings and uprights, as well as the arrangement of the cables for imparting the warp to the surfaces, was like that of the 1901 machine. . . .

While this machine was being assembled, we made measurements of one of the surfaces flown as a kite, and found that the pull on the kite strings, in proportion to the load carried, was less than that of the surfaces of the 1901 machine. This was in accordance with the estimates which we had made from calculations based on our own tables. These measurements were taken when the kite strings stood in a horizontal position, so that only the drift of the surface was measured.

The assembling of the machine was completed about the 19th of September, when we began making glides with it. . . . In the first flight we found that the machine was able to glide with a much smaller angle of descent than either of our former machines. The first glides made with it, but which were not entirely free, led us to think that the lateral control had been improved by the addition of the fixed vertical vanes in the rear. In these first tests the wing with the larger angle would rise, while the opposite wing was depressed.

This machine was assembled with the spars straight from tip to tip, but as these first tests showed the same trouble that we had had with the 1901 machine when the wings were straight, on the 22nd of September we altered the truss wire so as to arch the surfaces from tip to tip, making the tips at least four inches lower than the center. We also made the angle of the surfaces at the tips greater than the angle at the center of the machine. We found that the trouble experienced before with a cross wind turning up the wing it first struck had been overcome, and the trials seemed to indicate that with an arch to the surfaces laterally the opposite effect was obtained.

Later, when we began to make free flights with the machine, we found that when the wings were warped, first with the larger angle on one side and then on the other, the machine descended the hill rolling from side to side. But later in some of the flights, when the machine was allowed to slide a little to one side or the other as the result of one wing being at an almost imperceptibly lower height than the other, we found that the fixed vertical vane, instead of maintaining an equal speed at the two opposite wing tips, as we had expected, as a matter of fact did just the reverse, and caused one wing to be checked and the other one to be speeded up. This was due to the fact that when the machine began sliding laterally a pressure was created on the fixed vanes on that side which was toward the lower side of the machine and the side toward which the machine was sliding. The increased speed of the high wing gave it a still greater lift, and the decreased speed of the lower wing produced a lesser lift upon it, with the result that the lower wing dropped and the higher wing went still higher. When the wings were warped in an attempt to recover balance, with the low wing having a greater angle of incidence than the upper wing, a still greater drag was produced upon the low wing with a result that its speed was further decreased and the speed of the higher wing was increased. These flights ended usually with disaster to the machine in what is called today a 'tail spin.' ...

Our first change in the machine, as the result of our experiences in these flights just mentioned, was to remove one of the vertical vanes in the rear of the machine. By doing this we hoped to remove at least a part of the disturbing effect of the vanes when the machine was sliding slightly sidewise. We found that this only slightly mitigated the evil influence of vanes. After a good deal of thought the idea occurred to us that by making the vane in the rear adjustable, so that it could be turned, ... to entirely relieve the pressure on that side toward the low side of the machine, and to create a pressure on the side toward the high wing equal to or greater than the differences in the resistances of the high and low wings, due to their different angles of incidence, all of the good properties of a vane in the rear would be secured without any of its bad properties. But this was going to add one more burden to the operator. He would now not only have to think, and think quickly, in operating the front elevator for maintaining the longitudinal equilibrium, but he would also have to think so as to operate this rudder . . . to present its surface to the wind on that side which is toward the high wing, or the wing having the smaller angle of incidence.

While this change to make the vane adjustable was being made, the idea came to us of connecting the wires which operated the rudder to the cables which operated the wing warping, so that whenever the wings were warped the rudder was simultaneously adjusted, . . . to produce a pressure on that side of the rudder which was toward the wing having the smaller angle of incidence.

With the machine as now constituted we began a long series of gliding flights. The disastrous experiences which we had when the fixed vanes were used now seemed to be entirely avoidable. In fact, in the seven or eight hundred gliding flights that were made after the adjustable rudder was installed, not once did we encounter the difficulty we had experienced with the fixed vane. . . .

The fore-and-aft control of the 1902 machine had proved very effective, so that when at last we felt that the problem of lateral equilibrium had been entirely solved, we began to turn our thoughts to the construction of a machine to be driven with a motor. . . . To provide for the installation of motor, propellers, etc., we thought a more rigid structure at the center of the machine would be useful. As a result we decided to rigidly truss the upper and lower planes of this 1902 machine in all excepting the outermost panel at each end. This was accomplished by putting in fore-and-aft stay wires, running diagonally from the upper end of the forward upright posts to the lower end of the corresponding rear posts, and from the upper end of the rear posts to the lower end of the corresponding front upright posts. In this manner we provided a rigid structure in the three center panels. Before this modification was made, in warping the wings, the upper and lower surfaces were drawn into diagonal positions with reference to each other. With the new form of trussing, the front edges of the upper and lower planes were maintained parallel to each other, so that only the rear edges of the outer panels at either end of the machine could be adjusted up and down for the purpose of securing different angles of incidence at the two opposite tips.

All of the later flights made with this machine in 1902, as well as the early flights made with it in 1903, were made with the wings trussed in the manner just described.

The flights of 1902 demonstrated the efficiency of our system of control for both longitudinal and lateral stability. They also demonstrated that our tables of air pressure which we made in our wind tunnel would enable us to calculate in advance the performance of a machine. Before leaving our camp at Kitty Hawk we began the designing of a new and larger machine to be driven by motor."

As a tail-piece to this section, here is part of a letter written by Orville Wright to his sister Katharine, on October 23rd 1902:

"The past five days have been the most satisfactory for gliding that we have had. In two days we made over 250 glides, or more than we had made all together up to the time Lorin left. We have gained considerable proficiency in the handling of the machine now, so that we are able to take it out in any kind of weather. Day before yesterday we had a wind of 16 metres per second, or about 30 miles per hour, and glided in it without any trouble. That was the highest wind a gliding machine was ever in, so that we now hold all the records! The largest machine we handled in any kind of weather, made the longest distance glide (American), the longest time in the air, the smallest angle of descent, and the highest wind!"

II

The Non-Use of the Hargrave Box-Kite: 1902

In the July 1902 issue of *L'Aérophile* there appeared a review of J. Lecornu's now well-known book on kites, entitled *Les Cerfs-Volants*, which had just been published. Then, in the next issue (August) of *L'Aérophile*, the Editor published an illustration of one of Hargrave's designs for a box-kite float-glider (Fig 12), along with a brief paragraph describing it.

But neither Lecornu's book, which must have been fairly widely-read, and which was in the Aéro-Club's library, nor the references in *L'Aérophile*, had any effect at that time on European aviation as far as the box-kite was concerned. It is hard to say why this was, as one would imagine that someone among the European pioneers would have tried out the box-kite idea, full-size.

As a matter of fact, the box-kite, as a configuration, had been known among the more knowledgeable of the aeronautical fraternity since 1893, when Hargrave invented it, and published it.[1] But little notice had been taken of this kind of kite, except in England, where in 1899 Percy Pilcher was very struck by its quality of stability, and built a triplane glider which probably[2] incorporated the box-kite idea.

However, the general reader of aeronautical books may be forgiven for believing that Hargrave's box-kite had already made a considerable impression in Europe, owing to the misleading—and indeed mischievous—inclusions in Gabriel Voisin's autobiography, *Men, Women and 10,000 Kites*, both the 'persuasive' text and the illustrations of a Chanute-type glider with a box-kite tail (Fig 43) whose caption says it is seen being tested in 1899; but it was, in fact, not built by Gabriel Voisin until 1907.

[1] The Hargrave box-kite was first published to an international audience when drawings and descriptions of it were printed in *Proceedings of the International Conference on Aerial Navigation, held in Chicago 1893*, New York, 1894.

[2] It is not known what this machine looked like, as it was destroyed after Pilcher's death in 1899. See postscript on next page.

This is not the first time that we find Gabriel and the box-kite mentioned together. In 1953 he wrote the following in *Interavia* (Vol VIII, No 12, 1953):

"It was not until 1898 that we first learned about the Hargrave Kite from a magazine. Our first Hargrave, made of two equal-sized cubes, one behind the other, astounded us. . . ."

Gabriel goes on to say that they built bigger and better Hargrave kites; then tried gliding experiments—the type of machine was not stated—but these "produced no results in the gliding direction". What is so significant is that he was piloting Archdeacon's Wright-type glider in April 1904; yet we hear nothing of any Hargrave-influenced machine of Gabriel's design until the Voisin-Archdeacon and Voisin-Blériot float-gliders which were completed and tested in June and July 1905.

It is difficult to avoid the conclusion that, as with the case of the misdating in his autobiography, Gabriel did not, in fact, give any serious consideration to the box-kite configuration until some years after he claims.

Postscript. For an excellent discussion embodying new research, see JARRETT(P.). "Pilcher and the multiplane—a neglected aspect of a pioneer's work," in *The Aeronautical Journal*, vol. 76, May 1972.

THE YEAR 1903

12

Ferber's Article in *L'Aérophile* : February 1903

In the issue of *L'Aérophile* for February 1903, there appeared an illustrated article on Ferber, entitled "Expériences d'Aviation", ostensibly by a Monsieur de Rue. But 'de Rue' was actually Ferber himself, and this name was later to become well-known to the aviation world as both his 'nom de plume', and 'nom d'aviation', since he was a serving officer in the army and had to remain nominally anonymous.[1] This article proved a landmark in aviation history, and on two counts:

(1) it was the first clarion-call on behalf of aviation, directed at the aeronautical world, a world now almost exclusively devoted to balloons and airships;
(2) it was the first time that illustrations of the Wright glider configuration were placed clearly—however crudely—before the would-be European pioneers.

Ferber's article covered $4\frac{1}{2}$ pages, and included 6 illustrations; four of these illustrations were of his Wright-type machines. He opened almost aggressively:

"Captain Ferber, who commands the 17th Alpine Battery at Nice, represents—in the Aéro-Club—the group of militant aviators (le parti des aviateurs militants). What is original about him is that he does not claim to be an inventor—he is an aviator: that is to say, he mounts flying-machines as those who are chauffeurs mount automobiles. That does not denigrate him at all; for, according to Lilienthal's formula, 'to invent an aeroplane is nothing; to build one is something; but to fly one is everything.' Lilienthal, moreover, is the Captain's master, and he believes that from 1891—the date when this brave pioneer made his first experiment—men have known how to fly. If we do not yet witness everyone flying off their own bat,

---

[1] Rue was one of his family's estates in Canton Fribourg (Switzerland).

it is because they have not known how to see, how to listen to, or how to imitate, the unfortunate Lilienthal. To those who say of him that his experiments resembled parachuting, the Captain would point out that up to the present, parachutes have descended vertically and not come to earth diagonally at the will of the aeronaut. If one calls the Lilienthal machine a parachute, we would have the right to call the pigeon a parachute when it descends from the roof to the street. But just as one still says of a pigeon that it flies, even when descending, so must one also say of the Lilienthal machine, that it also flies."

This was indeed revolutionary talk; but it seems that the more stolid members of the Aéro-Club were not unduly moved by it, despite the description of Ferber (by himself) as representing the group of "militant aviators" in the Aéro-Club. It is quite clear that the members of the Club did not recognise any such group, since aviation was not to be taken seriously until April of this year 1903, when Chanute gave his epoch-making lecture, which is discussed later. Ferber went on to say that he practised what he preached, and wished to recruit pupils to increase the number of Lilienthal's disciples. Then, in a footnote to the words "Lilienthal's disciples", he makes this interesting statement:

"There are actually five of Lilienthal's disciples in the world: Messrs Chanute and Herring in Chicago, Messrs Orville and Wilbur Wright in Dayton, USA, and Captain Ferber in Nice. The Englishman Pilcher died in 1899. One sees that Europe is insufficiently represented (insuffisament répresentée)."

After this masterly understatement on the insufficient representation in Europe, he then states that, since 1898, he (Ferber) had tested five machines, four being of the Lilienthal type, and the fifth of what he called the "Chanute-Wright type; much more stable and, for that reason, probably the type of the future (probablement le type de l'avenir)". As already noted, it was a purely Wright-type machine, albeit a crude one (Fig 11). Ferber also mentioned that, in 1902, the Wrights had achieved a glide of 200 metres—it was actually $622\frac{1}{2}$ feet—information about which he had recently had from Chanute. As noted before, Ferber did not incorporate any wing-warping, or other form of control in roll in his Wright-type glider; nor did he say a word about lateral control of any kind in this article.

This article of Ferber's was not only the first clarion-call made direct to the would-be pioneers, urging them to revive aviation in Europe; but it established Ferber as a leader in the eyes of those who were soon to become his fellow-workers. In April 1903, Ernest Archdeacon was to say that Ferber was "well known in our particular sphere (bien connu dans le monde spécial) as the only man in France who is making gliding experiments in the manner of Chanute."

Later, *L'Aérophile* (March 1904) spoke of

"Captain Ferber, who was for a long time the only French disciple of Lilienthal, and who made the only gliding experiments which we could set against the American aviators."

13

Chanute's Lecture in Paris: April 2nd 1903

Important as Ferber's article was, it soon became eclipsed by the strong and wide impact on Europe of a lecture given by Octave Chanute to the Aéro-Club de France in Paris on April 2nd 1903,[1] and by the well illustrated articles—which appeared during the following months—reporting the lecture, or enlarging upon it, written both by Chanute himself, and others.

There can be no doubt that Chanute's 1903 Paris lecture and these articles, collectively formed the second—and major—channel through which flowed to Europe the Wright brothers' influence and inspiration, which powerfully reinforced Ferber's initial impact on Europe, and led directly to the birth of practical powered European aviation, or, shall we say, the rebirth of European aviation.

The results of the lecture and the 'follow-up' articles may be tabulated as follows:

(1) the Wrights' successes in gliding were vividly conveyed—by illustrations and descriptions—throughout Europe; and, in 1903, the Wrights' 'message' penetrated all the minds who were now to pioneer aviation on the Continent;

(2) Ferber's previously published choice of the Wright glider configuration, as the spear-head of the aviation revival, was powerfully reaffirmed;

(3) the Wright glider configuration—their modified No 3 of 1902—became, in the minds of the European pioneers, the leading type of heavier-than-air machine; and served as the pattern from which future development would grow;

(4) Ernest Archdeacon was established as the undisputed leader of the European revival of aviation, owing (a) to his report of the Chanute

---

[1] Chanute says that he gave several lectures, but the venue of the others has not been discovered.

lecture; (b) his dramatic exhortations to the French to "get going"; and (c) his forming the Aviation Committee of the Aéro-Club de France.

On January 3rd 1903, Chanute had sailed from Boston with his daughters for a four-month tour. This took them first to Egypt; then to Naples, Rome, Florence, Nice, Milan, Venice, Vienna, Berlin, Paris, and London. Wherever he went, Chanute made a point of contacting all those he knew who were actively interested in aviation, or potentially so. One of the main objects of his visit to France was to promote Continental interest in the aviation section of the St Louis Exposition, which was to be held in 1904.

But it was Chanute's short stay in Paris that was to prove of such vital importance in the development of practical flying.

Chanute said that he "had to give several talks"; but the momentous occasion was this illustrated lecture which he gave in the evening of Thursday April 2nd 1903, at a 'dîner-conférence' of the Aéro-Club de France at the Hotel de l'Automobile-Club, No 6 Place de la Concorde, Paris. Although the lecture is not available verbatim, we know fairly accurately what was said, thanks to the two principal illustrated accounts of it, one in *La Locomotion* (issue of April 11th 1903) written by Ernest Archdeacon,[1] and the other in *L'Aérophile* for April 1903, by an anonymous writer. As Archdeacon said:

> "there has been a great flutter in the 'aerial world' during the last few days, caused by the arrival in Paris—and his reception at the Aéro-Club—of M. Chanute, the American inventor who, if he is not extremely well-known among ordinary mortals, is a veritable celebrity among the 'aviators'. For several years, M. Chanute has worked with indefatigable ardour at this most difficult question of aerial navigation by heavier-than-air means."

Chanute spoke to a distinguished audience at the Aéro-Club, including such rich and talented patrons of aeronautics as Archdeacon and Deutsch de la Meurthe, and the professionals like Renard, Mallet, Lahm (an expatriate American), Tatin, Peyrey, Surcouf, Besançon, Borde, Tissandier, and La Vaulx. La Vaulx took the chair on this occasion.

Chanute—speaking in French—started by dealing with Lilienthal, and paying him tribute, before passing on to describe his own work with gliders, especially his biplane hang-glider of 1896. There is then little doubt that he sowed—however innocently—the seeds, not only of future dissension between himself and the Wrights, but of endless confusion about his and their aircraft, and about his supposed influence on them. The *Aérophile* writer said that the Wrights became Chanute's

[1] This article is reproduced complete in translation as Appendix B. See also Section 15.

"devoted collaborators"; Archdeacon, in *La Locomotion*, said that Chanute—after admitting he was no longer young—"took pains to train young, intelligent and daring pupils (élèves) capable of carrying on his researches by multiplying his glider experiments indefinitely. Principal among them is certainly Mr Wilbur Wright of Dayton . . ." *L'Aérophile* went even better: its anonymous writer reported:

> "In 1900, the Wright brothers, bicycle-makers of Dayton (Ohio), wrote to M. Chanute to request details of his experiments. They wished to repeat them with a purely sporting aim in view. M. Chanute willingly provided them with the information they required, and the Wright brothers then constructed, on this data, machines similar to those of M. Chanute (appareils semblables à ceux de M. Chanute) with which they soon experimented with real success. In 1901, new experiments by the Wrights, at which M. Chanute assisted, gave better results, . . . Finally, in 1902, the Wright brothers succeeded even better . . ."

If Chanute gave this 'information' in his lecture—and it seems certain that he did—it is very hard to acquit him of making mischief, even if one does make allowances for a "grand old man" who was being lionized abroad, and feeling paternal into the bargain. The truth, for everyone to see, is in a well-known letter which Wilbur Wright wrote to Chanute on May 13th 1900; in it Wilbur gave a long and explicit account of his own ideas about aviation; of the successful experiments with his warping kite of 1899; and of his gliding intentions, especially in regard to launching his new machine. At the very end of the letter came Wilbur's only two requests; (a) asking Chanute if he knew of "a suitable locality where I could depend on winds of about fifteen miles per hour"; and (b) saying that he would be grateful if Chanute could tell him where he could obtain "an account of Pilcher's experiments". Wilbur had no interest at all in the operation of Chanute's machines, since he and Orville had already invented their warping system, and were now planning their first full-size machine to test-fly it. At this moment, it is clear that the brothers were primarily interested in two points—weather conditions and launching techniques; hence their interest in Pilcher, who had had horses tow him off the ground. The Wrights had written to Chanute as young experimenters, telling an older and admired pioneer about their intentions.

When it comes to mention of 1901, and "new experiments by the Wrights, at which M. Chanute assisted", the old gentleman seems to have had even more to answer for. For Chanute only spent the inside of a week in the Wrights' camp at the Kill Devil Hills in August 1901, and only as a fascinated onlooker after a friendly invitation from Wilbur, dated May 12:

"It is scarcely necessary to say that it would give us the greatest pleasure to have you visit us while in camp, if you should find it possible to do so."

Chanute paid a second visit to the Wrights' camp at the Kill Devil Hills from October 5th to the 14th 1902, to watch two of his own gliders being tested by Herring in friendly 'competition' with the Wrights: both the Chanute machines failed miserably. Chanute admiringly watched and photographed the first glides of the Wrights' final and triumphant glider, the modified No 3, with its warp-and-rudder-linkage.

In his lecture, Chanute went on to give a genuine and glowing account of the Wright gliders and their performances: these machines—culminating in the modified No 3 of 1902, with its rear rudder—provided the main focus of interest in the whole lecture. That this statement is no exaggeration is shown by Archdeacon, who, after only a few introductory paragraphs, said:

"... their [ie the Wrights'] last machine appears actually to be the model of its kind (le modèle du genre). I will therefore describe only this machine, for fear of unduly prolonging my article."

When it came to the nature and development of the Wright gliders, either Chanute or his reporters, or both, made about as big a 'hash' of the whole story as can be imagined: it is today impossible to say exactly how this happened, or why. To begin with, *La Locomotion* (Archdeacon's account) and *L'Aérophile*, each reproduced the same four photographs of one glider; this was a Wright glider, but—unbelievably—it was the tailless No 2 of 1901, and not the modified No 3 of 1902, which Chanute had so extolled in his lecture. In this No 2, with which the French pioneers were now presented, the Wrights had used the old aerodynamic "tables and formulae" inherited from Lilienthal; these had proved so unsatisfactory that the brothers had had to abandon all they had learnt from their predecessors. "Having set out with absolute faith in the existing scientific data," wrote Wilbur, "we were driven to doubt one thing after another, till finally, after two years of experiment, we cast it all aside, and decided to rely entirely upon our own investigations". The end-result of this fresh start was the modified No 3 Wright glider of 1902, properly described and praised by Chanute. Yet now—to accompany and enshrine Chanute's literally epoch-making lecture—this vitally important glider was apparently unavailable in photographs. We still do not know what photographs Chanute exhibited at his lecture, or why he did not ensure that the correct photographs were available—and used—by the Paris periodicals. From the surviving correspondence, it seems certain that he not only had a good supply of prints from the Wrights' own negatives of this modified No 3 glider, but also prints from his own photographs, which he had taken in October 1902 at the Kill Devil Hills. Perhaps it was the fault of the authors of the articles, who did not realise the difference between the various Wright machines. But whatever the reason, these incorrect illustrations caused much confusion.

One is also led to wonder how much Chanute really understood of the Wrights' control system in general—when it came to involve the warp-rudder linkage—despite his presence in the Wright camp in October 1902, immediately after this linkage was incorporated in the machine: indeed, on October 8th, he had actually witnessed the first flight of the modified No 3. He certainly did not understand the *modus operandi* of the rudder, as we find him in July of 1903 asking Wilbur how it worked; so, in view of the fact that he saw and discussed the machine in October 1902, without discovering how it worked, he also probably remained ignorant of the *raison d'être* of the warp-rudder linkage. It would follow that the reporters of Chanute's lecture were also confused; and, even if they had by them the photographs of the Wright's No 3 glider, they may not have been able to distinguish this machine from the others, despite the presence of a tail in 1902! Today, perhaps, we may appear to be rather critical of those men in 1903—who were completely new to the business of looking at aeroplanes or photographs of them—when the presence or absence of a tail might not have been so evident, and when it was clearly too much to expect authors and editors alike to realise that the photographs they decided to use did not fit the descriptions given by the lecturer.

Then, to render confusion even worse confounded, all the photographs of the Wright glider published in the *La Locomotion* and *L'Aérophile* accounts *were described in their captions as being of the Chanute type!* In each journal, the caption "L'appareil Chanute" appeared under the first two photographs, along with the sub-captions "le Lancer", and "... vu par-dessous"; the caption "L'appareil Chanute" clearly referred also to the two other illustrations, since they simply bore the sub-captions "Le planement" and "La direction".

But there is still more confusion to come. For the *Aérophile* reporter, presumably following Chanute's words—since he had none other to follow—wrote:

"Finally, in 1902, the Wright brothers succeeded even better. Their machine was much larger than the original one of M. Chanute's; ... The Wright brothers had, in addition, made two important modifications as a result of their previous experiments. First, they placed an elevator out front, a device which acted much more efficiently for control of height than the former rear elevator: they replaced this rear elevator by a vertical rudder. Then—and this is the essential point—the pilot was placed in a prone position, instead of being upright, which reduced by four fifths his resistance to the air. M. Chanute had well understood the advantage of this position for the aviator but, out of prudence, had not risked advising it."

Confusion and misrepresentation could not have run more wild, however innocent the causes. As pointed out previously, the Wrights decided on a forward elevator early in their work, and had used it on all their gliders and powered machines from their first in 1900, until 1910, when they fitted their first rear elevator. They had no tail-surfaces in 1900 and 1901; then, in 1902, they added first a double fixed fin; then substituted a rudder, working in concert with the warping system. Chanute, on the other hand, never fitted either an elevator or a rudder to his machines, using only fixed tail-surfaces, with Herring's "elastic fastenings" to act as a would-be gust-damper. Chanute never employed pilot-operated control surfaces of any kind in his gliders, whereas the Wrights employed them from the first. The last remark by the reporter —about the prone-pilot position—is also highly misleading. The Wrights had decided on this position, on their own, after assessing the disadvantages of Lilienthal's hanging-position, and had used it in 1900 on their first glider. Wilbur even wrote a whole article on the subject in 1901 (see Section 8).

It was bad enough to find the Wrights so entangled with Chanute in the minds of the Europeans; but then Archdeacon was to add a few extra touches of confusion not included in *L'Aérophile*: again we find ourselves asking how much was stated, and how much implied, by Chanute himself in his talk. For, after noting with surprise and approbation Chanute's new word for a glide—'glissade'—Archdeacon described the Wrights' launching technique; they he says of their gliding in general:

"What is particularly remarkable about these experiments is that they have been conducted with such care (prudence), and so methodically, that, since the beginning, M. Chanute has not had to record a single accident other than a pair of torn trousers!... M. Chanute prides himself, and justly, on this absence of accidents in his experiments: it is, moreover, the only proud thing I know about him, for he is the most modest man in the world, and is always ready to attribute his own merits to others. Finally, here is what M. Chanute said in conclusion: 'Our experiments, methodically carried out, will allow us, little by little, to learn completely the art (métier) of the bird; an art which, without appearing to be, is extremely difficult. And so, in a year—or perhaps two—we shall know it thoroughly. Until that time, it is useless, and even dangerous, to saddle oneself with a motor, and I much prefer such untroublesome and simple a motor as gravity'."

Then Archdeacon adds:

"M. Chanute, as soon as he returns to America, will send us the complete plans of his latest machines, and so allow us to produce similar ones in France."

As I have said above, it is hard indeed to imagine that Chanute's listeners—in this case Archdeacon—were materially mis-reporting his words; and it is hard indeed to feel that Chanute did not make these quite outrageous claims. Every relevant document is available on the Chanute-Wright relations; also on the Wrights' design, construction and testing of their gliders during their three seasons. At no time had Chanute had anything whatever to do with the conduct of any of the Wright's activities: he was later at pains to explain to the Wrights that he had not conveyed the idea that they were his pupils. Nor did Chanute ever claim to have contributed anything of basic importance to the Wrights' flying.

Years later, in the sad quarrel that arose between the Wrights and Chanute, Wilbur felt he had to bring up this question of Chanute's remarks. The old gentleman answered:

"I have never given out the impression, either in writing or speech, that you had taken up aeronautics at my instance, or were, as you put it, pupils of mine. I have always written and spoken of you as original investigators and worthy of the highest praise."

With all due charity, one can only conclude that this fine old man, finding himself in 1903 in the romantic country of his birth, and surrounded by the warmth, respect, and indeed adulation of his audience, succumbed to the temptation of acting the part of inspirer, teacher and mentor of the Wrights. Chanute's greatest contribution to the development of aviation was as encourager, critic, and propagandist-in-chief; not as technical innovator or adviser.

The final point to consider is one of the most important; that of control in roll. For a long time it was assumed by historians that the Wrights' vital discovery of warp-drag, and its cure by simultaneous movements of the rudder—as incorporated in their modified No 3 glider of 1902—was not revealed to the French by Chanute in this lecture, because the Wrights wished it to be kept secret. But perusal of the two articles in question reveals immediately that Chanute did indeed reveal the simultaneous action of wing-warping and rudder, without, however, revealing either the aeronautical *raison-d'être* or the mechanism by which this action was effected. The reason for this lack of explanation is seen clearly in the Chanute-Wright correspondence: it was simply that at this time, Chanute did not know the answer to either question. Both *La Locomotion* and *L'Aérophile* stated the bare facts of the operation, as Chanute had obviously given them. Here is Archdeacon in *La Locomotion*:

"To regulate lateral equilibrium, he operates two cords which act on the right or left side of the wing by warping (gauchissement), and simultaneously by moving the rear vertical rudder."

So much for the fabric of Chanute's epoch-making lecture. Perhaps the most remarkable outcome—or rather lack of outcome—was that no European pioneer troubled to figure out the reasons for this clear description of the Wrights' system of lateral control. A principal cause of this neglect—a neglect of true lateral control which was to dog the Europeans until Wilbur Wright revealed its full significance before their eyes in August and September of 1908—was surely that, until a pilot actually got up into the air in a glider and flew it, not once or twice, but scores of times, he would never discover the necessity of proper lateral control, in its full significance, which involved primarily a combination of roll-control and yaw-control. The European pioneers, despite their expressions of determination to glide like the Wrights, before trying to power-fly, were not, in fact, to practice what they preached; and after the most tentative of gliding efforts, they were later to abandon proper glider experiments in favour of committing themselves to the air foolhardily and unsuccessfully in powered machines before acquiring even the rudiments of flight-control.

Here, to end this section, is the Comte de La Vaulx (*op. cit.*) on the effect of Chanute's lecture (page 274).

"For most of his listeners, except Ferber and his friends, it was a revelation, and even one which was a little disagreeable; when we spoke at times in France rather vaguely about the flights of the Wright brothers, we were far from doubting their remarkable progress; but Chanute was now admirably explicit about them and indicated their full significance. The French aviators at last felt that, despite their own enlightened views, they had been resting on the laurels of their elders a little too long, and that it was time for them to get seriously to work if they did not wish to be left behind. These facts particularly impressed themselves on the mind of one man, who was to play a great role in the evolution of French aviation—Ernest Archdeacon.... Anxious to keep for his country the glory of seeing born the first man-carrying aeroplane which would raise itself from the ground by its own power ... Ernest Archdeacon decided to shake our aviators out of their torpor (de secouer le torpeur de nos aviateurs), and put a stop to the indifference of French opinion concerning flying machines."

❧ 14 ☙

Ferber's Letter to Archdeacon
with its Exhortations and the Suggestion for a Gliding Competition: April 1903

One of the most important letters in aviation history does not survive complete, nor is its exact date of writing known. It was written by Ferber to Ernest Archdeacon, just after Chanute's lecture, and quoted by Archdeacon in *La Locomotion* (see next section). Ferber had not been able to attend Chanute's lecture on April 2nd—his military duties probably kept him away—but he must have read or heard a report of it, and then written straight away to Archdeacon. In this letter, occurred the following passages, the italics being Ferber's:

> "As I know that you are more of an aviator than a balloonist, I beg to write to you and ask you to use your influence with the Aéro-Club, and get them to found as soon as possible *a prize for distance for gliders* (un prix de distance pour appareils de planement). With our experience of the automobile, we know that it is racing which leads to the machines being improved; *and the aeroplane must not be allowed to reach successful achievement in America* (et il ne faut pas laisser l'aéroplane s'achever en Amérique)."

We have here, in April of 1903, the first clarion call to the would-be pioneers of Europe to rival and overtake the Wright brothers. Although included in a private letter to Archdeacon, this exhortation of Ferber's was soon to be broadcast when Archdeacon included it at the end of his report of Chanute's lecture in *La Locomotion*; and when it was republished in *L'Aérophile* (see next section). Such clarion calls and exhortations were to echo throughout Europe during the next three years. Ferber's was also the first proposal to promote glider competitions to stimulate the spirit of emulation. How enthusiastically Archdeacon received this suggestion of Ferber's will be seen in the next section.

15

Archdeacon's Call to Action: April 1903

Ernest Archdeacon[1] was a wealthy Paris lawyer, automobilist, and patron of flying: he is now to become the undisputed leader of European aviation, both as its patron and its prophet. Even before receiving Ferber's letter, he had already decided to sound a clarion call of his own. The Wrights had—from 3,000 miles away—exercised another powerful 'influence-impact' on Europe, this time on a most prominent Frenchman. After completing his write-up of the Chanute lecture for *La Locomotion*, Archdeacon had added a peroration of his own, which was to ring through French aviation circles for years, and become famous in aeronautical history (the italics and capital letters are Archdeacon's):

> "The result of all this, my very dear 'colleagues in locomotion' (mes très chers collègues en locomotion), is that the solution is approaching, and even approaching *very quickly*; also, that France, this great homeland of inventors, assuredly does not hold the lead *in the special science of* AVIATION, even when the majority of good minds are today convinced that this alone is the true way. Will the homeland of Montgolfier have the shame of allowing this ultimate discovery of aerial science—which is certainly imminent, and which will constitute the greatest scientific revolution that has been since the beginning of the world—to be realised abroad? Gentlemen scholars, to your compasses! You, the Maecenases; and you, too, of the Government; put your hands in your pockets—or else we are beaten!"

Then he quoted Ferber's letter, with its specific and practical suggestion that a glider competition should be held to stimulate their countrymen. As noted before, the full text of Ferber's letter does not seem to have survived, nor its exact date of writing. But Archdeacon said that he had received it "at the precise moment" of his sending to press his write-up

---

[1] Archdeacon was pronounced "Arsh-dek", *not* "Arsh-dek-ong".

of Chanute's lecture, which was published in the issue of *La Locomotion* for April 11th 1903.

This was a remarkable coincidence, which was to have a profound impact on European aviation; for Archdeacon held up sending off his article for the brief time it took to add his momentous postscript. He wrote:

> "At the precise moment of sending the present article to press, I received a very curious letter from Captain Ferber, well known in our specialised world as being the only man who is conducting gliding experiments in the manner of Chanute [*ie* of the Wrights] ... This letter is extraordinarily curious in that it demonstrates once more how certain ideas are in the air ... Written by Captain Ferber, *whom I have never had the pleasure of meeting*, it expresses ideas that one would have sworn he had agreed on with me in advance ..."

Archdeacon then quoted the passages from Ferber's letter given in the previous section, followed by these comments of his own:

> "It is really impossible to put it better, and I am immediately going to refer this project to the Technical Committee of the Aéro-Club. It is only a question of *working out the proper regulations*, and of finding a location. In view of the great interest in this project, this is a really good occasion for the 'Maecenases of locomotion' to manifest their generosity. Here it is a question of machines which cost 300 francs to construct; and even relatively modest prizes will be amply sufficient to produce a gigantic rivalry (une colossale émulation). As for me, I am quite ready to place myself at the head of the subscription list, with a sum of 3,000 francs ..."

It is clear from these statements that although Archdeacon was determined to lead a crusade to rival the Wrights, it was Ferber who suggested the practical course of promoting a gliding competition; but it was a competition that was never to materialise.

The Editor of *L'Aérophile* was so impressed with Archdeacon's stirring call to action in *La Locomotion*, that he reprinted the whole of both his peroration and postscript in his April issue—*L'Aérophile* appeared in mid-month—under the title, "A Glider competition, if you please! (Un concours d'appareils de planement, S.V.P.!)"

❧ 16 ❧

Archdeacon forms the Aviation Committee of the Aéro-Club: May 1903

When Ernest Archdeacon was later (1909) created a Chevalier of the Legion d'Honneur for his services to French aviation, *L'Aérophile* (August 1st 1909) wrote of him that:

> "Stirred by the first successes of the Wrights in America, he started an admirable and fruitful campaign on behalf of French aviation, by his words, by his pen, by his example, and by his liberality. He created the Aviation Committee of the Aéro-Club; stimulated both intellect and energy; and has become the promoter of our present-day successes."

For Archdeacon had been as good as his word, and had immediately approached the Aéro-Club after receiving Ferber's letter. At the next meeting (on May 6th) of the Club's Technical Committee on Aerial Locomotion, he asked for permission and support in setting out his proposals for organising the glider competition and its prizes, and repeated his offer to head the subscription list with 3,000 francs. His suggestions were received with acclaim, and a sub-committee was set up to prepare the ground-work: its official title was at first the Sub-Committee on Aviation Experiments (Sous-Commission des Expériences d'Aviation). As is usual even now, the Chairman (President) of the sub-committee was a respected "veteran"—in this case Colonel Renard—with Archdeacon, its creator and spokesman, as Secretary. As the Technical Committee on Aerial Locomotion was virtually uninterested in heavier-than-air flight, the new sub-committee was, from now on, to become the centre of all aviation activities in the Aéro-Club, and thus the whole of France, since all the noteworthy pioneers were welcomed by the Club as members, regardless of income or social status. Although the sub-committee was to operate from the first as an independent committee, it was not granted the status of a full committee—the Aviation Committee—until March 1906, by which time

Archdeacon had become its Chairman (President): but it has always been customary to refer to it as the Aviation Committee, even in the beginning. As mentioned above, the initial idea behind the Aviation Committee, *ie* to conduct a glider competition, was lost sight of, and no such competition was ever to be held.

17

Ferber's Gliding during the Year 1903

Ferber was to progress very slowly in 1903, and to hold himself up by over-ambitious schemes. He had already, in 1902, improved his 1902 Wright-type machine by giving it a better front elevator, and adding two primitive wing-tip rudders (Fig. 13): this machine was further tested in 1903 on the coast at Conquet, 15 miles west of Brest, on Cape Finistère. It is clear, from the continued absence of any attempt to fit a roll-control system, that he was hovering between a controllable machine without proper controls, and an inherently stable machine with a minimum of controls. This new move did not materially improve matters; but he still could write confidently:

> "Already, as a result of my experiments at Beuil in 1902, I had sufficient equilibrium in hand to think of passing immediately to the installation of a motor."

It is interesting to find Chanute, when on his grand tour, paying a visit to Ferber in March 1903, and writing home to Wilbur Wright on March 7th:

> "In Nice I met Captain Ferber, of whom I spoke to you, who has been trying experiments with a machine similar to yours. It is rudely made by a common carpenter, ... He says that he is not trying to invent a new system, but simply to experiment [with] the best that others have designed, and is much of the opinion that you are ahead of all others. He says that he is much inclined to go to America to take lessons from you, and that he wishes to purchase your 1902 machine, ..."

It was now becoming clear that, despite his agreement with Chanute on the need for long and careful testing, Ferber did not have it in his mental make-up to carry out such a programme. For late in 1902 he had gone ahead and built an enormous whirling arm at Californie, near Nice. He had then adapted one of his gliders to take a 6 hp Buchet engine driving two contra-rotating tractor propellers, and hung this machine on the arm. He made some tests in 1902, but most of the testing seems to have taken place in the Summer of 1903 (Fig 14). It is interesting to note that, in the caption of one of Ferber's photographs, the machine is described as "l'appareil Wright". These tests were, of course, a failure, and of little significance. Ferber seems now to have realised he was trying to proceed too fast, for his next step was to be in the direction of more gliding experiments, although still not extensive enough to attain any sort of proper control over his machine.

❧ 18 ❧

Chanute's first French Article: August 1903

During June and July 1903 there was continued talk in the Aéro-Club about the necessity for a glider competition. The Aviation Committee met together and discussed the matter; but nothing was done about it.

Then, in the August 1903 issue of *L'Aérophile*, there appeared the next big 'prod' to be administered to the European aviators: this was Chanute's own first article to appear in Europe, entitled "La Navigation aérienne aux Etats-Unis".[1]

This article is one of the most important documents in aviation history; for it was here that the Wright's modified No 3 glider was first presented visually to the world, through both general arrangement drawings (plan, side-elevation, and front-elevation), and one excellent photograph.

Chanute spent the first half of his article in describing and illustrating three of his own machines, the 'multi-wing', the 'oscillating-wing', and the biplane; all were hang-gliders, and none had any pilot-operated control-surfaces.

Then he turned to the Wrights. At the outset, he strongly implied that the brothers took up aviation in 1900 at his instigation, which was quite incorrect, as they had been directly inspired, as had Chanute himself, by Lilienthal. The Wrights' first test model—their warping kite—was made in 1899. Chanute wrote:

"The invitation to amateurs to repeat these experiments remained unacted upon till 1900, when Messrs Wilbur and Orville Wright of Dayton (Ohio), took up the question. They have accomplished such advance upon all previous practice, that the rest of this paper will be devoted to giving an account thereof. The improvements which they have introduced are the following:

---

[1] This article is reproduced complete in translation as Appendix C. Its illustrations are reproduced in Sections 4 and 5.

> 1st. Placing the horizontal rudder, or tail, at the front, a position which proves more efficient in acting upon the air;
> 2nd. Placing the operator prone on the machine, thus diminishing by 4/5 the resistance due to his body;
> 3rd. Warping the wings to steer to right or left."

Chanute then described the Wrights' testing grounds at the Kill Devil Hills; their preliminary work with their No 1 glider in 1900; their method of take-off; their No 2 glider of 1901; the aerodynamics of gliding; the Wrights' laboratory experiments of 1901-02; and their modified No 3 glider of 1902, in considerable detail, with plan, elevations, etc., and a fine photograph of the machine in flight, along with tables of its best glides. He wrote:

> "The two brothers glided alternately, and they soon attained almost complete mastery over the inconstancies of the wind. They met the wind gusts and steered as they willed. They did not venture to sweep much more than one quarter circle, so as not to lose the advantage of a head wind; but they constantly improved in the control of the machine, and in learning the art of the birds. Some 800 glides were made ... In point of fact the Messrs Wright are now gliding very nearly as well as the vulture, which generally descends one metre in ten in calm air."

Not only was this a vivid and glowing enough account to stimulate any would-be aviator; but Chanute deliberately introduced the question of a powered machine:

> "the time is evidently approaching when, the problems of equilibrium and control having been solved, it will be safe to apply a motor and a propeller". (On peut donc prévoir l'époque à laquelle l'équilibre et la direction étant résolus, il sera possible d'ajouter un moteur et en propulseur).

This statement, made now in August of 1903, in the monthly "bible" of the Aéro-Club de France—*and four months before the world's first powered flights*—is one of the most important in aviation history. For it was here "spelled out", in the words of the much-respected Chanute, that "the problems of equilibrium and control" must be solved before powered flight was attempted. It was the flouting of this prescription— after a derisory amount of glider testing—that held up properly successful powered flying in Europe until 1908-09.

It is noticeable in this article that, although Chanute rightly dwells on the importance of equilibrium and control, he does not show either in his general arrangement drawings of the Wright glider, or in his descriptions of it, how either the warping system works, or the rudder;

and does not mention here—as he did in his April lecture—that the warping and rudder action was linked and simultaneous. He knew that the Wrights were applying for a patent, and were not anxious to publicise their control system; and he may have felt he had already gone a bit too far in his lecture by revealing the simultaneous action.

It may be felt that the relevant statements were brief, and so they were; but to the keenly interested audiences of pioneers, it does seem extraordinary to us today that these brief but vital control clues, published for all to read in April of 1903 were, from now onwards, studiously neglected by every Continental pioneer. The reason, I believe, was that since no one was willing to undertake proper and thorough gliding tests, no one was at all aware of the practical problems involved; consequently, no one bothered to follow up these clues, clues that could have led to the Europeans understanding the significance of control in roll. Control in roll remained neglected in its full significance until after Wilbur Wright entered upon his spectacular 'season' of flying in France from August to December 1908.

Chanute closes his article with various words of advice; "a word of warning" for would-be experimenters to "exert the utmost possible prudence in conducting gliding experiments"; about selecting a suitable testing ground; then starting gradually, and from only a small elevation; and so on:

> "More important than all," he said, "do not try to beat previous records. This leads to taking risks and to producing accidents. . . . We must always remember that the important things to be secured are control of the machine and safe landings."

His last paragraph reads:

> "It is hoped that when many experimenters get to work, such progress shall be made as materially to advance the time when aviation shall become practical."

Charles Dollfus has this to say in *L'Homme, l'Air et l'Espace* (1965):

> "On April 2nd 1903, Octave Chanute gave a lecture to the Aéro-Club de France, which was published in the journal *L'Aérophile*, along with drawings of the machines, that revealed the use and method of construction of the Chanute and Wright gliders. This communication had a decisive importance: it orientated French aviation towards an attempt at imitating the Wright gliders."

Further Reminders of the Wrights: August 1903

In the same issue of *L'Aérophile* (August 1903)—which contained Chanute's long-awaited article—there appeared a half-page article by Archdeacon entitled "Towards Aviation (Vers l'Aviation)", in which the readers were reminded of the recently created Aviation Committee of the Aéro-Club; of the efforts to promote and subsidise a gliding competition; and of the necessity of finding a suitable terrain for gliding experiments, suggestions for which were called for. There was also included the now familiar exhortation:

> ". . . in a word, to stimulate—from one end of France to the other—the great rivalry which is nearly always necessary to quicken progress."

Then, in the September 1903 issue of *L'Aérophile*, there was printed the discussion which followed Wilbur Wright's second, and now famous, lecture given before the Western Society of Engineers in Chicago on June 24th 1903, and reprinted in the Society's Journal for August 1903. The reason why *L'Aérophile* did not print this outstanding lecture *in toto*—or even a summary of it—and why it only included the desultory discussion which followed, is explained by the Editor in these interesting words:

> "Our eminent collaborator, Victor Tatin, has kindly made for *L'Aérophile* a literal translation [of the discussion] which usefully completes the article by Monsieur O. Chanute on the same subject, that appeared in our last number."

However, the $2\frac{1}{2}$ pages, headed (in English) "Observations in Soaring Flight—Discussion", again brought the name of Wright forcibly before the eyes of the French pioneers. The points covered in this discussion were not of great interest. Wilbur was now deliberately evasive when subjects like control, or the possible application of power, were brought up, as he was afraid his secrets would leak out: but he gave full information about materials of construction and so on.

20

Chanute's second French Article: November 1903

On June 30th 1903, Chanute had written to Wilbur telling him that he (Chanute) was writing an article for the influential French journal, the *Revue Générale des Sciences*: he asked Wilbur, "should the warping of the wings be mentioned? Somebody may be hurt if it is not".

This seems to us now an absurd question, since Wilbur himself had given an explicit account of warping in his first Chicago lecture (1901)—which was available in Europe—and Chanute had referred to it in his recent (April 1903) Paris lecture, as well as in his article in August. Wilbur, now becoming more and more secretive—and with some justice—replied cautiously that he and his brother did not wish "any description of this feature" to be given "at present".

Chanute thereupon sent him a draft of his article (on July 12th) in which he wrote that the rear rudder of the Wright glider was operated by "twines leading to the hand(s) of the aviator". Wilbur promptly replied that this was not true, and suggested that Chanute leave out the clause. Chanute wrote back, saying:

> "How is the vertical tail operated? I fear that it will not do to strike out the clause altogether, and you did not desire that the warping of the wings should be mentioned at all."

To which Wilbur sent a highly equivocal answer, mentioning previous points in their correspondence, previous errors in the article which had been discussed, and ending with the statement that "the statement in regard to the 'twines' leading to the 'hands' is more serious, and we hope will be omitted".

Not unnaturally, Chanute immediately comes back with a somewhat tart note, sent "special delivery", on July 23rd:

> "I have yours of 22nd, but it does not answer the question 'How is the vertical tail operated?' You will understand that I forwarded

the paper to the *Revue* some time ago, . . . You will understand also that I desire to comply with your wishes, but that I cannot send off corrections without substituting something."

Wilbur replied next day with an explicit description of the manner in which the rudder wires were linked to the warping wires, for simultaneous and automatic operation; but he added that "this statement is not for publication". He goes on to say—and this is perfectly reasonable—that the patent laws of France and Germany could invalidate patent claims if devices claiming patent protection were previously described in publications. After discussing alternatives to revealing the true facts of the control system, Wilbur suggested removing the "entire last sentence in the upper paragraph on page 14". Unfortunately we do not know what this consisted of; but it will be seen below what in fact Chanute did include.

Before describing this important article of Chanute's, I feel it is appropriate here to express considerable surprise, not at Wilbur's attitude—which was at least understandable—but at Chanute's lack of true understanding of the Wright control system, a lack I have already noted. For Chanute was actually staying in the Wright camp in October 1903, after the crucial change from the two fixed fins to the linked single moveable rudder had been made. Chanute witnessed the machine making glides, and he even took photographs of it! I find it almost incomprehensible that such a dedicated and able man who was "taking records" of the machine's first flights—as Orville's diary records—and who was literally sleeping in the same hut as housed the glider, either did not have the curiosity to look, or to ask how the control system worked; it is equally impossible to see how Chanute failed to understand it when he was told. This is one of the major mysteries of aviation history.

Chanute's article appeared in the *Revue Générale des Sciences* for November 30th 1903. It was a long (10 page) contribution entitled "L'Aviation en Amérique", and included reproductions of 15 photographs, of which no less than 10 were of the Wright No 3 glider of 1902 —properly credited—with nearly half the text devoted to the Wrights. It is an admirably conceived and written article, lucid and informative: in many ways it echoes his earlier piece in *L'Aérophile*, but now there is evidence of far greater thought and balance of expression.

Although not as historically decisive as the one in *L'Aérophile*, this article is a minor aviation classic. Its main achievement—and indeed Chanute's probable intention—was to present in facts and photographs the Wrights' achievements in gliding, and to establish once and for all, in the minds of the Europeans, the Wright configuration as being the starting point of the sorely-needed revival of aviation on the Continent.

He started the article with the customary tribute to Lilienthal; then gave a clear summary of his own work with gliders, illustrated by an artist's sketch of his multi-wing machine, one photograph of his oscillating-wing machine, and three excellent shots of his biplane hang-glider, one of which showed it with side-curtains. Then, although carefully avoiding any specific reference to the Wrights as his pupils or collaborators, the old gentleman could not resist trying to perpetuate the two untruths which represented the next best thing to their being his pupils: (a) that the Wrights took up aviation as a result of his "invitation" in 1897 to "amateurs" to repeat and better his own experiments; and (b) that the Wright machine was a "similar machine" to his biplane, "with some important improvements (avec quelques améliorations importantes). These consisted of placing the 'tail' in front, where it proved itself more effective than at the back; and of placing the operator in a horizontal position instead of the upright position which had been adopted by Lilienthal, Pilcher, and myself". Not a word about the fact that the Wrights' aircraft was totally different in conception, and was fitted with sophisticated pilot-operated control-surfaces, etc. These statements—which the reader of the article had to interpret for himself in the light of descriptions farther on—show, I feel, that Chanute had indeed persuaded himself to believe that the Wrights descended aeronautically from him, and that he was determined to further that impression in Europe.

He then described the Wrights' three gliders, dwelling particularly on the modified No 3, with its rudder; he showed ten excellent photographs of the machine, including one of it being launched, three of it on the ground, and six of it airborne.

Whatever Chanute intended to omit, we do not know, as the English text of his second article does not survive; but, first, there is a confusing but specific reference to the Wrights' wing-warping, when Chanute says of the pilot: "he also steers to left or right by slightly displacing his weight, and by curving the surfaces when he is able to (en courbant les surfaces quand il le peut)". Here, I believe, is another indication that Chanute did not properly understand the Wrights' control system; for his reference to "slightly displacing his weight" obviously refers to the Wright pilot swinging his hip-cradle to operate the warp and rudder wires, not to any intention to shift his weight. It is also interesting to find the French translator of Chanute's English text using the word 'courbant' instead of the word Archdeacon used in April when reporting Chanute's lecture, which was 'gauchissement' (from 'gauchir'), which was later to become the misleading, but standard, word in French for warping. The Wrights, in their French patent, used the correct words, 'tordre' (to twist), and 'torsion'.

We next find the rear rudder inevitably referred to, which is also clearly to be seen in the photographs. Two of the photographs are of the machine turning, and one of them—taken immediately from the rear—shows very clearly the down-warp on the starboard wings, the up-warp on the port wings, and the rudder put over to port; but the machine is *banked to starboard*. Chanute did not know that here was a photograph showing the pilot, having initiated a bank to starboard, being now either in the act of holding off the bank, or initiating a levelling-out, by reversing his controls. Nevertheless, the simultaneous action of warp and rudder should have been clear to all, no matter how confusing the circumstances. Chanute's description of this photograph is as follows: "Fig. 11 shows the machine turning to the right. It will be noted that the rudder is turned to one side, and that the right-hand wings are much lower than the left-hand: the attitude resembles that of a bird when it is turning in a circle".

Incidentally, a further indication that Chanute did not know as much as he should, is to be seen in one of the 'airborne' photos: this shows the first form of the Wright No 3 glider when it had two fixed fins, and Chanute notes that the machine was photographed when it "possessed a double rear rudder (un double gouvernail arrière), one of which was later removed.[1]"

Chanute was enthusiastic in his praise of this modified Wright glider, giving his readers a detailed description of its construction; telling them that it made between 800 and 1,000 glides; also that the angle of glide had been 6-7 degrees; and, on one or two occasions, 5 degrees; and that the results obtained with it "mark a great advance on all earlier experiments ... the only thing that remains to be done is to acquire the bird's skill in its craft (d'acquérir de l'adresse dans le métier d'oiseau)".

Among its many interesting statements, Chanute's article contained a number which have a close bearing on subsequent aviation history in general, and of European aviation in particular. First, about his own work, reiterating what he had said before:

"our experiments had not had the aim of attempting a true flying-machine, but simply of studying the problem of equilibrium."

Then, in speaking of the Wrights' work in general, Chanute wrote:

"the theory of the Wright brothers is that one should control the machine all the time, instead of trusting to automatic devices to ensure the maintaining of equilibrium."

This position of apparent mutual exclusivity was to dog the steps of

---

[1] Oddly enough, when the Wrights returned to the Kill Devil Hills early in 1908, they first practised on this No 3 glider, to which they now fitted a double rear rudder.

early flying, with the Wrights insisting (at first) on inherent instability in their machines to allow of maximum controllability, and the Europeans—somewhat misunderstanding Chanute—insisting (at first) on maximum inherent stability with minimum controllability. It was a great pity that the Europeans did not determine first to achieve the Wright's mastery of gliding; and then, when they properly understood what flight-control involved, to design powered machines which incorporated stability, whilst also being fully controllable. If this course had been followed, there is little doubt that Europe could have had practical powered aeroplanes by 1905, or 1906 at the latest; instead of having literally to 'flounder' until 1909, when the first fully practical European machines appeared.

Chanute saw much of this problem clearly: for example, at the end of his article, he wrote (my italics):

"The problems to be resolved at first are those of equilibrium and of manoeuvre; and one will only arrive there by a persevering 'apprenticeship'. There is now room for extensive development, for improvements on the existing gliding machine, and the development of new types. But it must be done gradually ... *Up to the present, I have advised against the application of motors to gliding machines, because I believed that one must first become master of equilibrium and control.* The motor will introduce complications; but I think that enough progress has now been made to envisage the application of motors. The machine will at first have to be tested as a glider, with great care, as well as in the form of a small model: then, when one has acquired enough experience, the motor and propeller will be added to it."

The French had this article in their hands in November 1903. Within a month, they were to hear news which—whether they wholly believed it or not—should have shaken the Aéro-Club members out of their complacency, or 'torpor' as its President was later to describe it.

Looking back on this article of Chanute's, any historian of aviation must be acutely aware, not only of the pictorial beauty and "impact" of the in-flight photographs of the Wright glider, but of the close parallel with the equally beautiful—I use the word advisedly—photographs taken of Otto Lilienthal gliding (Fig 1), which started to appear in the world's periodicals in 1894, and had soon become an integral part of the mental lining of every man-jack among the pioneers of Europe. It was these superb photographs, rather than the written descriptions of Lilienthal's gliding, which produced the most profound impact on the men who were to pioneer powered flight: an impact of vision which brought home the very essence of a man flying through the air—the mythical 'homme-oiseau'—supported by wings, like a bird.

The photographs of the Wright glider which Chanute published in this article were of the same order, but even more expressive of the true nature of flight. There can be little doubt that to the more sensitive and imaginative European readers, these photographs, and the descriptions of the flights, conveyed the message of a highly sophisticated machine, properly controlled by its occupant, which could take to the incorporeal medium of the air, and master it. It was the visible manifestation of the 'airman's' ideal, rather than the 'chauffeur's'. To the former, the pilot was part of his machine, the wings—and later the engine—being but extensions of his body and spirit: his field of operation was three dimensional space, in which he was as adept at control as the eagle or the swift. To the 'chauffeur', the aeroplane pilot was but the driver of a winged automobile, transposed from the flat layer of the highway to a similarly 'flat' layer of air above it: to him, control and manoeuvre were matters of level and direction, of steering this winged automobile along an aerial highway which he felt should also have been flat, but found was deceptively not so.

But this ideal of flying through the air—the airman's ideal—which Lilienthal was first seen to symbolise, and which was now renewed by the Wrights, was only to be hesitantly acted upon in Europe for a short space of time; then abandoned at the dictates of lesser men. It was, *au fond*, this abandonment of the airman's ideal in favour of the chauffeur's, that let to the lamentable postponement of practical aviation in Europe.

21

The Wrights build their first Powered Flyer: 1903

This brief account is also taken from Orville Wright's deposition of 1920:

"The flights of 1902 demonstrated the efficiency of our system of control for both longitudinal and lateral stability. They also demonstrated that our tables of air pressure which we made in our wind tunnel would enable us to calculate in advance the performance of a machine. Before leaving our camp at Kitty Hawk we began the designing of a new and larger machine to be driven by motor. The wings of the new machine had a spread of 40 feet 6 inches and a chord of 6 feet 6 inches, having a total area of a little over 500 square feet.

Immediately after our return from Kitty Hawk in 1902 we wrote to a number of the best-known automobile manufacturers in an endeavour to secure a motor for the new machine. Not receiving favourable answers from any of these, we proceeded to design a motor of our own, from which we hoped to secure about 8 horsepower. When the motor was tested it gave more power than we had anticipated. It developed a little over 12 horsepower and weighed about 160 pounds, without magneto, water, or oil.

We next proceeded with the construction of the parts to be used in this first power machine, and while we were doing this we began an investigation of screw propellers. At first we hoped to be able to procure a theory of the reactions on a screw propeller from works on marine engineering, but we soon found, after examining the few books we were able to secure in the Dayton Public Library pertaining to marine engineering, that water screw propellers at that time were not based upon theory but almost entirely upon empirical data. We had thought that we could adopt the theory from the marine engineers, and then by using our tables of air pressures, instead of the tables of water pressures used in their calculations, that we could estimate in advance the performance of the propellers we would use. When we found we could not do this, we began the study of the screw propeller from an entirely theoretical standpoint, since we saw that with the small capital we possessed we would not be able to develop an efficient air propeller on the 'cut and try' plan. As a result of this study we developed a theory from which we designed the propellers which we used in this 1903 power machine.

These propellers had an efficiency of over 66 per cent, an efficiency, I believe, rarely exceeded by the marine engineers, and never approached by any of the aeronautical investigators up to that time....

We went to Kitty Hawk the latter part of September 1903, and after a few days spent in establishing camp and in erecting a building in which to assemble and house our new machine, we began the work of assembling.... While in general the structure of these wings was similar to that of the previous gliding machines which we had built, yet a number of changes in design were made, among which I may mention that of the ribs and the covering of the surfaces with cloth. Instead of using thin strips of ash, bent to the desired curvature, as had been used in the earlier machines, for the new machine the ribs were made by [sawing] a piece of ash, with a cross section of about three-eighths by one-half inch, ... from one end to within a few inches of the other, inserting blocks of wood between the two halves of the strip and gluing and nailing them in position.... Through this structure we secured at the same time great strength and lightness. Ribs of this type are used in practically all flying machines of today. The cloth was stretched over both the top and bottom sides of the spars and ribs.

These, I believe, were the first double-surfaced planes ever designed or built.... The control of this machine was the same as that of the 1902 machine. Like the 1902 machine in the later part of the season, the central portions remained fixed, while the outer portions of the wings were adjusted to different angles of incidence..."

22

The Wrights' first Powered Flights: December 17th 1903

The first powered Wright machine, the *Flyer I*, was a biplane on a skid undercarriage, of 40 feet 4 inches span, a wing area of 510 square feet and a camber of 1 in 20: it had a biplane elevator out front, and a double rudder behind, whose control cables were linked to the warping-cradle: the motor drove two geared-down pusher airscrews through a cycle-chain transmission in tubes, one being crossed to produce counter-rotation (Fig 16). The launching technique was as follows: the *Flyer's* skids were laid on a yoke which could run freely on two small tandem wheels along a 60-foot sectioned wooden rail, laid down into wind; the machine was tethered whilst the engine was run up, and then unleashed; then, when its speed produced sufficient lift, it rose from the yoke and flew. The Wrights did not use any assisted take-off device for their 1903 flights.

After minor but exasperating set-backs at the Kill Devil Hills, North Carolina—where the tests took place—and after brushing up their piloting on the 1902 glider, the first attempt was made on December 14th 1903, with Wilbur at the controls (he had won the toss of a coin): but owing to over-correction with the elevator, the *Flyer* ploughed into the sand immediately after take-off.[1]

It was on the morning of Thursday, December 17th 1903: between 10.30 a.m. and noon, that the first flights were made. After five local witnesses had arrived, Orville (whose turn it now was) took off at 10.35 into a 20-22 mph wind and flew for 12 seconds, covering 120 feet of ground, and over 500 feet in air distance (Fig 15). On the fourth and last flight, at noon, Wilbur flew for 59 seconds, covering 852 feet, and over half a mile in air distance. Their speed was about 30 mph.

These flights were the first in the history of the world in which a

---

[1] This abortive test was made downhill. The tests on December 17th were all made from level ground.

The Wrights' first Powered Flights: December 17th 1903

piloted machine had taken off under its own power; had made powered, controlled, and sustained flights; and had landed on ground as high as that from which it had taken off. It should, I feel, be repeated that no aeroplanes, other than the Wrights', could remain in the air for more than 20 seconds until November of 1906; and it was not until November of 1907 that a full minute's duration was achieved by a European machine (see Section 80).

From time to time it becomes fashionable in certain circles to deny that the Wrights well and truly flew on December 17th 1903. In fact, if these brief 'flight-hops' had been all the Wrights had ever attained, they would not have been rated very high, historically. But the brothers made rapid progress to proper flying, and so the 1903 events are seen as the clear beginnings of achievement.

Here is the statement made by the brothers in an article written jointly by them in *The Century Magazine* for September 1908: they have just described the four flights made on December 17th 1903; then they say of the first flight:

> "The first flight lasted only twelve seconds, a flight very modest compared with that of birds, but it was, nevertheless, the first in the history of the world in which a machine carrying a man had raised itself by its own power into the air in free flight, had sailed forward on a level course without reduction of speed, and had finally landed without being wrecked."

A seldom reproduced account of the flights of December 17th, written by the brothers themselves, is reproduced in full in Section 25.

To close this section I would like to quote Charles Dollfus, from his *L'Histoire de l'Aéronautique:*

> "THE FIRST HUMAN FLIGHT—December 17th 1903 is the great day in the history of aviation. The Wright brothers made, in succession, four sustained flights, with their powered aeroplane. This final 'consécration' took place at the Kill Devil sand-dune, . . . It assures an imperishable glory for the two first aviators and for the United States."

❧ 23 ❧

The Wrights and Chanute: 1903

The sequence of communications between the Wrights and Chanute is of considerable interest: there was complete trust between the three men, and Katharine Wright knew that Chanute had to be informed as soon as the news arrived from Kitty Hawk; it was received at Dayton at 5.25 p.m. She immediately wired him, the telegram being received in Chicago at 8.07 p.m. It read:

> "Boys report four successful flights today from level against twenty-one mile wind. Average speed through air thirty-one miles. Longest flight fifty-seven seconds."

Chanute wired back straight away (still on the 17th):

> "I am deeply grateful to you for your telegram of this date advising me of the first successful flights of your brothers. It fills me with pleasure. I am sorely tempted to make the achievement public, but will defer doing so in order that they may be the first to announce their success. I earnestly hope that they will do still better."

Next day he wired the brothers:

> "Immensely pleased at your success. When ready to make it public please advise me."

The Wrights arrived home on December 23rd. On the 27th, Chanute wrote to Wilbur at Dayton:

> "I have had no letter from you since I left your camp, but your sister kindly wired me the results of your test of Dec. 17. Did you write? The American Association for the Advancement of Science holds its winter meeting at St Louis, Dec. 28th to Jan 2nd. I have been asked for a paper on aerial navigation, and have made it very

general in character. It is fitting that you should be the first to give the Association the first scientific account of your performances. Will you do so? Please wire me at the Southern Hotel, St Louis, on Monday. I go down Sunday night."

To which Wilbur wired in reply on December 28th:

"We are giving no pictures nor descriptions of machine or methods at present."

Wilbur followed up this telegram with a long letter to Chanute, also dated December 28th:

"Your telegram of congratulations and the Christmas remembrances have reached us, and, of course, we are deeply gratified at your kindness.
 The axles for which we were waiting when you visited us, did not arrive for one whole week after your departure. We spent part of this time in installing a new system of operating the wing tips and rear rudder, as the old system did not seem quite satisfactory. We then spent three days putting the axles in place again, and giving the machine the final touches. When ready for trial a three days' storm kept us penned up so that another week was lost. We however made some indoor tests of the thrust of the propellers and found that we would have plenty of power as the transmission only cost 5 or 10 percent apparently, instead of the thirty percent you had estimated. The thrust of the screw came within three or four pounds of our calculations of what it would do in a fixed position. But as we were concluding these experiments a peculiar feeling led to an investigation which revealed the fact that one of the axles was giving way. Accordingly we removed both of them and Orville went home to make new ones. He was gone two weeks more, so that by the time everything was ready again, five weeks had elapsed since the trouble with the axles began. We accordingly determined to try the machine at the earliest opportunity instead of waiting for the conditions we desired. So on the 14th inst., although the wind was only 2 to 3 metres a second, thus making it necessary to use the hill in starting, we got the machine out and made the first trial. It rose from the track and soon reached a point as high as the starting point but as this was done too suddenly it lost speed somewhat so that it was no longer fully supported. In turning down to regain speed the rudder was moved too far, and the machine darted down and touched ground before it could be turned up again. The time was only $3\frac{1}{2}$ seconds and the distance a little more than a hundred feet. The landing was made with the propellers still going, and with the machine sidling somewhat. The lower struts of the front rudder frame sank into the sand and as it was braced only at the ends the side pressure of the sand broke one of them, and it in turn twisted off one of the upper struts. The main machine and the skids under it,

of which we were so fearful, stood the test perfectly, although the landing was made at a speed of more than twenty miles an hour.

Our next flights were on Thursday, Dec. 17th, on which occasion the flights were all made from a level spot about 200 feet west of our buildings. The wind had a velocity of 24 to 27 miles an hour according to the Kitty Hawk anemometer which was almost directly to windward of us, but our measurement made with the English anemometer at a height of 4 feet from ground was only $20\frac{1}{4}$ miles. The conditions were very unfavourable as we had a cold gusty north wind blowing almost a gale. Nevertheless as we had set our minds on being home by Christmas, we determined to go ahead. Four flights were made, the first lasting about 12 seconds and the last 59 seconds. The 'Junction Railroad' worked perfectly and a good start was obtained every time. The machine would run along the track about 40 feet propelled by the screws alone, as we did not feel it safe to have strangers touch the machine. It would then rise and fly directly against the wind at a speed of about 10 miles an hour. The first flight was of about 12 seconds' duration and the last 59 seconds. The controlling mechanisms operated more powerfully than in our old machine so that we nearly always turned the rudders more than was really necessary and thus kept up a somewhat undulating course especially in the first flights. Under the prevailing conditions we did not feel it safe to rise far from the ground and this was the cause of our flights being no longer than they were, for we did not have sufficient room to manoeuver in such a gusty gale. Consequently we were frequently on the point of touching the ground and once scratched it deeply but rose again and continued the flight. Those who understand the real significance of the conditions under which we worked will be surprised rather at the length than the shortness of the flights made with an unfamiliar machine after less than one minute's practice. The machine possesses greater capacity of being controlled than any of our former machines.

One of the most gratifying features of the trials was the fact that all our calculations were shown to have worked out with absolute exactness so far as we can see, though we have not yet made our final computations on the performance of the machine.

Orville and I alternated in the flights according to our usual custom.

With wishes for a Happy New Year, (&c.)"

This little-known account, straight from the horse's mouth, so to say, is by far the most interesting of all the Wrights' statements. It was written by an expert for the consumption of an expert, and contains many points of technical interest not to be found in the brothers' other, and more popular, descriptions. I often feel that this simple and straightforward statement, by one honest man to another, should be required reading for those who profess to doubt the Wrights' word that they were indeed airborne for fifty-nine seconds, a flight-duration not surpassed by any other inventor until November 9th of 1907, when Henri Farman flew at Issy in his Voisin biplane for 1 minute 14 seconds.

The winter meeting of the American Association for the Advancement of Science at St Louis, to which Chanute referred in his note to Wilbur, was addressed by him on December 30th 1903, the subject being "Aerial Navigation". In the course of this paper Chanute made the following statements about the Wrights:

"Like the failures of Maxim and of Ader, it does indicate that a better design must be sought for, and that the first requisites are that the machine shall be stable in the air, shall be quite under the control of its operator, and that he, paradoxical as it may appear, shall have acquired thorough experience in managing it before he attempts to fly with it.

This was the kind of practical efficiency acquired by the Wright brothers, whose flying machine was successfully tested on the seventeenth of December. For three years they experimented with gliding machines, as will be described farther on, and it was only after they had obtained thorough command of their movements in the air that they ventured to add a motor. How they accomplished this must be reserved for them to explain, as they are not yet ready to make known the construction of their machine nor its mode of operation. Too much praise can not be awarded to these gentlemen. Being accomplished mechanics, they designed and built the apparatus applying thereto a new and effective mode of control of their own. They learned its use at considerable personal risk of accident. They planned and built the motor, having found none in the market deemed suitable. They evolved a novel and superior form of propeller; and all this was done with their own hands, without financial help from anybody.

Meantime it is interesting to trace the evolution which has led to this result and the successive steps which have been taken by others. ... Three years later Messrs Wilbur and Orville Wright took up the problem afresh and have worked independently. These gentlemen have placed the rudder in front, where it proves more effective than in the rear, and have placed the operator horizontally on the machine, thus diminishing by four fifths the resistance of the man's body from that which obtained with their predecessors. In 1900, 1901, 1902, and 1903 they made thousands of glides without accidents and even succeeded in hovering in the air for a minute and more at a time. They had obtained almost complete mastery over their apparatus before they ventured to add the motor and propeller. This, in the judgment of the present writer, is the only course of training by which others may hope to accomplish success. It is a mistake to undertake too much at once and to design and build a full-sized flying machine *ab initio*, for the motor and propeller introduce complications which had best be avoided until in the vicissitudes of the winds bird-craft has been learned with gravity as a motive power.

Now that an initial success has been achieved with a flying machine, we can discern some of the uses of such apparatus, and also some of its limitations. It doubtless will require some time and a good deal of experimenting, not devoid of danger, to develop the machine to practical utility."

❧ 24 ❧

The first Impact on Europe of the Wrights' first Powered Flights : 1903

Contrary to general belief, the direct influence of the Wrights' first powered flights on December 17th 1903 was rapid, profound and far-reaching.

Now that the news of the Wrights' powered flying had come to Europe, it provoked within three months an almost violent reaction among the members of the Aéro-Club de France; it renewed their ambition to fly; it produced active discussion of ways and means; it led directly to Archdeacon officially announcing in March 1904 his highly 'prize-provided' Gliding Competitions; and also in that March, to Deutsch de la Meurthe's offer of his "Grand Prix d'Aviation"—25,000 francs—for the first man to fly a powered machine in a closed kilometre circle; a prize not to be won until January 13th 1908, nearly four years later, when Henri Farman succeeded in flying a circle (see Section 84).

The reports of the Wrights' flights, most of them totally absurd, were naturally treated as pipe-dreams almost everywhere in the world except in Chanute's circle in the USA, and in the community of would-be French pioneers who had been made growingly aware of the success of the Wright gliders in 1901-02, and which had in turn culminated in the publication of Chanute's second article in November 1903, with its properly accredited photographs. It was now clear to all exactly what the Wrights' final glider looked like; and what it had achieved. And Chanute had already prepared the way for the news of possible powered experiments at the end of his November article.

The first news to reach Europe of the Wrights' powered flights came within a day or two—through the daily press—in the form of garbled and almost meaningless reports from America. But the more thoughtful Frenchmen feared the worst, despite the obvious exaggerations of the reports; and their fears were scarcely alleviated by a short notice which the Editor of *L'Aérophile* just managed to insert in his December issue (the magazine often appeared late in the month): he headed it simply "L'Aviation en Amérique":

"According to the foreign daily press, the brothers Orville and Wilbur Wright whose trials in gliding flight we described in our April and August 1903 numbers, have on December 18 (*sic*), at Kitty Hawk, successfully tested an aeroplane equipped with propellers driven by a tricycle motor. Wilbur Wright by this means covered 5 kilometers.

The place chosen was a hill known as Kill Devil Hill (le Tue-Diable); it is about 30 meters high and is located in the dunes on the Atlantic coast. The aeroplane was carried to the summit, and Mr Wright took his place on it lying at full length. There was no special arrangement for launching. Mr Orville Wright pushed the machine, and, the motor turning at its maximum, the machine rose to a height of 20 meters (?). The experimenter was able to maintain directional control at will and keep up a speed of 12 kilometers an hour. At the end of 5 kilometers, he came down without difficulty.

Mr Chanute, interviewed by telegraph on the veracity of this news, replied to Mr Drzewiecki: 'Newspaper accounts considerably exaggerated, await details by letter'.

We shall wait before commenting, but may we recall to French aviators the cry of alarm uttered by Mr Ernest Archdeacon [*ie* in *La Locomotion* of April 11th 1903]: 'Will the homeland of the Montgolfiers have the shame of allowing that ultimate discovery of aerial science to be realized abroad'? and this recommendation of Captain Ferber: 'The aeroplane must not be allowed to be perfected in America'. There is still time, but let us not lose a minute."

It will be noticed that this account repeated the misleading reports from America,[1] which included the absurd statement—absurd enough to calm any fears in certain quarters—that the Wrights had made a flight of 5 kilometres! The date of the flights was also given incorrectly, as December 18th. Stefan Drzewiecki, was an Aéro-Club member who had telegraphed the trusted Octave Chanute for confirmation.

One of those who were to believe that the Wrights had done what they claimed, step by step, was Ferber. His attitude towards them was later to become somewhat ambivalent; but he was always honest about them, and always gave them their due if he thought the evidence merited it. He also kept in touch with Chanute, and realised that Chanute was absolutely truthful; and that when he (Chanute) made a report, it was to be trusted. In March 1904, in the first of his two long articles in the *Revue d'Artillerie*, Ferber was to write (see Section 35):

---

[1] This garbled account also seems to include items which might belong to the abortive test by Wilbur on December 14th.

"Although the results were less remarkable than was announced at first, the date December 17th 1903 non-the-less marks the day when a *piloted* flying machine *has really flown* (a réellement volé), and the honour of this memorable experiment falls to the name of Wright." [Ferber's italics.]

As a tail-piece to the European activities of 1903, we may record one of the more charming notes of historic irony. For in this same issue of *L'Aérophile*—December 1903—which recorded the Wrights' first powered flights on December 17th, there is the official report of a meeting of the Aéro-Club's Aviation Committee, which had taken place on December 2nd. This report reads, in part:

"M. Drzewiecki communicated a letter from M. Chanute, the American aviator, giving interesting details of the recent gliding experiments of the Wright brothers, his pupils and collaborators. The Wrights are such masters of their machines that they have succeeded in soaring in a stationary position.... The Wright brothers have not yet ventured to attempt making circles, a manoeuvre which must follow stationary soaring... These results are very encouraging."

THE YEAR 1904

❧ 25 ❧

The Wrights' Powered Flights of 1903:
the second (and major) Impact on Europe:
January 1904

The misleading reports which appeared in the Press concerning their brief flights on December 17th 1903, led the brothers to write an admirable account, which they sent to the Associated Press for world distribution. This account—a minor classic of aviation history—is virtually unknown to the public. Here it is in full:[1]

Statement by the Wright Brothers to the Associated Press, January 5th 1904
"It had not been our intention to make any detailed public statement concerning the private trials of our power 'Flyer' on the 17th of December last; but since the contents of a private telegram, announcing to our folks at home the success of our trials, was dishonestly communicated to the newspapermen at the Norfolk office, and led to the imposition upon the public, by persons who never saw the 'Flyer' or its flights, of a fictitious story incorrect in almost every detail; and since this story together with several pretended interviews or statements, which were fakes pure and simple, have been very widely disseminated, we feel impelled to make some correction. The real facts were as follows:
On the morning of December 17th, between the hours of 10.30 o'clock and noon, four flights were made, two by Orville Wright and two by Wilbur Wright. The starts were all made from a point on the level sand about two hundred feet west of our camp, which is located a quarter of a mile north of the Kill Devil sand hill, in Dare County, North Carolina. The wind at the time of the flights had a velocity of 27 miles an hour at ten o'clock, and 24 miles an hour at noon, as recorded by the anemometer at the Kitty Hawk Weather Bureau Station. This anemometer is thirty feet from the ground. Our own measurements, made with a hand anemometer at a height of four feet from the ground, showed a velocity of about 22 miles when the first flight was made, and 20½ miles at the time of the last

---

[1] The reference to the 'rudder' is, of course, to what came to be called the elevator. It was often referred to as the 'horizontal rudder'.

one. The flights were directly against the wind. Each time the machine started from the level ground by its own power alone with no assistance from gravity, or any other source whatever. After a run of about 40 feet along a monorail track, which held the machine eight inches from the ground, it rose from the track and under the direction of the operator climbed upward on an inclined course till a height of eight or ten feet from the ground was reached, after which the course was kept as near horizontal as the wind gusts and the limited skill of the operator would permit. Into the teeth of a December gale the 'Flyer' made its way forward with a speed of ten miles an hour over the ground and thirty to thirty-five miles an hour through the air. It had previously been decided that for reasons of personal safety these first trials should be made as close to the ground as possible. The height chosen was scarcely sufficient for maneuvering in so gusty a wind and with no previous acquaintance with the conduct of the machine and its controlling mechanisms. Consequently the first flight was short. The succeeding flights rapidly increased in length and at the fourth trial a flight of fifty-nine seconds was made, in which time the machine flew a little more than a half mile through the air, and a distance of 852 feet over the ground. The landing was due to a slight error of judgment on the part of the aviator. After passing over a little hummock of sand, in attempting to bring the machine down to the desired height, the operator turned the rudder too far; and the machine turned downward more quickly than had been expected. The reverse movement of the rudder was a fraction of a second too late to prevent the machine from touching the ground and thus ending the flight. The whole occurrence occupied little, if any, more than one second of time.

Only those who are acquainted with practical aeronautics can appreciate the difficulties of attempting the first trials of a flying machine in a twenty-five mile gale. As winter was already well set in, we should have postponed our trials to a more favourable season, but for the fact that we were determined, before returning home, to know whether the machine possessed sufficient power to fly, sufficient strength to withstand the shocks of landings, and sufficient capacity of control to make flight safe in boisterous winds, as well as in calm air. When these points had been definitely established, we at once packed our goods and returned home, knowing that the age of the flying machine had come at last.

From the beginning we have employed entirely new principles of control; and as all the experiments have been conducted at our own expense without assistance from any individual or institution, we do not feel ready at present to give out any pictures or detailed description of the machine."

A copy of this statement was sent to *L'Aérophile*, and published in full in its issue of January 1904, under the boldly printed title "La Machine

Volante des Frères Wright". The Editor then added his own comments, which were strangely mixed in sentiment:

"Mr Orville Wright, protesting against the ultra fantastic accounts of the experiments at Kitty Hawk, addresses to us a faithful and very interesting recital of the first aerial flights with a man-carrying 'heavier-than-air' [machine], moving under its own power. . . .
  Any commentary would detract from the value of this document, which is nevertheless marred, we must point out, by several obscurities. Among other things, Mr Orville Wright does not tell us the difference in the level between the starting and landing points. Notwithstanding, the experiment was a fine one and merits our warmest applause.
  Gliding flight, so vigorously launched in France by Mr Archdeacon, will not be long in bearing its own fruit. What do we lack? A few specialists trained in the tricks of the trade (quelques spécialistes rompus déjà aux tours de mains du métier)."

Although Ernest Archdeacon was an admirable man, to say that he had "vigorously launched" gliding flight in France was indeed somewhat odd! For it was to be exactly a year from his write-up of Chanute's lecture in April of 1903 to the first test of the Archdeacon Wright-type glider in April 1904: and there was no one else in the field as yet except Ferber.

The caution about the Wrights' statement expressed by *L'Aérophile*, but not the courage-sustaining whistle about 'tricks of the trade' at the end of the comments, was shared by the few other serious writers who felt the subject worth dealing with. They tended to play down the Wrights' flights of December 17th by treating them as extremely tentative—as indeed they were—but failing to realise what such tentative success obviously portended.

Four days after the Wrights released their press statement, Chanute wrote (on January 9th) to his old and distinguished friend in England, F. H. Wenham:

"To you, who first called attention to the possibility of artificial flight, 38 years ago, I send the first correct account which has been published of the Wright brothers.[1] It is a beginning and, if no accident occurs, may lead to practical results; but the uses will be limited."

Chanute must have been in both a gloomy and forgetful mood; forgetful because he knew perfectly well that Cayley had vividly drawn attention to the possibility of artificial flight as far back as 1809; and he himself, in his *Progress in Flying Machines* (1894) had written:

[1] Chanute enclosed the Wrights' statement to the Associated Press of January 5th 1904.

"It seems difficult, therefore, to forecast in advance the commercial results of a successful evolution of a flying machine. Nor is this necessary; for we may be sure that such an untrammelled mode of transit will develop a usefulness of its own, differing from and supplementing the existing modes of transportation. It certainly must advance civilisation in many ways, through the resulting access to all portions of the earth, and through the rapid communications which it will afford."

The aged Wenham, on his part, could only commiserate. He answered Chanute on April 7th 1904 saying:

"I see by your concluding paragraph that you have somewhat lost faith in the possibilities of flying machines proper, and their utility; but no one has yet arrived at the conditions required for flight, that is, horizontal continuous speed, and a certainty of obtaining equilibrium of stability. This deterred me in my early experiments. . . ."

Chanute had also written to Ferber, enclosing a copy of the Wrights' press statement: here is Ferber's answer, dated from Nice, January 27th 1904:

"I thank you heartily for your letters of Jan 1st and 7th which have interested me very much. When I learned on the 21st of December that Wright had succeeded with his motor, I was at first quite annoyed at not having been able to take this first step myself. But now, just think that this success of Wright is doing me lots of good, and is much to my advantage. I believe that people are now saying: 'Why, that Captain was not such a fool after all, as the other chap has met with success'. I would like to know whether Wright had already begun on his motor last June, or whether it was the news that I was on the point of experimenting with one, which determined him to apply a motor himself? Archdeacon is very active, and hence I believe that not fewer than 6 apparatuses of the 1902 Wright type are now being built in France. I believe we will see a great movement."

There is a certain pathos in Ferber's question as to whether the Wrights were spurred on to make their powered flights by hearing the news of his intended experiments. The European Wright-type glider programme he mentions was confined chiefly to himself and Ernest Archdeacon, and then later to Esnault-Pelterie.

But Ferber was also to write to Archdeacon at this time. The exact date of this letter is not known, but here is Ferber's letter in full; which was quoted in an article by Philippe Rey in *L'Automobile*, issue of February 27th 1904 (see also Section 31):

"I finally have a letter [*ie* from Chanute]. It is not as wonderful as they say.

The machine has 47.3 square meters of surface, is 12.9 meters across, weighs 338.61 kg. mounted on wheels, with 12 horsepower motor.

It started from a little slope on the plain and went for fifty-nine seconds against a wind of 10 meters a second. It travelled only 266 meters, four times in a row, on December 17. The cold drove away the experimenters, who will not begin again until next season. That gives us a respite of six months.

The experiment is not as grand as we thought, but it nevertheless represents a new fact. For the first time, a heavier-than-air machine (338 kilograms!) has flown in horizontal flight for 266 meters.

My machine, weighing about 225 kilograms, with 50 square meters and only 6 horsepower, ought to fly too. I am going to spend some more money to make sure; but, really, I think we ought to unite our efforts."

❈ 26 ❧

The first published Picture of the Wright Flyer I: January 1904

I am still unable to trace the complete history of the first published picture of the Wright Flyer (Fig. 17); but I know that it was published in the *New York Herald* of January 11th 1904; and that it appeared in the leading German aeronautical periodical, the *Illustrierte Aeronautische Mitteilungen* for March 1904. I cannot as yet discover whether it appeared in any of the French dailies or periodicals. But whatever its history, this illustration did not have much impact on the European pioneers. The reason, presumably, was that it looked almost exactly like the French thought it would look, following Chanute's illustrations of the Wrights' modified No 3 glider of 1902, and his remarks about the Wrights' intention to make a powered machine. The illustrator had simply taken the modified No 3 glider, and positioned two pusher propellers behind the trailing edges of the lower wings, without any indication of where the engine might be. This picture cannot be described as a 'pirated' picture, because it was clearly not made up from the statements of any reliable witness, or from an actual photograph of the Flyer: it was a simple essay in probabilities, based on known facts about the Wrights' glider.

❦ 27 ❦

Archdeacon's Wright-type Glider is announced: January 1904

It was also in the January (1904) issue of *L'Aérophile* that mention was first made of what was shortly to become the most significant focus of aviation attention in Europe, the Archdeacon Wright-type glider. As noted before, Ernest Archdeacon had become the outstanding driving force in European aviation, especially since his forming the Aviation Committee of the Aéro-Club. But the procrastination displayed in getting this first glider off the ground—a year after Chanute's lecture— is extraordinary.

L'Aérophile noted that on January 22nd, members of the Aviation Committee had visited the military establishment at Meudon (Chalais-Meudon) where they had "examiné avec intéret l'aéroplane de M. Archdeacon, type Wright 1902". Archdeacon himself was to describe it next month (see *L'Aérophile* for February) as:

"A machine which is, apart from subsequent modifications, exactly copied from that of the Wright brothers (exactement copié sur le modèle des frères Wright)."

With the advice of Colonel Renard, the glider had been built by the chief craftsman at Meudon, M. Dargent, "un habile modeleur". Notice was given that the glider would be taken to the Aéro-Club's "parc d'aérostation", near Saint Cloud, on February 20th, where it could be inspected by those interested, until the 29th of the month. Then, on March 1st, it was to be transported to Merlimont, near Berck-sur-Mer, where tests would start immediately, under the direction of Archdeacon.

One cannot help feeling that it was the disturbing news from America, in December, that had hastened Archdeacon into production with his first glider: one also cannot help wondering why—after his impassioned plea in April 1903—he had done nothing about glider-construction for so many months.

With this news, we will leave this important machine until a little later, as the chronological march of events is of particular interest at this point.

28

Archdeacon's continued Efforts to launch Gliding Competitions: January 1904

The Wright brothers, even before Europe received news of their first powered flights of December 17th 1903, were again haunting the Europeans with their unseen but potent presence. During a meeting of the Aéro-Club on January 7th 1904—reported in *L'Aérophile's* issue of February—Archdeacon turned again to his pet project, the promotion of gliding competitions. He informed his fellow members that the prize offered—

> "had the object of encouraging the efforts of those pursuing the path of aviation ... [he] concluded by appealing to his colleagues on the Committee, and praying them to be willing to promote it [the idea of gliding competitions] in order that France—which has always occupied the front rank in this field—shall not be outdistanced by the United States."

29

Tatin's Protest and Exhortation: February 4th 1904

At a dîner-conférence at the Aéro-Club on February 4th 1904, Victor Tatin—perhaps the most respected of all veteran French pioneers at that time—gave a lecture entitled "L'Analyse des Expériences d'Aviation", which was reported in the February 1904 issue of *L'Aérophile*. As with Archdeacon the previous year, Tatin was genuinely outraged at the idea that the final conquest of the air might be achieved abroad, and he took the opportunity of reminding his audience at the start that "l'aviation est bien une science française". He proceeded to examine some of the problems involved in powered aviation, then quoted the figures for his own famous compressed-air model of 1879, and those of others. Towards the end of his lecture, Tatin inevitably touched on the subject which then occupied so many minds, the Wright brothers. The report of his lecture included the following:

> "M. Tatin then said a few words about the experiments made last year by the Americans with man-carrying powered machines. The reports, very incomplete and often contradictory, leave him somewhat sceptical, and he would only accept the validity of the results announced with the greatest reserve. In any case, he said, the problem cannot be considered as completely solved by the mere fact of someone having flown for less than a minute; and, moreover, under conditions with which we are not very well acquainted."

The last paragraph of this *Aérophile* report also includes some revealing passages:

> "In closing, M. Tatin protested against the tendency we seem to have in France of slavishly copying (à copier vigoureusement) the gliding machines of the Americans.... And then, where do such copies lead us? Does this not seem like a confession of our incapacity to make anything original ourselves? Nevertheless, it would appear that we still have in France some men of genius capable of successfully

carrying out such work without putting ourselves in tow of foreigners (sans nous mettre à la remorque de l'étranger) who in any case do not seem to be on the best track (meilleure voie)."

Then, regardless—or perhaps not?—of Archdeacon's work, and of the latter's Wright-type glider; and also possibly ignorant of the fact that Archdeacon was to follow him on the rostrum in a minute or two with more mention of the Wright-type glider, Tatin made this impassioned appeal:

"Must we one day read in history that aviation, born in France, only became successful thanks to the Americans; and that the French only obtained results by slavishly (servilement) copying them? For us, that would indeed be glorious! Have we not already seen enough French inventions completed (terminées) by foreigners, such as the steam-engine, gas light, steam-ships, and many others? Alas, are we to add aviation to them? As we have engineers ready to go to work, I think it would be a disgrace to remain behind any longer, since there is certainly still time. It is in France that the first journey (voyage) by a flying machine must be made. We need only the determination. So let us get to work!"

The Editor of *L'Aérophile* then expressed agreement with Tatin's ideas, and pleasure at the applause which greeted this "brilliante causerie" by "le savant aviateur".

❦ 30 ❧

Archdeacon's Lecture
on the Wrights and on the Position of Aviation in France: February 4th 1904

Archdeacon followed Tatin to the rostrum at the same dîner-conférence on February 4th 1904, to tell of the Wright brothers and France, and to speak of his own proposed gliding competition. But before reporting him verbatim, the Editor of *L'Aérophile* (in the February 1904 issue) included the following among his introductory comments on Archdeacon's lecture:

> "M. Archdeacon has indeed chosen the best means to draw the attention of public opinion to aviation, in inaugurating in a practical manner his effective campaign by testing the machine [*ie* the Wright-type glider] which we already know about, the results of which are assured. But, in the spirit of their promoter, the first tests at Merlimont are above all intended to stimulate French aviators, to urge them to emulation; in brief, to promote individual study and research; to lead us, we hope—by the employment of new means and new apparatus, or by the better application of those we know about already—to the practical machine which everyone awaits, and which will allow man finally to annex to his domain the uninhabited vastness of the atmosphere (l'engin pratique, que tout le monde attend, et qui permettra à l'homme d'annexer définitivement à son domaine les immensités libres de l'atmosphère) . . ."

Archdeacon, when he rose to speak, introduced his address with these words:

> "For many years I have been possessed by an interest in the experiments undertaken by the late Otto Lilienthal, then by the Americans; and I have always dreamed of acclimatising (d'acclimater) aviation studies in France. I have already said, and repeated everywhere, that we must make haste, on pain of being overtaken by the Americans, especially by Chanute and the Wrights, who have pushed much farther ahead in this kind of study than we have. We are, however, exceptionally well placed to succeed, since we are the real fathers of

the light motor (les vrais pères des moteurs légers). What I predicted has now come about. Despite various contradictions, and a fair number of exaggerated reports published in the periodicals, it appears to have been established that the Wrights have recently succeeded in making a flight (planement[1]) of 266 metres, with a machine of 350 kg., mounted on wheels [sic] and fitted with a motor *from which it derived the whole power of sustentation (dans lequel il puise toute sa force de sustentation)* [the italics are Archdeacon's]. It is certain, Gentlemen, that the results obtained are considerable, and—I do not cease to repeat—we must hurry if we wish to catch up with the enormous advance (rattraper l'avance énorme) made over us by the Americans. As you all probably know, since April of last year, desiring to accomplish the impossible—to push aviation studies in France (voulant faire l'impossible pour pousser en France les études d'aviation)—I have subscribed a preliminary sum of 3,000 francs, and have founded, within the Aéro-Club, the Committee for Aviation Tests, of which you know the members. . . . To obtain anything from inventors in France, it is necessary to stimulate rivalry (pour obtenir quelque chose des inventeurs en France, il faut exciter l'emulation); and, to stimulate rivalry one must—and the automobile has well proved this—establish races or competitions. . . ."

Archdeacon then went on to speak of his proposed gliding competitions, and of the aerodynamics of gliding. Then he dealt with the construction of his own Wright-type glider, which was—as quoted before— "sauf modifications ultérieures, exactement copié sur le modèle des frères Wright". He continued:

"I would announce, at the same time, that it has been decided in principle for the competitions to take place in the outskirts of Berck, on the sand-dunes of Merlimont; and I do hope that some of you will take part. M. Drzewiecki, who is known to you all, has an aeroplane under construction, as also has M. Mallet; M. Girardot, the well-known automobile manufacturer, is about to start one; lastly, M. Balsan, who is shortly going to America, will bring back with him, when he returns, a machine specially built for him by the Wright brothers themselves."[2]

He then notes the progress of subscriptions, and concludes:

"So, as you see, we are going along as well as possible; and provided that we continue, we shall rapidly catch up with—and even overtake —the Americans. . . . But, to succeed in this difficult task, we must

---

[1] The word 'planement' was generally used at this time as a synonym for 'vol-plane', meaning a gliding flight, as was 'glissade', which had been introduced last year by Chanute. But the context immediately below makes it clear that Archdeacon was referring to a flight in which gravity played no part. The word 'vol' soon came to mean a powered flight.

[2] This was a pious hope, and never came to pass.

really have the help of all (if faut absolument l'aide de tous). I am therefore counting on you, Gentlemen, not only as possible participants in our competition, but as those to spread the word far and wide; to recruit 'proselytes'; and bring in subscribers. It is absolutely essential to ensure for France the glory of the ultimate conquest of the air, which the homeland of Montgolfier must not let fall into the hands of foreigners."

Here, then, as a result of the news of the Wrights' first powered flights in 1903, is yet another strong 'influence-impact' being exerted by the brothers on European aviation, in that the news of their flights gave a strong new impetus to the building of Wright-type gliders.

But perhaps the most astonishing feature of this lecture of Archdeacon's was that there was now, in February of 1904, a virtual admission from the leader of French aviation that the Wrights had in fact made sustained powered flights—flights with a machine in which "toute sa force de sustentation" was derived from the motor—accompanied by the almost complacent exhortation to his colleagues to "hurry", if they wished "to catch up" with the Americans.

With Ferber still only tentatively gliding on his still unsophisticated machine in far-away Nice, and with no recent news of him to hand; with Archdeacon's own copy of the Wright glider still untested; and with the Wrights—after three years of glider development—now taking to powered machines; one wonders at the strange state of self-delusion the Europeans had reached. It was almost as if Archdeacon and his friends were only half-conscious of the true state of affairs, and were not allowing their minds to meet their emotions, in case such a confrontation might lead to despair!

31

Ferber writes to Archdeacon about the Wrights: February 1904

The issue of the French journal *L'Automobile* for February 27th 1904 contained an article by Philippe Rey entitled "Gliding Competitions (Les concours de vol plané)", in which he described Archdeacon's progress in building his Wright-type glider. But what concerns us here is that Rey quoted in full, in this article, the letter from Ferber to Archdeacon which I have already quoted, which certainly could not have pleased the latter. The date of the letter was not given, but it could have been in mid or late January, or early February. I feel it is worth quoting again here, in full:

"I finally have a letter [Chanute to Ferber, Jan. 1 and Jan. 7, 1904; a copy of the Wrights' press statement of Jan. 5 was enclosed in the latter]. It is not as wonderful as they say.

The machine has 47.3 square meters of surface, is 12.9 meters across, weighs 338.61 kg. *mounted on wheels*, with 12 horsepower motor.

It started from a *little slope* on the plain and went for fifty-nine seconds against a wind of 10 meters a second. It traveled only 266 meters, four times in a row, on December 17. The cold drove away the experimenters, who will not begin again until next season. That gives us a respite of six months.

The experiment is not as grand as we thought, but it nevertheless represents a new fact. For the first time, a heavier-than-air machine (338 kilograms!) has flown in horizontal flight for 266 meters.

My machine, weighing about 225 kg., with 50 square meters and only 6 horsepower, ought to fly too. I am going to spend some more money to make sure; but, really, I think we ought to unite our efforts."

This somewhat mixed-up piece certainly did not come from the Wrights, and must have been put together from bits and pieces of fact and gossip. The flights of December 17th did not start from a little slope; only the abortive attempt made on December 14th started down a slope.

❈ 32 ❧

The Aéro-Club's Tribute to Ferber and Archdeacon: March 3rd 1904

The mounting influence of the Wrights was curiously reflected in the speech made by another great French pioneer, the Comte Henry de La Vaulx, when, as acting President of the Aéro-Club, he opened its General Assembly on March 3rd 1904. In the course of a somewhat flowery address, he referred to:

> "A new branch (nouvelle branche) of aeronautics, aviation—which had not yet received the official 'consecration' of our Society—has been able to win the 'freedom of the city' among our labours (conquérir droit de cité parmi nos travaux) thanks to the persevering studies of Captain Ferber and the impetuous generosity of M. Ernest Archdeacon."

We have here the two men who were directly influenced by the Wright brothers to precipitate practical European aviation, being publically acclaimed as those who "won the freedom of the city" for a "new branch of aeronautics".

The acting President was then followed by the Secretary-General of the Club, Georges Besançon—also Editor of *L'Aérophile*—who included these words in his speech:

> "A new phenomenon, and one of capital importance, merits our attention. I should like to speak of the active and fruitful campaign in favour of experiments in aviation, revived by a man of good heart and initiative, who is M. Ernest Archdeacon. This movement could have very great consequences (consequences énormes), of which you know the extent and depth. From the words of M. Archdeacon, and carried along by his example, the French aviators have become aware of their worth, and of their responsibilities ... M. Archdeacon can be proud of the magnificent 'élan' with which he has succeeded in inspiring aviation studies. The splendid success attained by his initiative provides the brilliant (éclatante) demonstration of its

Hargrave design for a powered Multiplane: 1902 The drawing shown here, which appeared in the August 1902 issue of the French journal *L'Aérophile*, is of a design by the great Australian inventor, Lawrence Hargrave. It is a multiplane box-kite machine with a smaller but similar tail-unit, and a tractor propeller. It may well have been this drawing which was later to help inspire the box-kite machines of the Voisin brothers.

Ferber Wright-type Glider (with rudders): 1903 The slightly improved Wright-type glider, built and tested by Ferber in 1903. Again the poverty of construction is all too evident, and the addition of the two wing-tip rudders is the only obvious modification, which would have had little or no beneficial effect on the machine's behaviour. Ferber at least had the good sense to adopt the Wrights' prone-pilot position.

14 Ferber powered Wright-type Glider: 1903 Despite the design and construction of the Ferber machines, this French Captain showed great energy and acumen in his pursuit of practical aviation. Long before he could build even a passably successful Wright-type glider, we find him here making a powered version of a Wright glider, with a small engine and enormous propellers, and two waving wing-tip rudders. At great expense he then built a huge whirling arm, near Nice, to test it in 1903. Needless to say the whole enterprise was a failure.

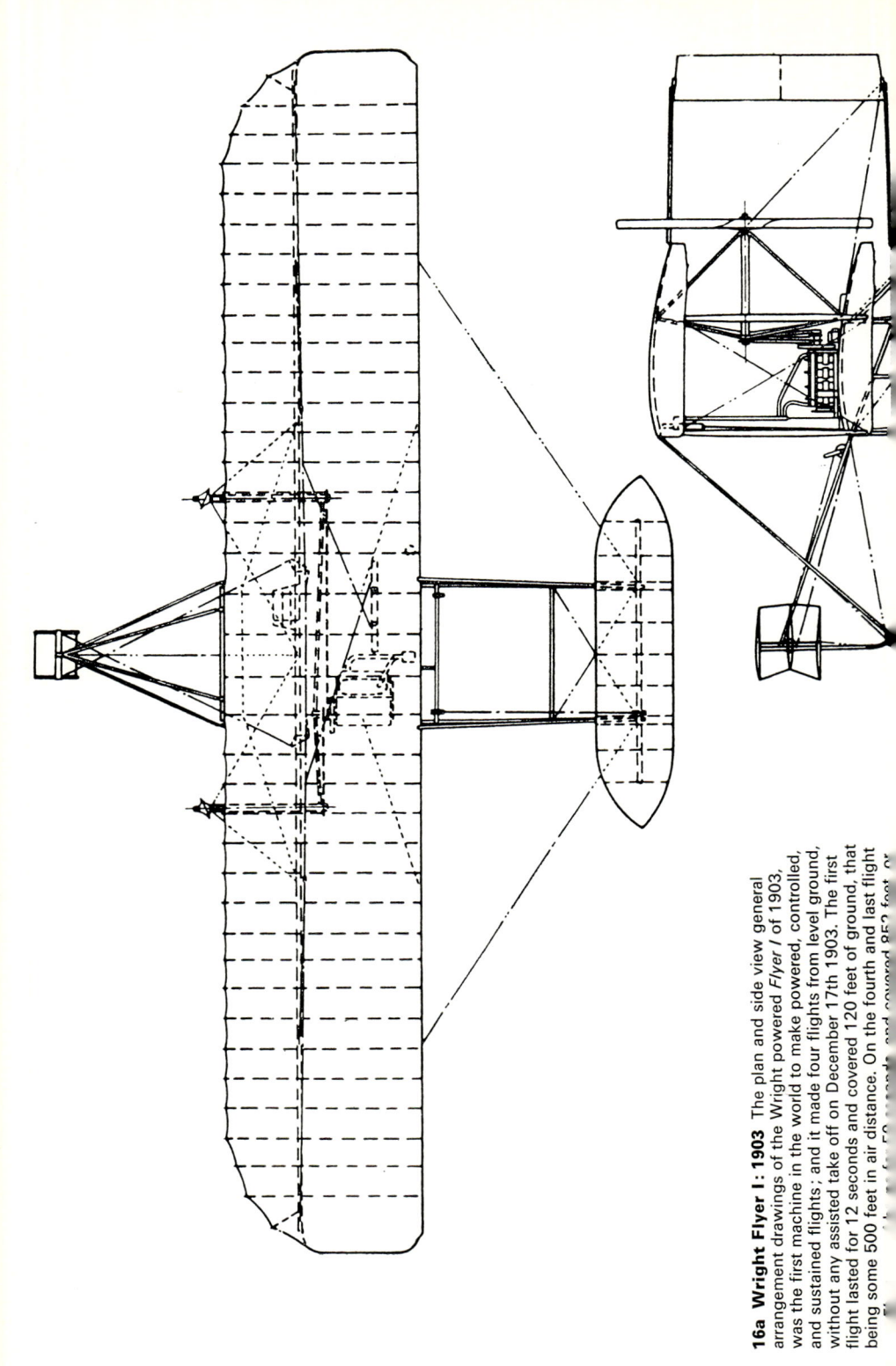

16a Wright Flyer I: 1903 The plan and side view general arrangement drawings of the Wright powered *Flyer I* of 1903, was the first machine in the world to make powered, controlled, and sustained flights; and it made four flights from level ground, without any assisted take off on December 17th 1903. The first flight lasted for 12 seconds and covered 120 feet of ground, that being some 500 feet in air distance. On the fourth and last flight

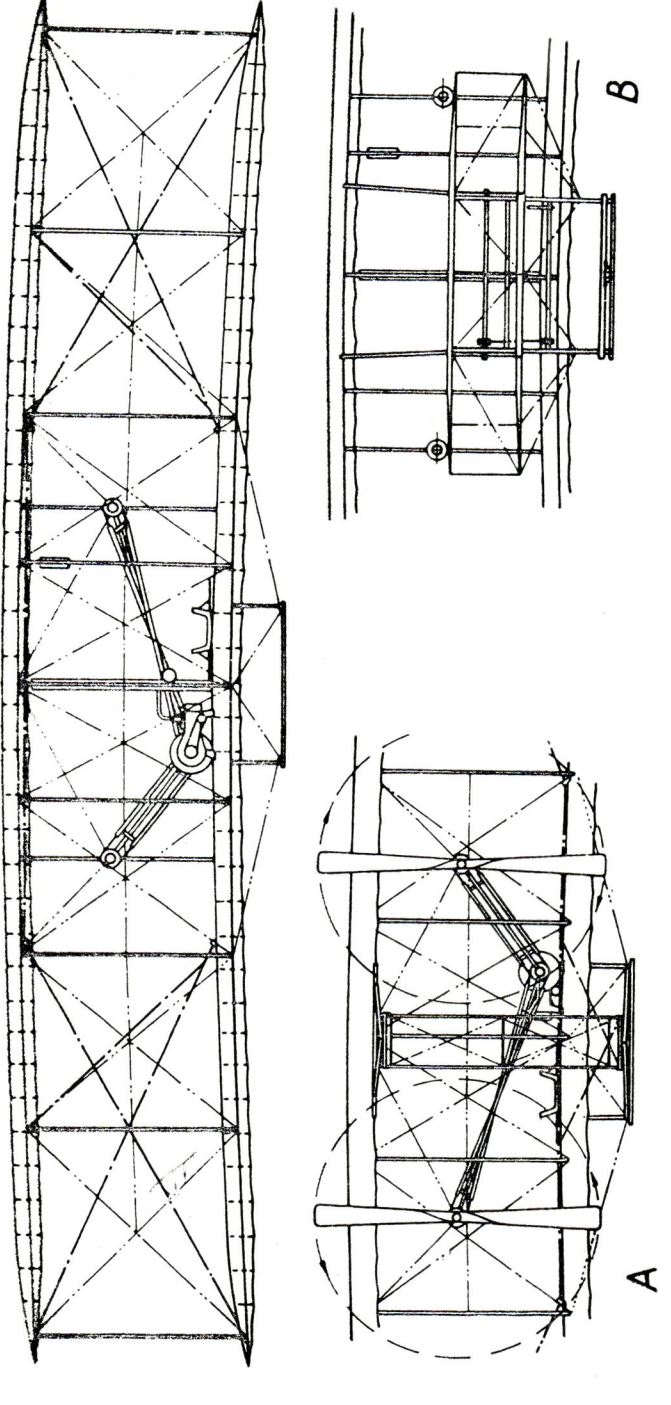

16b Wright Flyer I: 1903 The front and rear elevations of the Wright *Flyer I* of 1903. The machine was rigged with a very slight negative dihedral. It had a skid undercarriage, which lay across a yoke, which in turn (on two stripped cycle-hubs) ran along a wooden rail. After the engine was run up, the machine was released, and ran freely along the rail until it acquired enough lift to raise it from the yoke and become airborne. The 12 hp engine drove two counter-rotating propellers.

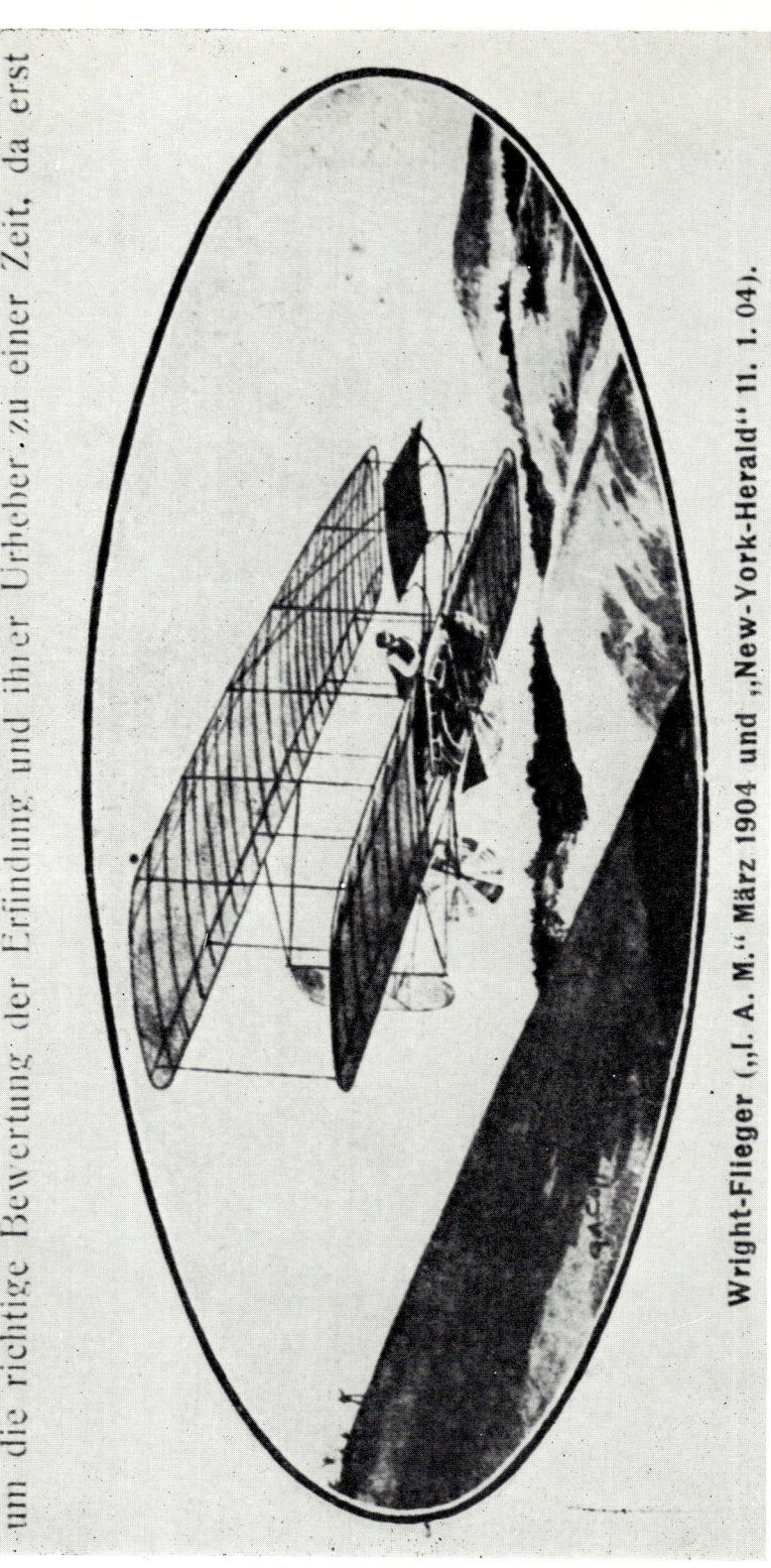

17 Wright Flyer I: 1903 This was the first pirated and published picture of the powered Wright *Flyer I*. It appeared originally in the *New York Herald* for January 11th 1904 and then in the March 1904 issue of the German journal *Illustrierte Aeronautische Mitteilungen*. It is of course, based on the published photographs of the No 3 Wright Glider (modified) of 1902, with the addition of two propellers. Except for the forward elevator of the *Flyer I* being a biplane structure, and the rear rudder a double unit, it is quite a

La pratique du vol plané

Nous avons déjà donné (*Aérophile* d'août 1903), sous la signature de l'homme [le] plus compétent en la matière, M. Chanute, une description complète et détaillée [d]es divers appareils de « vol plané » usités en Amérique. Mais peut-être n'avons-[n]ous pas assez insisté sur la pratique proprement dite des glissades aériennes, et [c']est là un point essentiel.

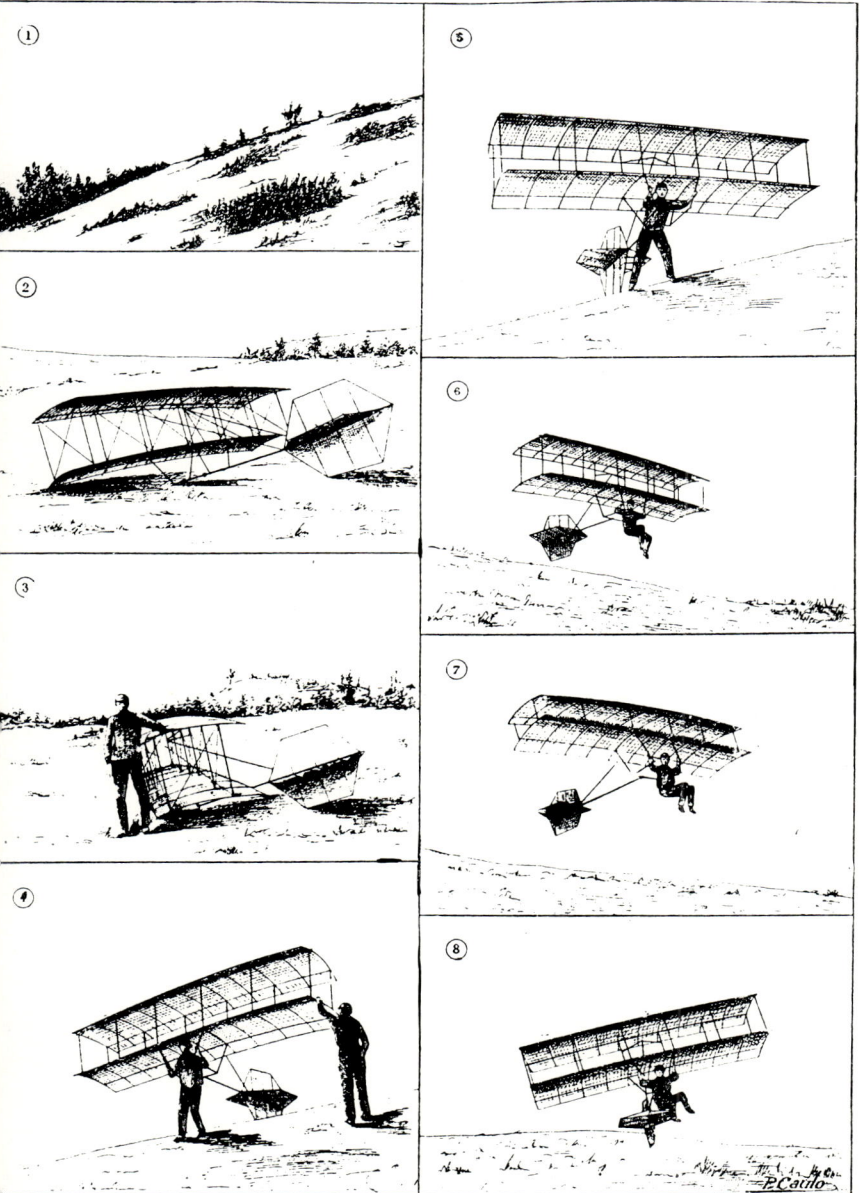

8 Chanute Hang-glider: 1896 A spread of drawings which appeared in *L'Aérophile* in its issue of March 1904, giving a sequence of drawings showing how to launch and pilot a Chanute glider.

19a/b Archdeacon Wright-type Glider: 1904 Inspired by the example of Captain Ferber, a ric(h) Parisian lawyer, Ernest Archdeacon, commissioned a Wright-type glider to be built, and had it tested (at) Berck-sur-Mer by a newcomer to aviation, Gabriel Voisin from Lyon. This was Voisin's debut, and afte(r) testing the Archdeacon glider he conceived the idea of designing and constructing machines on his ow(n) and thereafter was to become one of the leading European pioneers. The Archdeacon glider was again (a) poor copy of the Wrights' No 3 machine, but much better constructed than Ferber's, with at least a rigi(d) structure. But strangely enough there was no attempt to incorporate the Wrights' wing-warping, or indee(d) any other control in roll. The machine was a failure, but it provided an inspiration for others.

necessity. The Aéro-Club de France is happy to have furnished him with the efficacious and necessary assistance which he desired, in order to maintain French aviation at the level at which it should be preserved—that is to say, the highest."

⚙ 33 ⚙

L'Aérophile publishes an Article on How to Fly a Chanute Glider : March 1904

Ever aware of the practical side of flying, and of the urgent need to stimulate its readers to make practical experiments, *L'Aérophile* published, also in its March 1904 issue, the world's first application to flying of what one might call "how to do it in pictures"; but unfortunately the Editor picked the wrong aircraft. Written by G. Blanchet, and entitled "La pratique de vol-plané", the article presented in a sequence of 18 line-drawings (with descriptions), the suitable terrain, the launching, the flight, and the landing, of the classic Chanute biplane hangglider of 1896 (Fig 18). These drawings were made from a series of photographs provided by Chanute. The author of the article referred to the previous article in *L'Aérophile* by Chanute on gliders, but felt, rightly enough, that there had not been enough attention paid to the actual practice of gliding: hence his article.

This article was a purely negative contribution to aviation—as the author seems almost to admit—and was more suitable for the previous, rather than the present, generation of would-be experimenters. For M. Blanchet took care to point out that this picture-story of instruction concerned the Chanute machine, whereas, as he says, "the Wright machine, which has inspired M. Archdeacon, is later in date and more perfected (plus perfectionné)". He then unwittingly, but hopelessly, misleads the reader by adding, "but in the two cases, if the parts of the machine differ a little, the flight principle is the same (si les organes diffèrent un peu, le principe du vol reste identique)". To thus equate a hang-glider with a glider fully equipped with pilot-operated control-surfaces—M. Blanchet incidentally omits any mention of wing-warping, and refers only to elevator and rudder control—and tell the would-be pilot that the principle is the same, betrays a hopeless confusion in the minds of both the Editor, and the author of this article. The author closed his article with a detailed description of how to construct a Chanute glider, which was the last thing anybody wanted to know!

It is hard to imagine what Besançon (the Editor) thought he was about in publishing this article by Blanchet: it was almost as bad as telling his sophisticated Aéro-Club colleagues to go back to their youth and start flying kites again. For it should have been obvious to him and to all the better-informed members—especially after Chanute's explicit descriptions—that gliders must be properly pilot-controlled. All this, of course, should also be viewed within the context of the Wrights having already mastered controlled glider-flight two years earlier, in 1902, and having recently achieved tentative powered flying.

It is interesting to note that not a single European pioneer took to building and experimenting with Chanute-type gliders as a result of Blanchet's article; and no Chanute-type glider of any kind was built in Europe until 1907, when Gabriel Voisin produced his short-lived and sterile machine, as already described.

❧ 34 ❧

Further Efforts on behalf of Gliding
also Proposals for a Grand Prix: March 1904

Spurred on by the news from the Wright brothers, Archdeacon drove ahead with his idea of promoting glider competitions; and in the March issue of *L'Aérophile* his scheme was announced officially by the Editor, Besançon, along with the first list of subscribers, the article being entitled "La Souscription pour les Concours 'Vol Plané', et le Grand Prix d'Aviation".

Archdeacon himself headed the list with a generous contribution of 3,000 francs: there were twelve others listed, and the total amount received was a fraction under 5,600 francs. Archdeacon's idea was that such competitions would be

> "the best means of arousing fruitful rivalry among our aviators; to encourage their work, and to reward their success."

But even more significant was the latest news, also announced by Besançon in this article; that Henry Deutsch de la Meurthe and Archdeacon had each come forward with a cash prize of no less than 25,000 francs, for the first man to fly a kilometre circle in a powered aeroplane: this was to be the "Grand Prix d'Aviation". As Besançon said, this

> "Grand Prix d'Aviation—still in process of organisation—will crown in a deserving manner the complete solution of the problem [*ie* of powered flight]. But this great discovery is still, without doubt, rather distant. It necessarily remains to encourage and assist the investigators, to reward the more modest results, it is true, but immediately realisable and progressive stages toward complete success."

35

Ferber's First Article in the *Revue d'Artillerie* : March 1904

An important article by Ferber appeared in the March 1904 issue of the *Revue d'Artillerie*, and was entitled "Les progrès de l'aviation, depuis 1891, par le vol plané (The Progress of Aviation, by gliding, since 1821)". It is difficult to say how widely this article was circulated among the French aviation enthusiasts; it was probably circulated fairly widely, as Ferber was well known, and he himself would obviously tell his friends about it. It is a valuable historical document, and is itself well documented with dates and foot-notes; it described and illustrated the experiments of Lilienthal, Pilcher, Chanute, and the Wrights, before dealing with his own experiments to date; and contained no less than 40 excellent photographs and diagrams: unfortunately the *Revue* was printed on bad paper, which by today has considerably deteriorated.

Ferber's treatment of the Wright brothers is of great interest. As remarked before, I find it hard to understand how Ferber, who received —as we know—the Wright material from Chanute early in 1902, and who must have read one or more of the accounts of Chanute's famous 1903 speech, as well as the latter's *L'Aérophile* article, does not once mention the Wrights' wing-warping or problems of lateral control in general. This omission strengthens my belief that Ferber had—and was always to have—certain blind aeronautical spots, which explain why he progressed so slowly. Even though he later had the sound idea of adding a tailplane to his Wright-type glider, he never in his life achieved a successful glider, or a powered aircraft. He even goes so far as to say this of the Wrights:

"In 1900 and 1901, they had no vertical rudder; they changed direction by straightening up the elevator (ils changeaient de direction en dégauchissant le gouvernail horizontal). In 1902, in order to make more important movements (pour pouvoir faire des mouvements plus importants), they added at the back a vertical directional rudder, and commenced to describe quarter-circles."

A more astonishing passage it would be hard to find, especially when one considers what material he had to hand, and is known to have read, since early in 1902.

In this connection, and in support of my belief that Ferber did not understand even the rudiments of the Wrights' control system until he saw it in operation in 1908, we find in his book *L'Aviation* (1909) that he has slightly changed the above-quoted passage on the Wrights' gliding and it now reads:

> "In 1900 and 1901, they had no vertical rudder; they changed direction by warping the elevator and the lifting surfaces (en gauchissant le gouvernail horizontal et les surfaces portantes)."

It is still puzzling to find him using the verb "gauchir" of the elevator, as well as of the wings, since by the time he was writing this (1908-09) the word was only applied—aeronautically—to the Wright-type of wing-warping.

Returning to this Ferber article of March 1904, he then told his readers briefly of the Wrights' powered *Flyer I* of 1903, and—as quoted previously—announced without quibble that the date of December 17th 1903 marks the day when

> "for the first time a piloted flying machine had *really flown*, and the honour of this memorable experiment falls to the name of Wright."

Then, speaking further of powered flight, he says:

> "the tentative success of the Wright brothers seems to show that the solution is approaching."

Ferber went on to describe his own experiments in 1902 and 1903. Towards the end of his article he wrote:

> "We are often asked why it was necessary to copy the Americans. It is useless to copy them slavishly, and it would be annoying to fall foul of their patents (de tomber sous le coup de leur brevets). Today, with the light motors we possess, every kind of aeroplane should be able to fly (toute espèce d'aéroplane doit flotter). But we must not forget Lilienthal's principle, to 'know how to manage one's aeroplane is everything (savoir se mettre dans son aéroplane est tout)'. During three years, with rational types of machine, we have not been able to achieve a flight, because we have not known how to manage our machines. Today we would not fear to take our place in a Wright machine of 300 kg. because we have 'mounted' one of 100 kg.; but we would not perhaps risk taking off in a Tatin and Richet machine— which ought to go well enough when all is said and done—without

having passed through a long and delicate series of preliminary tests."

To close the article, came another of the familiar exhortations:

"Now, at this hour, there is not an instant to lose; and that is why we advise people temporarily (momentanément) to follow the track cleared by the Wrights."

✌ 36 ✌

Other French Experimenters: March 1904

Returning for a moment to the March issue of *L'Aérophile*, the article which was ostensibly on the tests of the Archdeacon Wright-type ("L'aéroplane Archdeacon et les expériences de Merlimont") was in fact divided into two: the first part was on Archdeacon, and the second dealt briefly with other experimental work being carried out in France. At the head of the experimenters, stated the author of the article, stood Ferber. The reader was reminded that

> "Captain Ferber, who for long had been the only French disciple of Lilienthal, and had made the only gliding experiments which we could set against the aerial glides of the American aviators (les seules expériences de vol plané que nous puissions opposer aux glissades aériennes des aviateurs américains)."

The following were listed as working on aeroplanes of one kind or another: MM. Drzewiecki, Mallet, Herbster, Bonnecasse, Girardot, Robart, and Lavezzari. Of these, Robart and Lavezzari built monoplane gliders which were tested briefly and came to nothing; and Robart —far away in 1908—was to build a sterile and unsuccessful biplane. None of these men was to occupy even a modest niche in the history of aviation.

But the *Aérophile* writer noted that two traditions were now starting in Europe, the biplane and the monoplane:

> "Not two schools, but two tendencies, are manifesting themselves among our aviators. The first of these, with Captain Ferber, the precursor, seems to incline towards machines with two superposed surfaces: the others, such as MM. Robart and Lavezzari, prefer, on the contrary, machines with only one surface."

He might have added the name of Levavasseur, who had started experimenting in 1903 with a monoplane, which was unsuccessful.

Levavasseur was, of course, later to become one of the giants of aviation. In this connection, it is interesting to look ahead for a moment to Ferber's London Lecture given in February 1908 (see Section 85), in which he was to make these statements:

"I ought to point out, finally, that the cellular form of the heavier-than-air machine, which is used at present, and which was introduced into France by my influence, is a bad solution because it presents great resistance to forward motion. It is not the French solution—the French solution being that which Pénaud indicated in 1868, and which he exemplified in model form with an indiarubber driving spring. His model is monoplanar [sic] and approaches very nearly to the form of a bird, there being a long tail for steadiness, and rudders behind."

Returning to the *Aérophile* article, we find the author—it was signed 'Philos'—like others before him, indulging in rhetoric and exhortation in his peroration:

"Thus is reduced to nothing the alarm of those who had had reason to fear seeing our efforts limited to a sterile imitation (stérile imitation) of the machines and experiments of the Americans. Our aviators are beginning to become conscious of their strength and their worth (de leur force et de leur valeur); they are already seeking the practical solution away from the trans-Atlantic methods and machines, which they feel certain they will be able to equal, and hope they will surpass. M. Archdeacon also has the right to congratulate himself on the success of his crusade. If he has been minded to start with American machines, it is only the better to stimulate the activity and emulation of the French investigators by setting before their eyes immediate results. He has attained his object, seeing that on all sides, ideas and new machines arise which—and this is his dearest wish and also ours—will retain, in the land which was the cradle of aviation, as it was of aerostation, the glory of finally conquering the air by means 'heavier than air'."

Alas, the author's hopes were not remotely to be realised. All the efforts with monoplane gliders were speedily to evaporate; their authors' names proved only to be "writ on water", and only one tentative monoplane glider was to be built in the early days of flying, *ie* the Wels-Etrich machine of 1907.

❧ 37 ❧

The Archdeacon Wright-type Glider and its Tests: April 1904

The Archdeacon Wright-type glider was tested during the first and second weeks of April 1904, at Merlimont, near Berck-sur-Mer, the pilots being Gabriel Voisin and Captain Ferber. The latter says that Voisin was pitched out of the glider and bruised; and that, as a result, he (Ferber) was telegraphed at Nice to come and show the "nouveaux aviateurs" how to handle the machine. This would appear to be true. During the fortnight of tests, this glider did not make many flights—the exact number has not survived—and had only minor success before Archdeacon decided it was not satisfactory, and had it taken back to Paris (Fig 19).

The machine was a Wright-type biplane based on Chanute's drawings of the modified No 3 Wright glider of 1902, but with certain important differences. Chanute's Wright figures were approximately correct, but Archdeacon had reduced the Wright span of some 32 feet to 24 feet 7¼ inches, the wing area from 301-305 square feet to about 237 square feet, and the empty weight from 112-117 lb to 75 lb. We do not know why these changes were made. The curvature seems to have been considerably increased in the Archdeacon machine, which caused trouble. But most important of all, there was no warping, or other attempt at control in roll. But Archdeacon described his machine as, "apart from subsequent modifications, exactly copied from that of the Wright brothers".

La Vaulx, in commenting on this machine, says that owing to the Wrights' having patents pending on their machine, Chanute was vague about its "transverse stabilising, and turning, by means of wing-warping". That is true, and the drawings did not include the mechanism. But La Vaulx must have forgotten that not only was the wing-warping mentioned in Chanute's own *L'Aérophile* article, but the vital point of the simultaneous use of warping and rudder had already been mentioned twice in the original write-ups of Chanute's lecture, in *L'Aérophile* and *La Locomotion*. Yet Archdeacon not only failed to incorporate means for roll-control in the glider, but—so far as I can find—never mentioned a word about the subject in his own descriptions of the machine. This "anti-control" aspect of European aviation is dealt with later.

The tests of the Archdeacon glider were written up in *L'Aérophile* for April 1904, over the initials "A. de M" (Albert de Masfrand), but much of the article in fact comprised direct quotations from Archdeacon himself.

Before noting his comments, a point of prime importance in this article was the accompanying photographs of the machine in flight. These, and other shots of it published now in April 1904, were the first photographs—since those of Ferber's primitive Wright-type machine published in *L'Aérophile* of February 1903, and of course since Lilienthal and Pilcher—to show a European glider in flight. Thus the Wrights' influence-impact was further increased visually.

The *Aérophile* article on this historically important glider was curiously defensive in tone. The writer says:

"The aim of these first experiments was simple: that is, to study the qualities and defects of the machine in practice, and to become familiar with its difficult handling. To those who may be tempted to be surprised at the tardiness of these tests, one must answer that the best and quickest results are obtained, not by hurry, but (on the contrary) with method, and by resigning oneself to begin at the beginning.... To hope to surpass, at the first 'go'—with a new machine and insufficiently trained operators—results which were obtained by the masters of the subject, the Wright brothers, over long months, even years, of groping, would have been foolish. M. Archdeacon saw the temptation offered to his natural impatience, and he avoided it; he cannot be sufficiently congratulated.... As to the results themselves, without yet being comparable with those of the American aviators, they were none the less very encouraging, and showed good promise for the definitive tests."

Archdeacon is then quoted as disapproving of various aspects of the previously much-vaunted workmanship, especially the use of screws; also mentioned was the uncomfortable support for the pilot's body, and the weakness of the bracing wires. He says that Gabriel Voisin had acquired "all the necessary mastery in the very difficult management of the aeroplane", and pays tribute to Ferber for his advice. Archdeacon goes on to comment on the good stability of the machine when flown as a kite; the necessity of perfect co-ordination of the four men who launched it—the Wrights employed two or four—the success of the take-offs, but the lack of success in its "progression"; the defective form of the wing-ribs, which had too great a curvature at the back; and so on. He reported that the longest glides were only about 20 metres (65.5

feet); then that he intended to construct a new glider—to cost less than 500 francs—which would incorporate all the improvements which the present one had shown to be necessary: this was to be carried out with the technical help of Victor Tatin. Archdeacon says:

> "Our new machine, which will soon be ready, will allow us—I am certain—to do as well as the Wright brothers."

Archdeacon closes with one of those all too familiar perorations, which smack more and more of the courage-sustaining whistle; they must surely have made each successive utterer feel more foolish as time went on:

> "One might therefore consider that the tentative phase, which is inevitable at the beginning of such an enterprise, is at an end. We are now entering the active period of tests, which will commence at Berck at the end of April. In a few weeks, French aviators will have nothing more to envy in their transatlantic rivals. They will even perhaps surpass them, to the greater profit of the 'aerial ideal'."

It seems almost unkind to record that no tests were to take place at Berck at the end of April; that, indeed, no new Archdeacon glider was to be tested until March of next year (1905), and then only to end disastrously; that no European glider of this epoch was ever even to approach the efficiency of the Wrights' last glider of 1902; and that at the time when the above remarks appeared in *L'Aérophile*, the Wrights were designing their second powered aeroplane!

It is, perhaps, of value to quote here the wise words of the Comte de La Vaulx (*op. cit.*), writing later about the colleagues he knew so well:

> "Archdeacon thought that the most rapid means of catching up with the Americans—if they really were ahead—consisted of making use of their own work . . . and reproducing their experiments in seeking to improve them. Considering the superiority of the French automobile industry, which was in a position to supply the lightest motors of that time, he did not doubt that when the moment came, the application of a motor to a perfected glider (au planeur bien au point) would be relatively easy, which would allow us to arrive well in front. The extraordinary 'tour de force' of the Wrights, in creating with their own hands the motor they lacked—and which the American industry was not in a position to supply—frustrated this calculation. This was not foreseeable; and Archdeacon's idea, which furthermore astonished an important group of French aviators—who intended to arrive at results by their own means and following the lines laid down by their national predecessors—was rational, if one recalled the necessity of succeeding quickly, indeed more quickly than others."

As tail-piece to this section, it is perhaps relevant to point out how curious it seems to us now that a man of Archdeacon's enterprise and energy—helped as he was by some very able assistants—had not taken a longer and harder look at the superb photographs of the modified Wright No 3 glider of 1902, which had been freely available in Paris since mid-1903; also that he had not realised that Chanute's description and drawings in *L'Aérophile* were not necessarily of perfect accuracy. This, too, could easily have been gauged by the very photographs Chanute himself used. Another clue which should have put Archdeacon, Ferber and all the others, on the alert, was that although Chanute in his April lecture had made explicit references to the Wrights' wing-warping, along with the simultaneous use of warping and rudder, neither the warp nor the rudder controls were shown on the measured drawings which Chanute later published in his article; but it was on these drawings that Archdeacon, and later Esnault-Pelterie, based their Wright-type gliders. The same caution should have been shown in adopting the type, and amount, of wing-curvature. Chanute actually gave the erroneous figure of 1/20 for the Wrights' 1902 machine; whereas the brothers had varied it between 1/24 and 1/30, and found the latter figure the best. By close scrutiny of the Wright photographs, these and a number of other points might at least have been guessed at, and then explored, before making what turned out to be unworkable so-called 'copies' of the Wright machine. There was certainly something very deficient about the powers of the French at that time to observe, analyse, and assess.

Finally, we are surely entitled to wonder why, after tests only aggregating a few minutes, the Archdeacon glider was abandoned; and why no successor was to be tested until March of 1905.

In short, there seems again here to be lack of sufficient "aero-talent" (as I have called it elsewhere), and indeed a lack of proper professionalism of outlook into the bargain.

⁂ 38 ⁂

Gabriel Voisin
enters professional Aviation :
1904

Gabriel Voisin, one of the great figures of French aviation, first appeared on the stage of history when he helped to pilot Archdeacon's Wright-type glider in April 1904, as has already been noted. Gabriel was to become an embittered man in his old age, and to make many extraordinary claims and misleading statements. In his autobiography, he attempts to make out that he was not only interested in aviation in the late 1890's—which is of course possible—but actively at work at that time. Two illustrations in his autobiography are entitled "Voisin the designer—Neuville-sur-Saône (Rhône) 1899", showing him with a Chanute-type glider, which is dated 1903 in the text. Unfortunately for his story, this glider has nothing to do with either 1899 or 1903: it is a well-known machine which was built by Gabriel in 1907 (Fig 43), and is described and illustrated in *L'Aérophile* for June 1907. Presumably the illustrations were included in his book in the hope that the readers would be convinced he had been active in aviation at the time the Wrights were starting; but there can be little excuse for including photographs in his autobiography of an aircraft which was not built until over seven years after the date attached to them.[1]

That Gabriel and his brother had been interested in aviation as teenagers (he was born in 1880) no one would contest; he and his brother Charles—who was later to join him in the famous firm of "Voisin Frères"—were probably making kites, and taking a general interest in flying by the first years of this century, like many other boys at the time. For example, Gabriel seems to have secured an interview with Clément Ader during these early years, probably in 1900. But his practical involvement with aviation on a 'national' level seems to date from January 1904. The Voisins were then living in Lyon, and at the

---

[1] It is interesting to note that in 1953, no such claims were advanced; in an article in *Interavia* (Vol VIII, No 12, 1953) entitled "Grandfather taught us how", Gabriel says that his efforts at this time "produced no results in the gliding direction".

end of that January, Captain Ferber gave a lecture on aviation in the city's Palais de la Bourse. He records in his London lecture of February 1908 (see Section 85);

"After a popular lecture which I delivered at Lyons, in 1904, a bright and intelligent young man came upon the platform and said to me 'I have understood the method which you teach, and I mean to devote myself to it.' The next day he left for Paris, and entered the service of M. Archdeacon as aviator, thus becoming the first man in the world to earn his living as an aeroman.[1] They were at Berck sur-Mer with a machine, and in considerable difficulties. On the receipt of a telegraphic request, I went there and showed Voisin what I had learnt in three years' work with the apparatus of Chanute and Wright. He understood, and we succeeded in making flights of about 20 metres length."

This story is borne out by the Comte de La Vaulx, who wrote (loc. cit):

"Luckily, he (Archdeacon) had to hand a newly discovered aviator. This was quite a young man, a 'Lyonnais', who had been fired with enthusiasm by a lecture in his home town, by Ferber; and who had resolved to devote himself to aviation research. His name was Gabriel Voisin, architectural student at the École des Beaux-Arts. Gabriel Voisin ... was recommended by Ferber to Archdeacon; and his technical knowledge allowed him very usefully to assist in the study and construction of the famous glider. ... He was also employed to make the tests, a mission which he aspired to, and which he fulfilled with remarkable skill and courage."

The first mention of Gabriel Voisin in the Aéro-Club annals was reported in the March (1904) issue of L'Aérophile, which recorded the

"interesting work of the Monsieurs Voisin in their aviation experiments, carried out in 1903, with Lilienthal and Chanute-type machines."

This presumably referred to the immediate results of Ferber's lecture in January.

Thus Gabriel Voisin was launched into 'official' European aviation circles in 1904, and was to take a vital place among the first generation of the practical "full-size" French pioneers. His role in this crucial period of aviation history has always been the subject of speculation, doubt, and controversy. It has never been satisfactorily settled as to how much Gabriel himself contributed to the design of his famous series of biplanes —which started with the two float-gliders of 1905—and established himself in the field of the powered biplane with the Voisin-Delagrange and the Voisin-Farman of 1907.

[1] Voisin says he was introduced to Archdeacon by Colonel Renard.

❦ 39 ❧

Esnault-Pelterie
tests his Wright-type Glider
and first fits Ailerons : May and October 1904

Inspired by Archdeacon to emulate the Wrights, an able engineer named Robert Esnault-Pelterie also built and tested in May 1904 what he falsely claimed was an exact copy ("absolument semblable") of the 1902 Wright glider, except for details; he found it was unsatisfactory and decided to improve it. He clearly could not make the warping work properly, and, like the other Europeans, did not understand the significance of Chanute's description of the simultaneous use of warping and rudder. Esnault-Pelterie then asserted that warping was structurally dangerous. In October 1904 he tested a rebuilt version, using—for the first time in history—ailerons, or rather primitive elevons, instead of the warping and front elevator (Fig 20). This machine, despite its ingenuity, was a crude affair and a failure, and Esnault-Pelterie—like his friends—would not take the trouble to experiment, modify and improve.

The deplorable influence of Esnault-Pelterie on French aviation did not take effect until 1905, when his lecture was given, and published later in the year: a full discussion of this pioneer and his 1904 glider will therefore be found in Section 51.

a/b Esnault-Pelterie Wright-type Glider: 1904 Another newcomer to aviation was the engineer Robert Esnault-Pelterie, who made a copy of a Wright glider in 1904, and then announced that he had abandoned the Wrights' wing-warping because it was too dangerous structurally; whereupon he fitted the first ailerons to a full-scale aircraft, a most creditable move. But, like the other Frenchmen, he would not persevere with his tests, and even after the ailerons — which were in fact elevons — had been fitted, the machine proved a failure.

21 Wright Flyer II: 1904 In 1904 the Wright brothers built and flew their powered *Flyer II*, using the Huffman Prairie, near Dayton, for its tests. This machine followed the general lines of the first *Flyer*, but there was now a new and greatly improved engine of 15-16 hp. This *Flyer* was flown mostly near the ground, so that the brothers could familiarise themselves with the controls. With this machine they made the first circle in history, and performed flights of up to five minutes' duration.

22 Ferber tailed Wright-type Glider: 1904 Also in 1904, Captain Ferber was moving ahead, with improved construction and with the ambition of building a stable machine, as opposed to the Wrights' inherent instability. In this year he made a vital move for European aviation by adding, for the first time — as seen here — a fixed stabilising tailplane to his new glider. The two 'waving' items at the wing-tips were rudders.

3 Archdeacon second Wright-type Glider: 1905 In 1905, Archdeacon had a second glider built and, luckily for those concerned, had it towed off the ground by an automobile but without a pilot: it crashed and would have killed anyone in it. But there was an important feature of this glider which was observed and copied, *ie* a fixed tailplane of the Ferber kind, and fins as well as the Wright-type forward elevator.

4 Wright Flyer (published 1905) Another pirated picture of a Wright Flyer appeared in the *Illustrierte Aeronautische Mitteilungen* for March 1905; but it told the world very little that they did not know already. Again it was based on a photograph of the No 3 Glider (modified), seen from the rear, and with two propellers revolving behind the wings. However, like the first, this gave a fair impression of what the machine might look like, but it was of no service to the European pioneers.

25 Voisin-Archdeacon Float-glider: 1905 In 1905, after many years of familiarity among the aviatic fraternity, the Hargrave box-kite entered full-scale aviation, and we find — shown here — the first Gabriel Voisin's machines, which he built for Archdeacon as the latter's third machine. This float-glide which was towed off the Seine by a motor-boat, was not successful; but it shows the Wright features allie with the box-kite structure for main wings and tail-unit. The wings had little curvature.

26 Voisin-Blériot Float-glider: 1905 Soon after the completion and testing of the Voisin-Archdeac float-glider, and also in 1905, came the second Voisin machine, the Voisin-Blériot float-glider. Blériot, wh had been on the edge of aviation for some time, now entered in full stature as the commissioner of th machine. It was basically similar to the other float-glider, but had the outer side-curtains of the box-ki wings given a pronounced dihedral angle. This glider was also a failure.

Ferber powered Glider: 1905 Captain Ferber was now determined to power-fly; and in 1905 he fitted a small engine and a propeller to one of his gliders, and set off from his launching towers. It was a creditable effort, but was not successful.

Wright Flyer : 1905 General arrangement drawings.

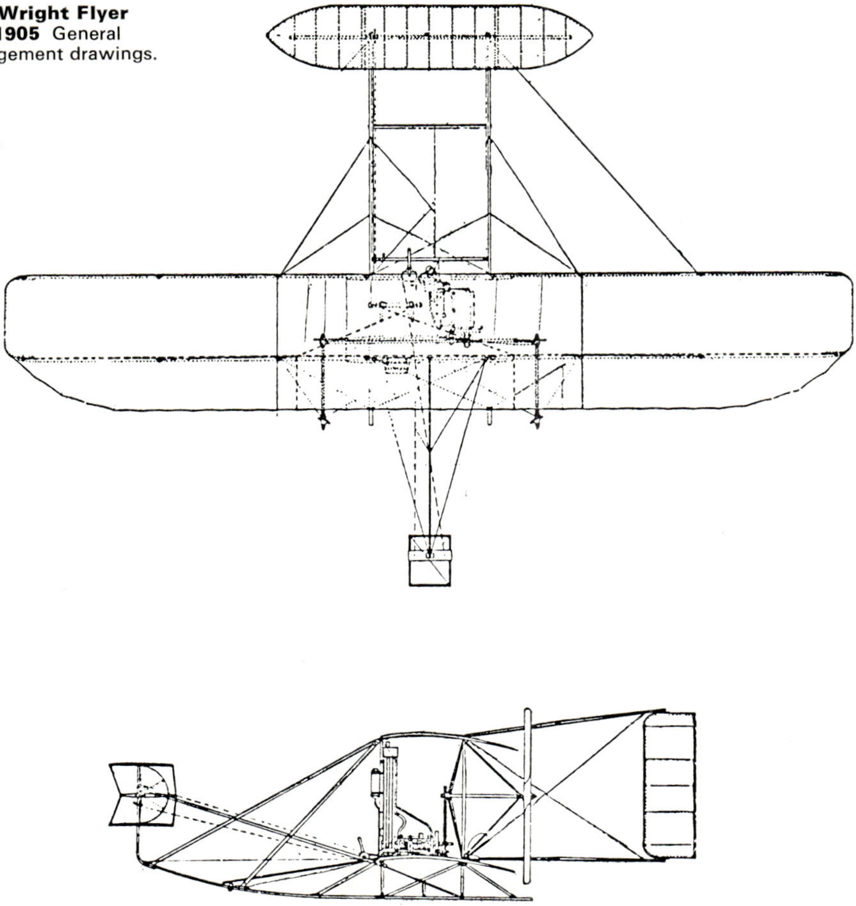

28b/c Wright Flyer III: 1905 This graceful and successful Wright *Flyer III* of 1905 again followed t
lines of the previous two, but the *III* had its forward elevator further away from the wings, as were the tw
rear rudders. It was powered by the same engine as the *Flyer II,* and could turn, bank, circle, and fly figu
of eight, and survive any number of landings on its robust skids. It was the world's first practical aeropla
This represented the culmination of the Wrights' experimental work, and they knew now that they ha
fully successful and marketable aeroplane. But their passion for secrecy dictated that they should not rev
the machine in public until an order was received. This machine, like the *II,* was tested at the Huffm
Prairie, and could fly for over half an hour.

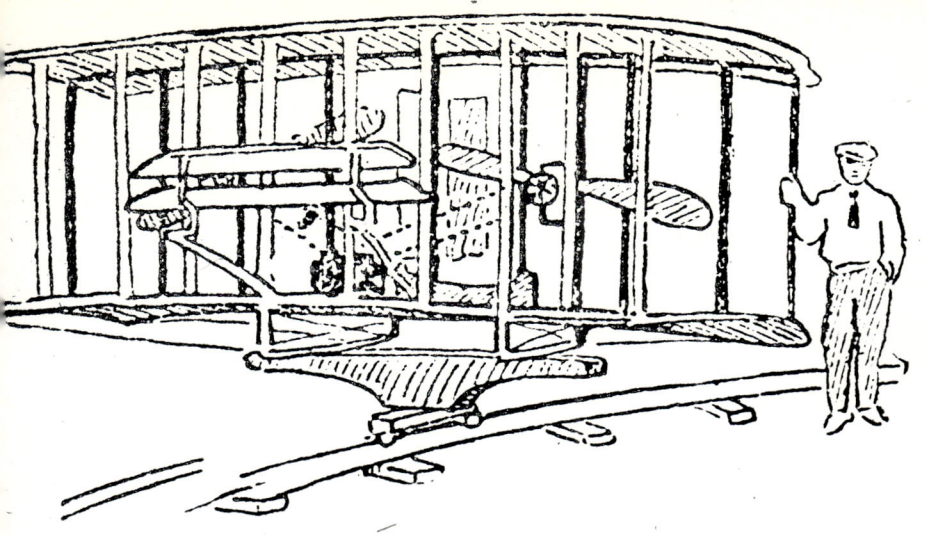

Wright Flyer: 1905 After news of the Wrights' triumphant 1905 season had reached France, great ~~eff~~orts were made to catch sight of the Wrights' machine. This pirated sketch appeared in the issue of the ~~Par~~is *L'Auto* for December 24th 1905; although crude, it showed two features of great interest to the ~~Eur~~opeans; the biplane forward elevator, and the yoke and rail take-off system. It was by this sketch that ~~the~~ Europeans were persuaded to adopt the biplane elevator on their own machines.

Cody aileroned Kite-glider: 1905 Off the beaten track, at Farnborough, England, the American ~~pio~~neer S. F. Cody, who had already invented an efficient man-lifting kite system, now produced a kite-glider ~~wh~~ich, although not very successful, incorporated one striking feature, the two ailerons. These came to ~~Far~~nborough *via* Esnault-Pelterie.

L'AÉROPHILE

Directeur-Fondateur : GEORGES BESANÇON

13ᵉ Année — N° 12   Décembre 1905

Les Frères Wright et leur Aéroplane à moteur

L'origine et les pièces du débat. — Exposé des faits avancés par les Wright. — Objections et possibilités. — Premiers résultats de l'enquête.

Dans son numéro de mars 1903 où il rendait compte de la conférence sur « L'aviation en Amérique », faite à l'Aéro-Club de France, par l'ingénieur américain O. Chanute, l'*Aérophile* fit connaître, le premier, chez nous, l'existence des frères Orville et Wilbur Wright, la nature et l'intérêt de leurs expériences. Cette révélation provoqua en France, un bienfaisant renouveau des études d'aviation sous l'impulsion énergique, généreuse et opiniâtre de notre collaborateur et ami Ernest Archdeacon et des personnalités compétentes et dévouées groupées autour de lui dans la Commission d'aviation de l'Aéro-Club de France.

Orville Wright   Wilbur Wright

Un peu plus tard, dans un article capital qu'il voulut bien écrire spécialement pour nos lecteurs (Voir *Aérophile* d'août 1903), M. Chanute exposa dans leurs détails, les résultats obtenus depuis quelques années, aux Etats-Unis soit par lui-même, soit par ses élèves les Herring, les Wright. Il y décrivait les machines employées, faisait connaître avec les dessins et les photographies permettant de les reproduire, leur fonctionnement, les procédés d'expériences ; il discutait enfin, avec une compétence remarquable, les côtés théo-

31 Title-page of L'Aérophile: December 1905 The furore created by news of the Wrights' 19 season with their *Flyer III* led to major articles being written on them, and finally to the Fren aeronautical press admitting that the Americans had been the first to conquer the air.

40

The Wrights' Work in 1904
and the Arrival of the News of it in Europe: 1904

The second Wright powered *Flyer* was completed in May 1904 and—through the kindness of their friend Torrence Huffman—an "aerodrome" was set up at the Huffman Prairie, a 90 acre pasture at Simms Station, about 8 miles east of Dayton. The *Flyer II* had approximately the same dimensions as the first, but had less wing curvature (1 in 20 to 1 in 25), and a new engine of 15-16 hp: the pilot still lay prone, and the warp and rudder controls were still linked. From May 23rd to December 9th, about 80 short flights were made which enabled the brothers to obtain practice in controlling and manoeuvring a powered machine (Fig 21). Some 100 starts were made in all, but various minor setbacks had to be overcome before consistent and productive flights were achieved. Their total airborne time was about 45 minutes: the longest flight lasted for 5 minutes 4 seconds, and covered about $2\frac{3}{4}$ miles. On September 7th they introduced for the first time their weight- and derrick-assisted take-off device, to make them independent of the weather, in view of the small area (90 acres) of the pasture. The most important event was their first circle—by Wilbur on September 20th—after which this became a necessary commonplace, as they did not want to overfly other people's property.[1] This first circle was the subject of a detailed eyewitness report made and published by Amos I. Root, which was the first report in history of a powered aeroplane flight,[2] part of which is reproduced below.

There was one control problem still outstanding, which was not to be solved until 1905; and that was a tendency to stall in tight turns.

Very little news of the Wrights' powered flying in 1904 filtered across

---

[1] Charles Dollfus (*op. cit.*) says of the 1904 season, that "these flights, taken as a whole, show that the problem of mechanical flight was entirely resolved (le problème du vol méchanique était entièrement résolu)".
[2] The whole of Root's statement was reprinted in my Science Museum book, *The Aeroplane: an historical Survey* (HMSO 1960).

to Europe during the year, and I can find only one reference in *L'Aérophile* in 1904, in the September issue. It was a short paragraph at the end of the issue, headed "New tests of the powered Wright aeroplane (aéroplane automobile)", and containing the familiar exhortations. It also assumed that the 1904 machine, *Flyer II*, was the same as the *Flyer I* of 1903. The paragraph reads:

"It is reported from Dayton (Ohio) that the Wright brothers appear to have made, on May 26th last, a new experiment with their flying machine. The machine apparently rose to about 4 metres and flew 30 metres against a high wind, stopping only because of engine failure. The first tests of the same machine, in December 1903, were more successful than that of May 26th, when it succeeded—we are assured—in flying 800 feet (260 metres) over the ground, in 59 seconds, only landing because of a wrong movement of the rudder [*ie* elevator]. We today fall very much short of these results (nous sommes aujourd'hui fort au-dessous de ces résultats). However that may be, it is certain that the progress of aviation is being keenly pursued in America. It is for the aviators of France to unite their efforts to those of M. Archdeacon—whose fine campaign to encourage similar experiments is well known—and not allow themselves to be overtaken."

Here are some excerpts from Mr Amos Root's article, which described the Wrights' first circle on September 20th 1904, published in *Gleanings in Bee Culture* for January 1st 1905, of which Mr Root was the Editor:

"God in His great mercy has permitted me to be, at least somewhat, instrumental in ushering in and introducing to the great wide world an invention that may outrank electric cars, the automobiles, and all other methods of travel, and one which may fairly take a place beside the telephone and wireless telegraphy. Am I claiming a good deal? Well, I will tell my story, and you shall be the judge. . . . I found them in a pasture lot of 87 acres, a little over half a mile long and nearly as broad. The few people who occasionally got a glimpse of the experiments, evidently considered it only another Darius Green, but I recognized at once they were really *scientific explorers* who were serving the world in much the same way that Columbus did when he discovered America, and just the same way that Edison, Marconi, and a host of others have done all along through the ages. . . . They have been prudent and cautious. I told you there was not another machine equal to such a task as I have mentioned, *on the face of the earth*; and, furthermore, just now as I dictate, there is probably not another man besides these two who has learned the trick of controlling it. In making this last trip of rounding the circle, the machine was kept near the ground, except in making the turns. . . . I was surprised at the speed, and I was astonished at the wonderful lifting power of this comparatively small apparatus. When I saw it pick up the fifty pounds of iron so readily I asked if I might ride in place of the iron. ... The operator takes his place lying flat on his face. This position offers less resistance to the wind. The engine is started and got up to speed. The machine is held until ready to start by a sort of trap to be sprung when all is ready; then with a tremendous flapping and snapping of the four-cylinder engine, the huge machine springs aloft. When it first turned that circle, and came near the starting-point, I was right in front of it; and I said then, and I believe still, it was one of the grandest sights, if not the grandest sight, of my life. Imagine a locomotive that has left its track, and is climbing up in the air right toward you—a locomotive without any wheels, we will say, but with white wings instead, we will *further* say—a locomotive made of aluminium. Well, now, imagine this white locomotive, with wings that spread 20 feet each way, coming right toward you with a tremendous flap of its propellers, and you will have something like what I saw. The younger brother bade me move to one side for fear it might come down suddenly; but I tell you, friends, the sensation that one feels in such a crisis is something hard to describe. ... When Columbus discovered America he did not know what the outcome would be, and no one at that time knew; and I doubt if the wildest enthusiast caught a glimpse of what really did come from his discovery. In a like manner these two brothers have probably not even a faint glimpse of what their discovery is going to bring to the children of men. No one living can give a guess of what is coming along this line, much better than any one living could conjecture the final outcome of Columbus' experiment when he pushed off through the trackless waters. Possibly we may be able to fly *over* the north pole, even if we should *not* succeed in tacking the 'stars and stripes' to its uppermost end."

Finally, here is part of Wilbur Wright's account of the brothers' work in 1904, taken from the "First Rebuttal Deposition of Wilbur Wright" (February-March 1912) in the case of Wright *vs* Herring-Curtiss, Vol 1, pages 493 ff:

"While our patent application was pursuing its slow course through the Patent Office, we built a second machine and flew it in a field near the city of Dayton, Ohio, in the summer and autumn of 1904. When we had familiarized ourselves with the operation of the machine in more or less straight flights, we decided to try a complete circle. At first we did not know just how much movement to give in order to make a circle of a given size. On the first three trials we found that we had started a circle on too large a radius to keep within the boundaries of the small field in which we were operating. Accordingly, a landing was made each time, without accident, merely to avoid passing beyond the boundaries of the field. On the fourth trial, made on the 20th of September [1904], a complete circle was made, and the machine was brought safely to rest after having passed the starting point. Thereafter we repeatedly made circles, and on the 9th of November made four circles of the field in a flight lasting a few seconds over five minutes. In all these flights the warping wires and the wires controlling the vertical tail were interconnected, as in the patent."

❧ 41 ❧

The Wrights' Influence on Flight-Control Principles and Practice : 1904 and on

Although Ferber implied that he understood the Wrights' wing-warping system, and decided not to use it at first, it was not until there appeared in *L'Aérophile* for June 1905, a large illustration—and a small one—of Esnault-Pelterie's elevon-equipped Wright-type glider (Fig 20), that any of the European pioneers of the period gave any serious thought to the proper problem of control in roll.

It is to the credit of Esnault-Pelterie, who was an excellent engineer, that he immediately realised at least part of the importance of the Wrights' doctrine of control in roll, *ie* its function of preserving "lateral balance", and that he incorporated it in his 1904 glider. As described in Section 51 Esnault-Pelterie wrongly decided that the incorporation of wing-warping would be dangerous structurally; and he therefore, for the first time in history, incorporated ailerons for roll-control on a full-size aeroplane: at the same time, he had the ingenious—but not new—idea of combining the functions of the ailerons with that of the Wrights' forward elevator, and so was also the first in history to apply elevons to a full-size aircraft.[1] But when he came to build his first powered machine in 1907, he quietly decided that wing-warping was not really structurally dangerous, and proceeded to adopt it!

Esnault-Pelterie's elevons were in fact to appear to most subsequent pioneers as only ailerons; and, as such, they led directly to all aileron development thereafter, on both sides of the Atlantic. This development can be accurately plotted, and the lines of its communication described in detail.

In passing, it is interesting to note that Esnault-Pelterie either did not understand—or never paid enough attention to—the available sources of information in France, to realise that roll-control had to be combined with control in yaw by the vertical rudder, to effect the safe and proper

[1] What the patent authorities call the "paper conception" of ailerons and elevons was, of course, nothing new, the first of such being Boulton's design of 1868.

lateral control of an aeroplane; for it was in the January 1906 issue of *L'Aérophile* that there appeared the detailed description of the Wrights' control system, together with its aerodynamic raison d'être.

Copies of *L'Aérophile* were also well distributed among the international aeronautically-minded community; and it is known that one of the subscribers was Colonel Capper, the officer in charge of the Balloon Factory, the embryo Royal Aircraft Establishment, at Farnborough. It was Capper, who, having seen Esnault-Pelterie's elevons in *L'Aérophile*, 'briefed' Cody in 1905, when the latter built his kite-glider at Farnborough; he included two slender ailerons on outriggers beneath its wings: these surfaces—far too small to be effective—were directly derived from the Wrights' wing-warping *via* Esnault-Pelterie's elevons (Fig 30).

The third use of ailerons was on Blériot's No IV (Fig 35) his unsuccessful biplane of 1906, where two very small between-wing surfaces were fitted; but this aeroplane could never fly in any of its various modifications.

The fourth application in history of ailerons was by Alberto Santos-Dumont on his modified 'canard' biplane *14-bis* of 1906, when two octagonal surfaces were fitted in November 1906, one in each outer box-kite cell (Fig 38). These ailerons were, like all their predecessors, looked upon and used solely as "balancing planes", as they were often called at the time; *ie corrective* control surfaces, for maintaining 'balance', and not as dynamic initiatory surfaces.

The fifth application of ailerons—this time they were also elevons—was on Blériot's tandem-wing *Libellule*, his aircraft No VI of 1907 (Fig 39): these were swivelling surfaces at the wing-tips. It was a large illustration of these elevons which was featured in the December 1907 issue of the *American Magazine of Aeronautics*, which was undoubtedly the immediate line of aileron influence travelling across the Atlantic to America. Everyone in the Aerial Experiment Association—which was based at Hammondsport on Lake Keuka in New York State—would read this magazine as a matter of course, as it was the premier magazine devoted to aviation in the USA. It is, incidentally, known from documents that this magazine was read within the AEA.[1] At this time the full significance of these surfaces did not strike them, and—as with their confrères in France—they treated ailerons only as corrective devices. One of the AEA members, Lieutenant Selfridge—later to be killed flying with Orville Wright—wrote a letter (on AEA notepaper) to the Wrights, dated January 15th 1908, asking for aerodynamic information, and was referred to Wilbur's Chicago lectures, which revealed the functions of wing-warping in roll-control.

[1] A copy was found in Selfridge's library.

The sixth application of ailerons was in England in 1908, when S. F. Cody fitted between-wing ailerons to his *British Army Aeroplane No 1* for a short while prior to its first flight on October 16th 1908 (Fig 48). Cody had abandoned ailerons before this flight, but subsequently used ailerons on all his machines. Cody's ailerons, as mentioned above, were the result of his being briefed by Colonel Capper. (See Postscript below).

The eighth application of ailerons, and the first flap-type (*ie* hinged to the wings), was on the *Blériot VIII-bis* monoplane of 1908 (Fig 49).

The ninth application—and the Aerial Experimental Associations' first—was on their *June Bug* of 1908, on which four small surfaces were fitted to the tips of the biplane wings.

The first really effective ailerons to be applied to any aircraft were fitted by Henri Farman to his already much-modified Voisin Biplane in October of 1908, after he had watched Wilbur Wright flying at Auvours (Fig 54).

On the subject of the Europeans' limited concept of roll-control, as a purely corrective function, it is interesting to note the early terms used in France and England for ailerons: "ailerons de stabilisation", and "equilibreurs latéraux"; and the English "balancing planes", and "righting planes".

Charles Dollfus, in his *Les Avions* (1962), confirms this limited conception, which lasted right up until Wilbur Wright demonstrated the full and proper use of roll-control, and its co-ordination with control in yaw, when he flew in France from August to December 1908. Dollfus writes:

> "lateral control—wing-warping—was used (by the Wrights) not like the primitive ailerons of the French machines, *ie* to redress the balance of the aeroplane (pour redresser l'avion), but as a means of controlling turns (un moyen de commande des virages)."

Postscript. As recently discovered by Mr Kenneth Molson, the seventh man to use ailerons (or elevons) was the American L. J. Lesh, who fitted them to one of his gliders in the latter part of 1907.

42

Ferber's tailed Glider: October 1904

Ferber, who had abandoned the front elevator on his glider (having inexplicably failed to make his surfaces rigid), again fitted one and made some tentative glides. He also used an ingenious testing device before launching out in free flight, consisting of an inclined rope between wooden towers with a trolley running down it, from which the machine was hung. Then, in October 1904, he made a vitally important and influential move. He completely rebuilt his glider and created a new type, with both forward elevator and fixed horizontal tailplane, as well as the two triangular wing-tip rudders (Fig 22).

This machine marks the beginnings of the full-size stable aeroplane. With this novel glider, Ferber now established a new configuration, a new dogma, which was to dominate European aviation for nearly a decade—the dogma of the longitudinally and laterally stable aeroplane consisting of Wright-type main wings set at a dihedral angle, and forward elevator, with a fixed horizontal rear tailplane: the rudder was later to retreat to the rear and become part of the tail-unit. Ferber, also in this year 1904, made the first tentative flight with a passenger in a heavier-than-air machine, when he took his mechanic Marius Burdin for a short glide at Chalais Meudon.

But, it is important to note, the influence of this vitally important tailed glider of Ferber's did not start 'operating' on his contemporaries until a photograph of it was published in the February 1905 issue of *L'Aérophile* (see Section 46).

☙ 43 ❧

Prizes are offered to stimulate European Aviation: October 1904

The Aéro-Club de France, in October 1904, published details of the first European prizes to be offered in order to encourage aviation. They were as follows:

(1) *Coupe Ernest Archdeacon.* A silver trophy, worth 2,000 francs; also a prize of 1,500 francs. Donated by Archdeacon. The trophy was to go to he who piloted an aeroplane which covered a minimum of 25 metres (82 feet), the money prize to he who covered a minimum of 100 metres (328 feet). (Both were won by Santos-Dumont on his "14-*bis*" (modified), the first on October 23rd 1906; the second on November 12th 1906).

(2) *Grand Prix d'Aviation Deutsch-Archdeacon.* 50,000 francs. Donated by MM. Henry Deutsch de la Meurthe and Ernest Archdeacon. For a piloted aeroplane flight of 1 kilometre ($\frac{5}{8}$ mile) in a circular course, without touching the ground. (Won by H. Farman on January 13th 1908).

(3) *Prix pour record de distance.* Donated by the Aéro-Club. (a) 1,500 francs: for the pilot of an aeroplane which covered a minimum distance of 100 metres (328 feet); (b) 100 francs and a silver medal; for each of the first 10 pilots who covered 60 metres (197 feet). (The first (a) was won by Santos-Dumont, also on his "14-*bis*" (mod), on November 12th 1906).

(4) *Prizes for aeroplane models.* Silver and bronze medals. Donated by the Aéro-Club de France, for the winners in the "Concours d'aéroplanes non-montés", held in February 1905.

* * * * *

It is interesting to note that a time-limit was set for the Grand Prix for five years, from October 1st 1904: it was won, as noted above, on January 13th 1908, by Henri Farman.

From the prizes offered, it will be seen that Archdeacon's proposed glider competition came to nothing. Historically speaking, this is to be regretted; since, if such a competition had been held, it might have encouraged the proper investigation of flight-control, and led to a more rapid achievement of powered aviation in Europe.

44

Aeronautical Comment in Britain: 1904

In Britain, and elsewhere on the Continent outside France—where of course the concern was vital and lasting—there was the to-be-expected burst of interest at the very end of 1903, and in the first month or two of 1904, following reports of the Wrights' first flights on December 17th 1903. Then the interest in aviation died away to almost nothing. By the attitude of the Aeronautical Society in London, it is doubtful whether in Britain aviation was treated as a serious subject at all at this time. And one of the many curiosities of British aeronautics of this epoch was a paper given by the President (Major Baden-Powell) entitled "The Development of the Aeroplane", which was printed in the October 1904 issue of the *Aeronautical Journal*: it did not contain a single word about either the Wrights or about the active French pioneers! Baden-Powell has, unfortunately, emerged in aeronautical history as a man with a talent for well-expressed aeronautical prophecy, but a painful ignorance of aeronautics, considering his position as President of the Aeronautical Society: it is certainly conducive to raised eyebrows to find him saying in this paper: "as regards balance, I am inclined to believe that it is not of such supreme importance..." This sort of naivete occurs so often in Baden-Powell's statements on aviation that one comes to realise perhaps why our great Society was in such a parlous state at that time.

Even the enterprising *Automotor Journal* did not shine very brightly in 1904, so far as aeronautics was concerned: but in its issue of October 15th 1904, it began to recover its old form: "What are Brothers Wright doing? An appeal". This was its head-line, followed by the introductory remark that "It is now almost exactly a year since Messrs. Orville and Wilbur Wright tried their first experiments with their motor-driven aeroplane..." The writer goes on to warmly praise the Wrights' 1903 results, but concludes with this plaintive appeal:

"But we should like to know what they have been doing. They need not be ashamed if their anticipations have not been fully realised. They have done such splendid work in the past that even news of temporary failure would be received by everyone with the most considerate sympathy. On the other hand, they may have succeeded and succeeded splendidly, and are keeping the knowledge of the fact to themselves. But we should like to know. In the name of the interest which everyone feels in the conquest of the air, do, O Brothers Wright, let it be known what *has* happened!"

It is hard not to sympathise with *Automotor*. But the Wrights were feeling their way toward a perfected machine, and did not court publicity of any kind in 1904. Their *Flyer II* of that year was, as they well knew, a good machine; but it was still only a tentative machine, and not one to advertise far and wide as the conqueror of the air.

45

The Wrights' first Impact on Britain: 1904-05

There was one Englishman—and only one—who was in a position to "know the score", so to speak, and had correctly appraised the Wright brothers and their work. This was Lieutenant Colonel J. E. Capper, then "Officer Commanding the Balloon Companies", who was later (in 1906) to become the head of the Farnborough research establishment of the day.

At the request of Colonel Templer, the then head of the establishment, Colonel Capper was sent to see the aeronautical exhibits at the St Louis Exposition of 1904. Among other things he did in America was to make personal contact with the Wrights. He made an excellent impression on the brothers; similarly they, and their work, made an excellent impression on Capper, thereby setting up yet another Wright influence-stream in European aviation.

Capper visited the Wrights on October 23rd 1904. In the course of his report to the War Office (P.R.O. AIR/1/1608) he wrote:

> "Both these gentlemen impressed me most favourably; they have worked up step by step, they are in themselves both well educated men, and capable mechanics, and I do not think are likely to claim more than they can perform.... The work they are doing is of very great importance, as it means that if carried to a successful issue, we may shortly have as accessories of warfare, scouting machines which will go at great pace, and be independent of obstacles of ground, whilst offering from their elevated position unrivalled opportunities of ascertaining what is occurring in the heart of an enemy's country."

Capper was one of the very few men to whom the Wrights showed photographs of their machine in flight: they also gave him performance figures, and even allowed him to see their engine. Dr P. B. Walker, author of the official history of the RAE Farnborough[1] shows that, when

[1] See Dr Walker's book in the Bibliography (*Early Aviation at Farnborough*, 1971, 1973) where the full story of this visit is told for the first time.

it came to the British Government sanctioning Capper to let Cody build the "British Army Aeroplane No 1", it was Capper who gave the basic ideas to Cody; and Capper followed the Wrights (see Section 82). This accounts for the first Cody machine of 1908 being accepted and described at the time as a Wright-type, as indeed is obvious from its configuration (Fig 48).

In the Proceedings of the RE (Royal Engineers) Committee of the War Office for 1905, Capper's further report—after an exchange of letters between the War Office and Dayton—included the following, dated January 5th 1905:

> "There appears little doubt but that the machine has done all that the Wright brothers claim for it, and that it is within their power to construct a similar machine capable of going much longer distances and carrying one or more passengers. That a machine of this nature may be of very great value for war is undoubted—but we should require not one but 50 or 100 machines.... That the Wrights are ahead of the rest of the world in this matter seems to be absolutely certain, and by securing their services we should obtain a lead over other nations.... I am inclined to think that the sum asked will be too great for acceptance, and that we must do our utmost to build successful machines ourselves and learn their use."

Postscript. In view of certain statements made in the past, it should be said here that, although Capper was sent officially to see the St. Louis Exposition, it was entirely due to his own initiative that he paid visits to the Wrights and others in the USA.

THE YEAR 1905

46

The first Illustration of Ferber's tailed Glider

to be published: February 1905

One of the few events of importance in a year which was to witness an ever-growing procrastination in European aviation, was the publication in *L'Aérophile* of February 1905, of a large illustration of Ferber's tailed Wright-type glider of 1904 (Fig 22). This was the first time the pioneers had "seen" this vitally significant machine, and it was later to bear valuable fruit. The illustration occurred in one of *L'Aérophile's* regular write-ups of well-known men in aeronautics, which in this case included not only the customary photographic portrait, and pen portrait, of Ferber, but an excellent photograph of the machine in flight, seen from the rear. It also dramatically revealed Ferber's weaknesses, with his non-rigid triangular rudders, called 'focs"—little more than thick flags—attached to the outer rear wing-struts, and specifically described as "gouvernails de direction." It was to be a very long time (*ie* until 1908) before he put a rudder on the tail-unit, despite a host of precedents in European model-making, and the obvious aerodynamic reasons for placing it there.

❊ 47 ❊

The Aéro-Club's Model Competition: February 1905

Considering the past exhortations by leaders of the Aéro-Club de France to their members to go ahead and master full-scale gliding, and then the powered aeroplane, one cannot help feeling that there was some symbolism of frustration about the elaborate arrangements and coverage for the display and competition of models held by the Aéro-Club in the Paris Galerie des Machines on February 11th-13th 1905.

Ferber was its "Secrétaire Rapporteur", and his illustrated report in the March issue of *L'Aérophile* ran to no less than eight pages, and was headed "Le premier concours d'aviation de l'Aéro-Club", The competition was officially called "Le concours d'appareils d'aviation non-montés", and included both a static show and a flying competition in which the models were launched from a 41-metre-high launching tower (pylone de lancement) erected in the huge building. It is fair to say that no model in this "concours" was to lead to any significant development, and that the meticulous recording and checking by all the officials concerned was mostly time wasted. The whole affair had, as its main aim, the encouragement of full-scale aviation; but no one proved to have been encouraged by it. What is more, the only model-makers whose names were to become familiar—although not of great importance—were José Weiss and Paulhan. Hubert Latham exhibited an airscrew, but no details of it are available.

48

The second Archdeacon Wright-type Glider: March 1905

Archdeacon's long-awaited new glider was still of the Wright-type with forward elevator; but he—like Ferber, and most probably influenced by him—had added a fixed tailplane bearing two fixed fins (Fig 23). It made its one and only test on March 26th 1905 at Issy-les-Moulineaux; it was towed off the ground unmanned, and in ballast, by an automobile, and crashed a few seconds later. This method of launching was suggested to Archdeacon by Pilcher's 'horse-drawn' method. Esnault-Pelterie had been the first to attempt this method of automobile launching in October of 1904. Chanute, on October 5th 1904, at the St Louis Exposition, had demonstrated a method of glider launching by a cable wound in by an electrically-driven winch.

It is interesting to note that Gabriel Voisin claims to have been responsible both for the design of this new Archdeacon glider and for the method of launching (*op. cit.* pp. 126 ff): but his autobiography is the only known source for these claims. It is significant that it bore a close resemblance to Archdeacon's first machine—which Gabriel certainly did not design—and Archdeacon was technically minded enough to think up the various improvements shown on this new glider.

This glider deserved a better fate, as it was a more sophisticated machine than Ferber's and augured well, provided control in roll was added. It was drawn along a wooden slip-way, and at first made a steady ascent: then part of the tail-unit came away and the glider crashed.

As before, Archdeacon did not persevere, and soon turned to the box-kite solution, as will be seen later (Section 52). The monoplane tail-unit would have been a far better avenue of experimentation than what was to come, and this case of the second Archdeacon glider is typical of the European refusal to "stay with" a configuration and make it work, instead of jumping from one shape to another. George Blanchet, who wrote the short piece—with an excellent large illustration—in *L'Aerophile* (April 1905), rightly said that this experiment "demonstrated the capital importance of tailplanes (plans de queue) from the point of view of stability."

As a tailpiece to this section, it ought to be said that Gabriel Voisin, in his autobiography, seems to have become confused about this second Archdeacon glider. He says that when his wing "cell" was completed,

> "Archdeacon decided ... that we should make a mechanical test on this cell. I modified an old engine which had been in the Turgan workshops and I fitted it to my glider. The power plant consisted of two tractor airscrews, belt-driven by the engine. This arrangement, which worked very well, was never tried outside the assembly shop."

This mysterious operation, performed for no stated reason, seems to have gone unrecorded and unphotographed.

But the oddest point is Gabriel's reference to a 'cell', always at that time meaning a box-kite type cell, which the second Archdeacon glider did not possess, but the subsequent Voisin-Archdeacon float-glider did.

It was still in connection with this second Archdeacon glider that Gabriel wrote (*op. cit.* p. 129):

> "I have read, in one of those lamentable 'Histories of Aviation' published in France, that for this structure I had been 'inspired' by the Wright machine. Now it should be noted that from 1903 to 2nd [*ie* the 8th] August 1908, no information about the Wright powered aircraft had ever been made public."

As elsewhere in his book, he neglects the fact that the Wrights' main early influence worked through their gliders and their European progeny. He was also forgetting the two pirated drawings, with descriptions, which appeared in Europe in January 1904 (Section 26) and March 1905 (Section 50), one or both of which Gabriel would certainly have seen or heard about.

… 49 …

Archdeacon challenges the Wrights: March 1905

It is not known why, on March 10th 1905, Archdeacon decided to approach the Wright brothers direct. It is perhaps strange that, since he had avowedly copied the Wrights' 1902 glider in his machine of 1904, and had failed with it, he had not contacted them before. This letter, dated March 10th 1905, was not only his first and last personal letter to the brothers, but it brought into the open for the first time Archdeacon's growing feeling of suspicion of—one might already say enmity for—the Wrights; this feeling was engendered, I believe, by his extreme patriotism, and his long-drawn-out efforts to spur his countrymen to emulation; laudable efforts which were being continually thwarted. His letter read as follows (the italics being Archdeacon's):

"Gentlemen, My name has certainly reached you as among those who are endeavouring to acclimatise, in France, studies relative to aviation.
 The results which you have already obtained—if I can believe the accounts of them *which have been made by you yourselves*, and published in different French journals—are absolutely remarkable; so remarkable, even, that they give rise to a certain incredulity in my country.
 This incredulity stems from two reasons: first of all, the long time which has elapsed, without new results being known, since your memorable experiments of December 1903; secondly, the mystery which it is claimed you supposedly surround yourselves with, to give you the time to secure patents.
 On this point, I have been a little astonished, in that, on the one hand, I do not believe in patents; and on the other, if you truly have new devices in your machines, you will have ample time to obtain such patents.
 However that may be, this question excites me to the extent that I will willingly make the journey to America to see your machines, only if I am *allowed to see them and see them functioning*; unless you yourselves think fit to come to France to show us what you have done

after you have dismantled one of your machines and put it on board the steamer.

There is no need for me to say that you would be received by us with enthusiasm.

I would add two things which may interest you: that if you are our masters in aviation, we are certainly yours in the matter of light motors, and that you will find with us motors weighing 2 kg. per horsepower, which you certainly have not got in America. I would add that perhaps you know that there is the Deutsch-Archdeacon prize of 50,000 francs for the first experimenter who flies a closed kilometre circle.

So please let me know *if one may see you* experiment in America; or if you would be disposed, should the occasion arise, to come and give us lessons in France."

It is known that the Wrights answered this letter, but the text of their reply has not survived.

50

Another Picture of a Wright Aeroplane is published: March 1905

Another small item, worthy of mention here I think, is an illustration of what was presumably meant to be the Wright *Flyer II*, which appeared in the *Illustrierte Aeronautische Mitteilungen* for March 1905. It is simply a drawing based on a well-known photograph of the Wright modified No 3 glider of 1902, seen from the rear, with the addition of two pusher propellers (Fig 24).

☙ 51 ❧

Esnault-Pelterie's Article on his Wright-type Glider : June 1905

On January 5th 1905, Robert Esnault-Pelterie gave a lecture to the Aéro-Club in Paris on the tests with his Wright-type glider in May and October of the previous year: the lecture was "trés applaudie". The printed article was promised for next month's *Aérophile*—ie February—but it did not appear until the June 1905 issue. It was entitled "Expériences d'aviation executées en 1904, en vérification de celles des frères Wright."

It started off by saying that certain French periodicals, particularly *L'Aérophile*, had published the results of the Wrights' gliding experiments; that angles of descent of as little as 6 degrees had been claimed; and the additional claim made that they could turn to left or right with ease. Esnault-Pelterie clearly showed himself on the offensive from the start:

> "A great stride seemed to have been realised in this very arduous and delicate question of the conquest of the air. We confess that results so magnificent, especially coming from the other side of the Atlantic, have left us a little sceptical. But in scientific matters, scepticism has no value. When an experiment seems surprising, there is a very simple means of resolving the doubts, and that is to make a repeat-experiment. It was with this aim that we constructed an aeroplane scrupulously following the directions (indications) of the Wright brothers, directions and diagrams which moreover were published in *L'Aérophile*. Our machine was exactly like (absolument semblable) that of the American experimenters, as much as to the general dimensions as to the curvature of the ribs and the disposition of the controls. Only some questions of construction and detail were different."

This Wright-type glider had a span of 10.20 metres; a chord of 1.50 metres and a gap of 1.50 metres (Chanute gave a span of 9.75 and a chord of 1.42 metres). The machine had a front elevator and a rear rudder, but the question of whether it had warping at this stage is in doubt (see below). The pilot was slung by straps so he could swing backwards or forwards to shift the centre of gravity, an idea totally alien to the Wrights; it is also impossible to understand how he kept himself in any one position, and was capable of operating the controls simultaneously.

The wing curvature was apparently 1 in 20 at one third chord, which Chanute erroneously gave in his article for the Wright glider. The Wrights actually varied their curvature from 1 in 24 to 1 in 30, the last giving the best results.

Where the controls were concerned, it is curious to find Esnault-Pelterie, in his detailed description of the first state of the machine, omitting all mention of the Wright's wing-warping. He wrote:

"The disposition of the controls was exactly that indicated by the Americans: the elevator was out front, supported by the skids; the vertical rudder was in the rear."

This is all that was said about the control system: there was not a word about control in roll; not a word about warping. One can only conclude that there was no roll-control at all incorporated in the first version of the Esnault-Pelterie glider.

The drawings and photographs in Esnault-Pelterie's article show only a crude glider, although not as crude as Ferber's. It was, of course, far from being even a good copy of the Wrights' No 3 (modified) glider of 1902, which Chanute had highlighted.

With Esnault-Pelterie, as with Archdeacon, it is hard to imagine why —if they were really dedicated to the matter in hand—they did not take a closer and more careful look at the many superb photographs of the Wrights' machine which had been readily available in Paris since mid-1903: the most cursory comparison would have revealed mistakes and crudities in the copies, which might at least have been partially corrected; it would also have revealed the differences in proportions and "feel" in the machine. Such a comparison would also have shown up the differences between the scale-drawings that Chanute published, and the real 'live' Wright glider. The whole attitude and approach of Esnault-Pelterie revealed—as also in the case of Archdeacon—the typical approach of the amateur. One may with justice say that at least part of the aeronautical disease that so sadly was to beset Europe was due to a lamentable lack of professionalism. The Wright brothers, in everything they did, proved themselves professionals to their finger-tips; most of the Europeans, on the other hand—until 1907—proved themselves amateurs in every aviation activity they undertook. These remarks may seem over-severe to those who feel some particular affinity to the early

French pioneers; but an examination of their words and deeds can only lead to such conclusions. It is all the more astonishing when one considers by contrast the extreme professionalism of everyone in France who was concerned, for example, with the automobile.

This glider underwent tests of "very brief duration" in May and was a complete failure; "defects appeared which could unfortunately not be corrected." Chief blame was thrown on the wing-curvature—which Esnault-Pelterie said was too pronounced—and the machine tended to pitch forward onto its nose. He came to the conclusion that

> "It was now proved that no suitable experiment (expérience convenable) could be attempted with this machine. The tests were then postponed until the Autumn, to be made with the machine modified in the light of experience."

The clear implication of this was that the machine had been an exact copy of the Wright glider; and since it was a failure, the Wright machine had also been a failure, and the brothers' claims were false.

The machine was modified thereafter, and was next given a series of tests in October 1904. The span had been reduced to 9.60 metres, with also a reduction of wing area to 27 square metres. The most important modifications were the alteration of the wing curvature to 1 in 50, and the abolition of the forward elevator: the "backward and forward movements of the experimenter were sufficient to assure the maintenance of longitudinal equilibrium". From his subsequent remarks, Esnault-Pelterie seems to have incorporated wing-warping on this modified machine. For he wrote of warping:

> "The twisting (torsion) of the surfaces, recommended by the Wrights —which we had tried—gave sufficiently good results for maintaining the transverse equilibrium; but we considered this system dangerous. It was possible, in our view, to cause magnified tensions on the [operating] wires. Consequently we feared breakages in the air. . . . We therefore believed we should abandon [wing-] twisting."

In the third—and final—series of tests, also made in October, Esnault-Pelterie fitted two rectangular surfaces forward of the wings to act both for control in roll and for control in pitch. These were the first ailerons and the first elevons in history to be fitted to a full-size aeroplane. He had reported, as a result of the second series of tests, which were made without the elevator, that "the front elevator is an indispensable device, contrary to what we had thought at first" (Fig 20). He wrote:

> "Nevertheless, in order to be able to effect lateral equilibrium, we then fitted two horizontal rudders in front, separately, each placed at the end of the wings. These two rudders were each connected to

a small control-wheel (petit volant de direction) within reach of the operator's two hands. Therefore, when these two rudders [*ie* elevons] were operated simultaneously, they controlled the fore-and-aft stability; when, on the other hand, they were operated 'contra-wise', they controlled the lateral stability."

It will be noticed that Esnault-Pelterie, like all the other European pioneers, considered any roll-control system as exclusively concerned with maintaining lateral 'equilibrium'; it was never conceived as an aid to turning, a manoeuvre which they thought of as solely the concern of the vertical rudder, as in a ship.

This attitude—owing to lack of gliding experience—was to persist right up to August of 1908, when Wilbur Wright started flying in France, and demonstrated the true nature of lateral control (*ie* roll-control in association with yaw-control). This was the principal reason for the Comte de La Vaulx referring to Wilbur's machine as the aircraft that "has just revolutionised the aviators' world".

In the third series of tests (still in October 1904), made with the yet-again-modified glider, the launching of the machine was carried out—as with the Archdeacon glider in 1904—by being towed by an automobile, the glider resting on a small two-wheeled trolley.

Yet again, as with every other French—or indeed European—pioneer, the tests were far too perfunctory and amateurish to establish anything of value. In his final 'conclusions', Esnault-Pelterie wrote:

"The ideas which one was able to formulate about the heavier-than-air problem, before being able to study the question in practice, were promptly modified by a number of experiments."

Esnault-Pelterie found that lightness of structure was not the all-important factor he had believed it to be, and that the wing-curvature was the most vital problem. He also thought, at the start, that 'equilibrium' could easily be obtained by placing the centre of gravity only a little below the centre of pressure, and found this error to be of even greater importance than the first mistake.

He then retreated to the favourite inherent stability idea of Chanute and the Europeans, that lateral 'equilibrium' could be best obtained automatically, *via* a dihedral angle:

"Lateral 'equilibrium', perhaps, could be obtained automatically by an appropriate form of the wings; but we believe that vertical 'equilibrium' [*ie* in the horizontal plane, involving control in pitch] should remain in the pilot's hand, as should also the (machine's) direction in the horizontal plane [*ie* as regards control in yaw]. It is nevertheless certain that all the difficulties involved can only be resolved by long practice. It is therefore necessary to arm ourselves with patience. Methodical experiments will alone be able to produce the answers to such questions. As soon as the summer comes, we hope to be able to devote ourselves to tests which will constitute the logical continuation (of those we have done), and which naturally, above all, will be tests of equilibrium."

Such further tests were never to be made.

This misleading article clearly implies that since Esnault-Pelterie's "absolutely similar" copy of the Wright-glider was not satisfactory, the Wrights' claims for their glider were therefore suspect, and could not be trusted. This is the typical assumption of a man who, so to say, 'dives' into a new craft—with no experience, but much enthusiasm—and tries to copy the achievements of the professional practitioners; fails; and then calls in question those achievements, without having attained any of their expertise of construction and pilotage.

If there is a slight saving grace in this article of Esnault-Pelterie's, it is contained in the last paragraph:

"It is nevertheless certain that all the difficulties involved can only be solved by long practice. It is therefore necessary to arm ourselves with patience. Methodical experiments will alone be able to produce the answers to such questions."

Esnault-Pelterie, despite his introduction of ailerons and elevons was, ironically enough, to abandon them when he introduced his own powered machines in 1907; he then incorporated the system which—here in 1905—he had condemned, *ie* the Wrights' wing-warping!

⚘ 52 ⚘

The Voisin-Archdeacon and the Voisin-Blériot Float-Gliders : June, July 1905

A milestone in aviation history came with the testing on the Seine near Billancourt, of two similar machines called today the Voisin-Archdeacon float-glider and the Voisin-Blériot float-glider, representing the marriage of the Wright glider configuration and the Hargrave box-kite (see next Section). Each was piloted, and towed off the water by a motor-boat, in June the *Rapière*, in July the *Antoinette*.

The intention behind these tests was described in *L'Aérophile* for July 1905 as follows:

"In this experiment (*ie* with the Archdeacon machine in June) the first of its kind, M. Archdeacon had a double aim: to substitute for the various methods of launching in use for aeroplanes, a method which allows the launching—as does the launching by automobiles— to be independent of the whims of the wind; and which, besides, allows experiments to be made without danger to the aviator, and without risking irreparable damage to the machine."

It is interesting to note that very few subsequent experiments in aviation history were carried out with the machines being towed over water.

The Voisin-Archdeacon machine (Fig 25) had main wings like a three-cell box-kite, with a very slight curvature, and a large tail-unit like a two-cell box-kite. The Wright forward elevator was retained. With Gabriel Voisin as pilot, the first two tests were made on June 8th 1905, the first with some success, the second ending in a mild crash. It was finally tested, with more substantial floats, on July 18th, again with moderate success.

On July 18th there was also tested the Voisin-Blériot (Fig 26), which had a shorter span, and a more pronounced curvature; the lower mainplanes had even less span, and the outer "side-curtains" connecting the wing-tips therefore formed a dihedral angle. This machine was tested only once (on July 18th) when after only a few seconds in the air, it went out of control and crashed, Voisin having a lucky escape from drowning.

The Voisin-Blériot machine heralded the arrival in the public eye of Louis Blériot, an engineer who had had great success in marketing his own designs for autombile lamps. He was, of course, to become one of history's greatest names in aviation. In the *L'Aérophile* report, he is described as one of the "aviateurs militants de l'Aéro-Club de France."

It is interesting to find Ferber, in his London lecture of February 1908 (see Section 85), writing as follows about the Voisin-Archdeacon float-glider, and the same could be said of the Blériot version:

> "He then made for M. Archdeacon a large aeroplane of the Hargrave kite pattern, but fitted with an elevating rudder, the usefulness of which I had shown him, having learned it myself from Wright."

It was the Voisin-Archdeacon machine which formed the archetype of the famous Voisin biplanes. It was, on the whole, a pity that this course of events took place, and that Voisin did not follow and improve on Ferber. This first float-glider flew at a 10 degree angle of attack, and Voisin—always a champion of inherent stability—seemed to lean more toward the idiom of a tethered kite moving through the air, almost as if it were being towed from the ground, than toward a stabilised aeroplane moving at an acute angle through the air, propelled by an airscrew. This machine underwent a further—and fruitless—series of tests on Lake Geneva in September 1905.

❧ 53 ❧

Hargrave's Box-Kite enters full-size Aviation: June, July 1905

It has always been a matter for speculation as to why the Hargrave box-kite abruptly entered full-size aviation in 1905 with the construction of the Voisin-Archdeacon and Voisin-Blériot float-gliders. There had till then been no tentative designs or model aeroplanes incorporating the box-kite.

I would hazard the suggestion that either Ernest Archdeacon, Gabriel Voisin, or Blériot were—early in 1905—leafing through back numbers of *L'Aérophile*, and that one of them came across the July and August numbers of 1902; the first of these contained a large clear photograph of a Hargrave box-kite, and the second had an illustration of Hargrave's own sketch for his proposed box-kite type of powered float seaplane (Fig 12).

The occasion of the first, in July, was a revue of J. Lecornu's now well-known work on kites, entitled *Les Cerfs-Volants*. Then, in the next month's issue (August) appeared a brief paragraph on Hargrave's design for a box-kite float glider.

Gabriel Voisin records[1] that he first heard of Hargrave's box-kite in 1898:

"It was not until 1898 that we first learned about the Hargrave Kite from a magazine. Our first Hargrave, made of two equal-sized cubes, one behind the other, astounded us."

Gabriel goes on to say that they built bigger and better Hargrave kites; then tried gliding experiments—the type of machine was not stated—but these "produced no results in the gliding direction". What is so peculiar is that he was piloting Archdeacon's Wright-type glider in April 1904, yet we hear nothing of any Hargrave-influenced machine of Gabriel's design until the Voisin-Archdeacon and Voisin-Blériot float-

---

[1] "Grandfather taught us how", by Gabriel Voisin, in *Interavia*, Vol VIII, No 12, 1953.

gliders which were completed and tested in June and July 1905. His possible retrospective feeling of 'guilt' at not having thought of the idea before, might lie behind the dating of 1899 of two illustrations of his semi-box-kite Chanute-type glider, which appeared in the English edition of his autobiography; whereas in fact the glider in question was built by Gabriel in 1907, as already noted.

But the fact remains that, whether or not I am correct in my surmise about its origin, the box-kite remained "unacted-upon" until 1905, despite its having been invented by Hargrave in 1893.

If I am correct in guessing that it was one of the three men I mentioned above who chanced to pick up the July and August *Aérophile* numbers of 1902—or a bound volume—it would not have been as curious as it might appear. Very few members of the Aéro-Club were interested in aviation in 1902, and it would not be likely that members would go back before about 1903 in any desultory search they might have been making. The idea of "research" was, by and large, alien to the pioneers of that day.

Another possibility, of course, is that Lecornu's book itself was being browsed through by Gabriel Voisin, or one of the others, in 1905: and that the illustration of Hargrave's box-kite triggered his interest. But whoever it might have been, would not have seen there the most significant illustration of all, which was Hargrave's design for his multiplane box-kite float machine, which *L'Aérophile* showed in its August 1902 issue, with its rear component smaller than the front portion. Also significant is the fact that the Hargrave machine and the two Voisins are on floats.

Considering how matters finally arranged themselves, intriguing food for speculation is provided if one imagines the Hargrave box-kite as being generally known and understood by, shall we say, about 1900; and there was no reason why it should not thus have been known and understood at that time.

✄ 54 ✄

Ferber's Work in 1905
and his second Article in the *Revue d'Artillerie*: August 1905

After the publication in *L'Aérophile* for February 1905 of the photograph of his tailed Wright-type glider, Ferber's direct impact on the history of aviation was virtually over. He was to remain the much respected doyen and spear-head of the European revival, but his remaining work was to have virtually no influence on anyone. His writings, however, were to prove valuable historically, and as a revelation of his own attitude and that of others.

He knew he was right in adding (in 1904), and then retaining, a horizontal tailplane to make for longitudinal stability, and dihedral on the wings: the forward elevator remained as before. In 1905 he continued to experiment at Chalais-Meudon with this machine; but, as before, he tried running before he could walk, and—instead of learning to control his glider properly—he went on to fit two seats in it, and produce what he called an "aéro-toboggan". Ferber commented on the Swiss sport of tobogganing, which he said was "un exercise passionnant"; but, he continued:

> "a thousand times more thrilling (passionnant), was the 'aéro-toboggan'; for, with it, one does not simply descend the 'slope', but in addition one describes on this 'slope' the undulations of a switchback."

The fact that Ferber now conducted the world's first passenger-carrying in a heavier-than-air aircraft with this two place glider—the passenger was his mechanic Burdin—should not divert attention from the fact that this modest feat was not really worth doing, so long as he continued to neglect the flight-control of the machine: for example, he did not even make the move of putting the rudder on the tail-unit until 1908.

Then in May 1905, Ferber added a 6 hp Peugeot engine, and on the 27th of the month, made his first motorised glide: the engine drove a small tractor propeller rotating within the elevator outrigger (Fig 27).

The power was not enough to sustain the machine, but it considerably reduced the angle of glide. Both glider and powered versions were "sling-launched" by Ferber's ingenious overhead cable system.

He said he found the effect of torque very troublesome—which would in fact have been negligible—and it may have been to obviate the resulting bank and side-slip of the machine that he tried the abortive extra vertical rudder which he was to recommend—half-heartedly—in his book *L'Aviation* (1909). There was no attempt to relate this phenomenon to lateral control by means of warping or ailerons, but he did consider contra-rotating propellers, as he had tested them on his machine in 1903, when it was slung from the whirling arm at Nice.

This attempt to plunge into powered flight was to advance Ferber scarcely at all. Next year (1906) he was to build a new powered machine (his No VIII) of which he had great hopes; but it was destroyed by a gale before it was tested. At last, in 1908, his No IX—an improved version of the VIII—was tested; but it was never to remain airborne for more than some 40 seconds. He then abandoned designing his own machines, and bought a standard Voisin, in which he was killed on September 22nd 1909 at Boulogne; taxying at speed, he hit a ditch (see page 257).

* * * * *

Ferber published his second article in the *Revue d'Artillerie* in August 1905, which was as bulky as a small book; it occupied 58 pages, with 48 illustrations included. Entitled "Les progrès de l'aviation par le vol plané", it was in effect a miniature history of gliding from Lilienthal to date, and thus the first of its kind in Europe. Its main interest in the present context is in the detailed account of his own recent work (which has been noticed in a previous section), including his ingenious method of launching, and his treatment of the Wrights.

Ferber reveals some strange aspects of himself in this paper: he seems to realise perfectly well that to ride the air is not—and can never be—a matter of "do-it-yourself-in-one-easy-lesson", and criticises those who feel this way:

> 'It is not possible to learn to walk, dance, skate, or ride a bicycle—let alone fly—in a minute; but that is what these aviators wish to do. No; one must slowly create the necessary reflexes; one must, as M. Chanute puts it so well in our language, learn little by little the art of the bird (son métier d'oiseau); and that is why Lilienthal's method—which I like to call the method of 'step by step, leap by leap, flight by flight'—is so fruitful; for after every check, one starts afresh.''

Ferber eulogises Lilienthal and his methods, and is convinced that success in flying only comes after long and arduous experience and practice:

> "It is thus that, taking solely to the Wrights' type in 1902 (m'étant rallié en 1902 seulement au type de Wright), I am two years behind him [*ie* Wilbur Wright] and I have not yet been able to catch up, whilst I retain a similar advance over my pupils. The reason for this is that to achieve practicality is always a long, difficult, and costly business."

Ferber seems, as ever, quite oblivious of the true nature of flying, and what it involves; and studiously avoids consideration or discussion of the overall picture of flight-control, and of lateral control in particular, of which his reading should have made him aware. One could understand his confusion of thought better if only he had at least brought up the subject of lateral control, discussed the question of wing-warping and its relation to the rudder, and then admitted that he did not understand what the Wrights were talking about.

But throughout his flying career—and the paradox of his thought is most evident here—he was to evince a most curious mixture, and confusion, of ideas which he was never satisfactorily to resolve. For example, he realised that the pilot's reflexes must be trained; but he misapprehended the proper role of control surfaces, and therefore conceived such training to be a much more formidable task than it really should be. Through ignorance of the true nature of control, Ferber came to over-emphasize the role of training; then—in order to shorten and simplify the prospect of pilot-training—he came to over-emphasise the role of inherent stability; and then, by the same token, to under-emphasise the role of control. He was never to understand the essentials of control in three dimensions, or the co-ordination of control. His instinct then seems bent towards simplicity at all costs, and this led him—as it did both Archdeacon and Gabriel Voisin—to embrace the concept of the motorised kite. Ferber wrote:

> "In short, an aeroplane is a kite in which the pull of the cord is replaced by the pull of the propeller.... Similarly, the kite is an aeroplane at anchor."

These statements admirably sum up the "anti-control" philosophy of the 'chauffeurs', who hoped that the whole job of flying an aeroplane would be done for them, except simply to raise it into the air with the elevator, and steer it about the sky with the rudder. Ferber later says of the aeroplane:

> "However, the aeroplane has two grave defects, which cause many people to say that it cannot claim to be the flying-machine of the future. It appears indeed rather unstable; and, above all, it must—

in order to start 'floating'—receive *instantaneously* a great initial speed it must have a launching device. It is difficult to reply to these objections. As to the first, one might propose the invention of better forms, and perhaps of automatic controls: but where the second is concerned, it is certain that special constructions must be built at departure-stations."

With all the data Ferber now had at his disposal, these statements are not only symptomatic of the chauffeur attitude to flying, but—perhaps —of a serious lack of imagination. He says later that the Wrights machine took off "sur un monorail", yet he states categorically that aeroplanes must be given a great initial speed *instantaneously* (his italics). His remarks about the lack of stability, and the hope of automatic controls, are of course typical thought-processes of the aerial chauffeur.

When he arrived at his historical sections, Ferber first paid glowing tributes to the Wrights, starting with the classic, "All honour to whom honour is due (à tout seigneur tout honneur)"; then referred to their flights of 1903 and of 1904—of which he received news from Chanute— and particularly to the Wrights' first circle of September 20th 1904 then to further circles of up to 4 kilometres in circumference, and a flight duration of over five minutes. This was the first news from a respected source of these 1904 achievements which reached the French pioneers. This news of Ferber's—his article was further brought to the attention of the pioneers by a notice in *L'Aérophile* for October 1905—was undoubtedly the "latest" news to which Archdeacon referred with scepticism in his address to the first conference of the Fédération Aéronautique Internationale in September 1905.

Ferber next voiced what was later to become a popular question: why, if the Wrights had achieved what they claimed, did they not compete for the Deutsch-Archdeacon prize of 50,000 francs, offered for what was thought to be the much more modest feat of a kilometre circle?

This led Ferber to explain that the Wrights were being secretive about their work in order to preserve and patent their inventions: then to this oft-to-be-repeated belief:

"There are actually no secrets in aviation: the mathematical theory has been well laid down by Pénaud, Tatin, and especially Renard and as to patents, since there has been on average one patent for an aerostatic machine every week for a century, it will be realised that for a long time past, everything has already been gone over (literally 're-seen' and 're-published'—'revu et réédité) from every side."

Ferber then dealt with the other pioneers, starting with Chanute Archdeacon and Esnault-Pelterie; the latter (his name three times misspelt!) was noted, and a close-up of his aileroned 1904 glider reproduced, with the remark that Esnault-Pelterie was "perhaps" wrong to substitute two forward control-surfaces for the one central one, but without a word about lateral control. Ferber then says a few words about various other pioneers then working in aviation, only one of whom—Levavasseur—was to make a real mark in history. Then, after a long account of the Aéro-Club's model competition, Ferber deals with launching devices and methods, and then gives a detailed account of his own recent experiments, ending with some interesting speculations about the future of aviation, including a shrewd assessment of its role in warfare.

This article, like his others, leaves one—irritatingly—with the feeling that if only Ferber had had a little more imagination, a little more 'aero-talent', he would have been an outstanding pioneer and a most valuable historian.

Ferber's article was rightly felt to deserve a much wider audience than it would receive in an obscure military journal; so *L'Aérophile* drew attention to it in the issue of October 1905. In the course of a brief summary, it mentioned Ferber's powered glider of 1905, and its origin. Ferber, it says:

"has carried out the first glide (vol plané) with a piloted powered machine (27 May 1905) which has been achieved in Europe, in the manner of the Wright brothers (à l'instar des frères Wright)."

But as a historically significant person, Ferber was still to play a part, in that it was to him that the Wrights wrote of their successes in 1905.

55

The Wrights' Work during the Year:
the first practical Aeroplane in History: 1905

Although the four brief flights of 1903 have naturally invested the Wrights' first *Flyer* with the greatest fame, their *Flyer III* of 1905—seldom dealt with in aviation histories—should stand equally with it; for the 1905 machine was the first practical powered aeroplane of history (Fig 28). It was of the same general arrangement as the others, but noticeable differences appeared in the placing of the elevator further forward—to improve longitudinal control—and the rudder further back. The span was 40 feet 6 inches; the wing area was slightly reduced, to 503 square feet; the curvature increased to 1 in 20; and new sets of propellers were used: but the excellent 1904 engine was retained. The prone pilot position was retained, and also—at the start of the season—the warp and rudder linkage. Its speed was approximately 35 mph. Like all the Wright aircraft, it was built inherently unstable, and had to be "flown" all the time by the pilot. In addition, the rudder outrigger was sprung to allow it to hinge upwards if it chanced to drag along the ground on take-off or landing.

The 1905 season at the Huffman Prairie lasted from June 23rd to October 16th, during which some 50 flights were made. But now the Wrights were concerned with reliability and endurance. They were airborne this season for just over 3 hours. In September, the trouble they were having in tight turns was diagnosed as a tendency of the lowered wings to slow up and stall; and the cure seen to be in putting down the nose to gain speed when turning. It was while seeking this diagnosis and cure that the brothers took the important step of unlinking the warp and rudder controls, and providing for their separate, or combined, operation in any desired degree.

With this *Flyer III* now perfected, the Wrights made many excellent flights, the five most outstanding being:

The Wright's Work during the Year: 1905

1905	Pilot	Duration	Km	Miles
September 26	W	18 minutes 9 seconds	18	11
September 29	O	19 minutes 55 seconds	20	12
October 3	O	25 minutes 5 seconds	25	15
October 4	O	33 minutes 17 seconds	33	21
October 5	W	38 minutes 3 seconds	39	24

W = Wilbur; O = Orville

"About two years ago," wrote Wilbur on December 4th 1905, "we succeeded in making the first free flight through the air with a motor-driven aeroplane. Since that time we have been busily engaged in developing the invention to the point of practicability. Though the difficulty of the task was increased not a little by the necessity of avoiding the eyes of the curious during the necessary preliminary flights, we have finally carried the machine through the experimental stage, and are now ready to offer it for sale as a secret practical invention. . . . The trial flights were made at Simms Station [Huffman Prairie] . . . and were witnessed by a dozen or more families living in the neighbourhood, as well as by a number of prominent citizens of Dayton who were present at our invitation."

My description of this machine as the world's first practical powered aeroplane is, I feel, justified by the sturdiness of its structure—which withstood constant take-offs and landings—its ability to bank, turn, circle and perform figures of eight; and its reliability in remaining airborne (with no trouble) for over half an hour.[1] It is now preserved in a specially built hall in Carillon Park at Dayton (Ohio).

It was in 1905 that the Wrights first offered their invention to the US and British governments. In January, on the experience they had had with the *Flyer II* of 1904, an offer was made to the US War Department, which was refused outright without even an attempt at investigation—"a flat turn down", as the exasperated Wilbur called it. They then offered it to the British War Office. The British dilatoriness proved just as bad as the US refusal. So, with persuasion from Chanute, the brothers again (in October, still of 1905) offered it to the US War Department, with all the added confidence now provided by the performance of their *Flyer III*. Again it was turned down; again the War Department assumed the Wrights were asking for financial *assistance*—whereas they made it quite clear they were offering a

[1] After writing this I find that Dollfus (*op. cit*) also says of the *Flyer III* that it was "le premier aéroplane pratique".

finished product—and for the second time the Department turned it down, in this case making one of the classic statements of aviation history: "the device must have been brought to the stage of practical operation without expense to the United States".

* * * * *

Mention has been made of Gabriel Voisin's autobiography, and its anti-Wright sentiments. Gabriel, by the time he came to publish his book in 1961, had built up what can only be described as a bitter hatred for the Wright brothers, a hatred which led him to make a number of highly fanciful statements. I have quoted a few of them in Section 1. Now, as we come to the end of the year 1905, let us look at some of his statements about the brothers, which may be said to apply to this period, after which they had decided to give up flying for the time being:

"the Wright of 1905 was not a straight line flight aircraft, but a machine using the undulations of small air currents." (page 224)

"It [*ie* the Wright *Flyer III* of 1905] could have flown, and it did fly, with the engine stopped, by means of favourable up-currents" (page 229)

". . . our conclusions which are categorical. *The Wrights never had in America a machine able to work without the help of upcurrents*. All they had was a poorly motorised glider." (page 221)

"In fifty years many things have become clear. In 1926, for instance Thoret, in an aircraft with a Hanriot engine, during a flight, at Blidah (Algeria) found that his machine would fly perfectly with the engine stopped. Thoret had discovered soaring flight which, after development, now allows flights of twenty-four hours to be made and distances of over 180 miles to be covered. The Wright mystery is elucidated. But the extraordinary thing is that the Wrights never understood clearly this explanation of their occasional successes." (page 238)

"The speed of the wind was carefully given for each test and it is easy to see that they never flew with less than 8.50 metres showing on the anemometer, and that all their attempts above this wind-speed were abortive!" (page 239)

"With the passage of time and thanks to the publication of the 'Wright papers' we now know why the Wrights surrounded their work with so much mystery. From 1903 until 1908, the Wrights only used motorised gliders, quite incapable of leaving the ground under their own power and absolutely incapable of flying in 'still air'. When they came to France, they could only fly with the aid of

French engines, made in France." (page 161)

"Orville and Wilbur Wright had never used in America (always without official observation) anything other than rather feebly motorised gliders. *They were only able to fly without the help of up-currents from the moment when they were able to use French Bariquand et Marre engines.*" (page 214).

The question of assisted take-off is dealt with in Section 94, and that of the French engines in Section 98. Here I should like to make a few comments on Gabriel's quite amazing proposition that the Wright machine of 1905 was "a poorly motorised glider", that it was not a "straight line flight aircraft", that "it did fly, with the engine stopped, by means of favourable up-currents", and that it was "absolutely incapable of flying in still air".

It is really a mark of how far 'overboard' Gabriel has gone in his hatred of the Wrights to suggest that a 700-pound powered aeroplane proceeded to "float about" on air-currents over the Huffman Prairie with its engine stopped for over half an hour; or that it could not fly in still air, when it went round and round and round the Prairie, and hence flew against the wind, with the wind, and athwart the wind on every journey it made. These wild statements were presumably made in an equally wild effort to do everything in his power to dissociate the Wrights from powered flight, so that he could claim that his own machines made the first powered flights. He therefore had to deny that the Wright engines were powerful enough even to keep the machines airborne! If his passions had not run away with him, he could not possibly have read, as he implies he did, the Wright Papers without being forced to abandon his fantasies. When, for example, he refers to the Papers, and says that the Wrights never flew with less than 8.5 metres showing on the anemometer, and that "all their attempts above this wind-speed were abortive", what did he suppose happened to the aeroplane when it turned round at the end of the field to come back on its endless circling of the Prairie? For more than half its airborne life, the *Flyer III* of 1905 was flying either with a side-wind, or a tail-wind.

Incidentally, had Gabriel really examined the Wright papers in any true sense of enquiry, he must have wondered why, in the numerous photographs of the Wright *Flyer III* in the air, it was impossible to see the two large propellers at rest, if the engine had been stopped!

It is sad, indeed, to find this grand old man allowing his imagination and his pen to run amok like this, as also in other contexts where the Wright brothers were concerned.

56

The Wrights send News of their Success with Flyer III to Ferber: October 1905

A document of great importance in aviation history is a letter written by the Wright brothers to Ferber, dated October 9th 1905. Ferber had always believed in what the Wrights said, and kept in touch with them from time to time. He had written to them in June 1905, asking if he could buy one of their machines. They answered as follows:

"Dear Sir,   October 9th 1905

At the time we received your letter we were just getting ready to resume our experiments, and we thought that in a short time we would be able to answer your inquiries in regard to the practicability of our flyer. We have been delayed longer than we expected. While our experiments last season had led us to expect much, nevertheless, until we had really made flights of much longer duration than those of five minutes, we could hardly consider that our flyer was practical for the purposes it will be called upon to serve in the future.

But our experiments of the past month have shown that we can now build machines that are really practical and suitable for many purposes, such as military scouting, etc. On the 3rd of October we made a flight of 24,535 metres in 25 minutes and 5 seconds. This flight was stopped through the heating of a bearing in the transmission, on which we had no oil cup. October 4th we made a distance of 33,456 metres in 33 minutes and 17 seconds. The transmission bearing heated again, but we succeeded in returning to the starting point before we were compelled to turn off the power and alight. On October 5th our flight had a duration of 38 minutes and 3 seconds, covering a distance of over 39 kilometres. Landing was caused by the exhaustion of the supply of fuel. An oil cup cured the trouble with the bearing which had terminated the previous flights. Witnesses to these flights have become so enthusiastic that they have been unable to hold their tongues, and as a result our experiments have become so public that we are compelled to discontinue them for the present, or at least until we find a less public place to carry them on.

The past several years have been given almost entirely to the development of our flyer, and but little time has been given to the consideration of what we would do with it when we had it perfected. But it is our present intention to first offer it to the governments for war purposes, and if you think your government would be interested, we would be glad to communicate with it.

We are prepared to furnish machines on contract, to be accepted only after trial trips of at least 40 kilometres, the machine to carry an operator and supplies of fuel etc., sufficient for a flight of 160 kilometres. We would be willing to make contracts in which the minimum distance of the trial trip would be more than 40 kilometres, but, of course, the price of the machine in that case would be greater. We are also ready to construct machines carrying more than one man.

Respectfully yours
(Signed) *Wilbur and Orville Wright*"

Thus Ferber was the first man in Europe to receive proper news of the Wrights' triumphant 1905 Season. He later wrote of this letter (in his *L'Aviation*):

"I was the first in the entire world to know—long before the others—this sensational news. . . . I wished to let my country profit, first of all the army, naturally. . . . I made a report to my official chiefs, who did not believe a word of it, and treated me as a mild lunatic (un doux illuminé). . . . The two main reasons for doubt at this moment were (a) if men had really flown through the air, one would have known about it; and (b) how could a simple Captain of artillery know something of which the American journalists were ignorant, men with whom it was a point of honour to be the best informed in the world!"

Poor Ferber! He was not the first, and was to be far from the last, to meet the stone wall of official incredulity. But worse was to come. The Wrights wrote again, on November 4th, saying the price of their machine was a million francs, payable after it had made a demonstration flight of 50 kilometres. Ferber said this price was prohibitive, but conceded that the buyers were thereby protected from deceit.

Meanwhile, Ferber sent further enquiries about the Wrights to Chanute, and the French consul at Chicago. In early November he also asked Lahm—the respected expatriate American member of the Aéro-Club de France—to make enquiries in America. But a whole month passed, and Ferber "lost all hope of interesting the War Minister through official channels (par la voie hiérarchique)".

Ferber then submitted his evidence of the Wrights' 1905 flights to Archdeacon, who "to my great surprise, had extraordinary doubts, which he formulated in an article in *Les Sports* of December 3rd 1905: and he left, saying that in no case would he head a purchasing committee [for the Wright machine]".

In the midst of these happenings, Georges Besançon, the Editor of *L'Aérophile*, received a letter from the Wrights similar to that written to Ferber, and published it in *L'Auto* for November 30th 1905 (see next Section). Ferber thereupon gave Besançon all the facts he had about the Wrights.

The journal *Les Sports* ranged itself alongside Archdeacon, and supported the "anti-Wright faction": in Ferber's words, it "unleashed the campaign of doubt and (accusations) of bluff, which was followed by the majority, people having in general more of a facility for denial than for belief".

So much for Ferber's brave efforts to convince his countrymen that the air had, in sober and melancholy fact, been conquered in America.

✄ 57 ✄

Archdeacon attacks the Wrights

and again issues Exhortations: October 1905

The newly-founded Fédération Aéronautique Internationale held its first Conference in Paris from October 12th to 15th 1905. One of the highlights of the occasion was Archdeacon's after-dinner speech to the delegates on Saturday October 14th, the main object of which was to tell the distinguished audience about his Voisin-Archdeacon float-glider experiments of June and July. But in a lengthy preamble, and not so lengthy peroration, he made some historically important statements. Soon after he started, he got on to the subject of the Wrights:

"Whatever the respect I feel for the Wrights—whose first experiments without a motor are undeniable and of the greatest interest—it is impossible for me to accept as historical truth (vérité historique) the report of their latest tests, which have not been witnessed, and about which they have voluntarily maintained the most complete obscurity. Perhaps, among the eminent American representatives we have the good fortune to have with us here, there is one who will be able to furnish the unpublished details on these sensational experiments . . . It is only 'yesterday' that aviation had become physically possible, and not much time has been lost; but, on the other hand, it is now the moment to get going; we are at a turning point in the history of science; the goal is in view, and we must all prepare for the final rush (au rush final) to try and arrive first. . . . But if the question of the motor has today been resolved, the question of stability certainly has not. It is necessary, above all, to study this stability, and discover, at all events, a practical method of carrying out the training (apprentissage) of the pilot without him breaking his bones . . ."

Archdeacon then goes on to say that he had devised this method of towing a glider on floats:

"My aeroplane, which you can see from the photographs that I have passed round, differs considerably from the original Wright type that

I felt I had to copy in my first machine, in order to give me my aviator's education. ... the elevator of the Wright brothers has been retained ..."

It is noteworthy that Archdeacon claims the float-glider as "my aeroplane", when the larger part of the design was probably due to Gabriel Voisin, along with the idea of water-towing. Archdeacon then speaks of his "aviator's education", when he said in the write-up of his 1904 machine that he did not pilot it; nor did he; nor did he pilot the float-glider in this year 1905, or any other glider so far as is known. The explanation is probably to be found in the fact that in those days, anyone actively concerned with aviation was liable to be called, or call himself, an "aviator".

Archdeacon then said that he had applied a modified Hargrave box-kite configuration to the machine, but had increased the number of side-curtains which, although they would increase the drag, would considerably improve the glider's lateral stability. He also said that the elevator could equally well be placed on the tail-unit.

He went on to give details of the machine and its performance over the Seine, and to say that further tests were to be made. These were carried out on the Lake of Geneva in September of this year 1905, but proved virtually fruitless.

At the end of his lecture, Archdeacon relapsed into that naiveté of hope and exhortation which—with the ever-growing "torpeur" of European aviators—is now familiar to us:

"The ex-Director at Chalais-Meudon, the late Colonel Renard, has published a series of theoretical studies on aviation, which possess the immense merit of showing researchers—which they can rely on entirely—the true way in which they ought to conduct their researches. If they follow this way strictly, without being halted by obstacles, or going astray en route, they will achieve their object. The problem has actually been so well propounded, that it is more than half resolved. Let two or three men of real ability tackle it, and its final solution will only be a question of months, perhaps of days (Que deux ou trois hommes de valeur s'y attaquent et la solution définitive ne sera plus qu'une question de mois, de jours peut-être). As for me, in the matter of science, I am a resolute internationalist: and wherever the final solution comes from, I shall always applaud it with equal enthusiasm. What is certain, is that the country from which it emanates will see its name inscribed for ever in letters of diamond on the tablets of world history (qui est certain, c'est que le pays, d'ou elle sortira, verra son nom inscrit à tout jamais en lettres de diamant sur les tablettes de l'historie du monde)"

I have quoted Archdeacon's most vivid pronouncements in his original words, as they are highly significant, coming, as they do, from the acknowledged spokesman and leader of European aviation. His lecture was not published until the January issue of *L'Aérophile* included it.

These quotations fall into perspective when one realises that just over a week before Archdeacon gave this lecture in Paris on October 14th 1905, the Wrights had made their record flight on the last but one test in the triumphant 1905 season, when they had achieved the world's first practical powered aeroplane: on this day (October 5th) Wilbur had flown for 38 minutes and 3 seconds, covering over 24 miles, and landing only when the fuel supply ran out. Among some fifteen witnesses was Bishop Wright—the Wrights' father—who noted in his diary: "In the afternoon, I saw Wilbur fly twenty-four miles in thirty-eight minutes and four seconds, one flight." The Dayton *News* and *Post* carried the story.

This 1905 flight of Wilbur's *was not to be exceeded by a European pilot until October 2nd 1908*, when Henri Farman flew for 44 minutes 32 seconds. It was not until June 5th 1909, that Farman's duration was bettered by another European, the Wrights having meanwhile flown for much greater distances.

The historically important aspects of this Archdeacon lecture, apart from further evidence of his antagonism toward the Wrights, are:

(a) his remark about stability being the most necessary subject to study;

(b) the absence of any remark (in the whole lecture) on lateral control;

(c) his espousal of the box-kite configuration with its maximum inherent stability, large angle of attack (10 degrees), and great drag; which meant

(d) his final abandonment of true controllability, and the crystalising of his 'chauffeur's' view of the aeroplane as a passive creature simply to be steered about the sky;

(e) his extraordinarily naive belief that his "two or three men of real ability" could tackle the problem of flight and arrive at a final solution in some months, or even days.

Like Ferber, Archdeacon has emerged as a man who—despite his passionate advocacy of aviation, and his determination that it should be achieved in the "homeland of Montgolfier"—had really no mature conception at all of what the proper flying of an aeroplane involved; for him—despite the superb photographs of the Wright gliders in flight which he had been the first to publish—flying had become a matter of motorising a kite.

There has never been any doubt about which names should be written in diamond letters; and one cannot help feeling that even Archdeacon had secretly realised the 'score', especially as he implicitly trusted Chanute, and Chanute had publicly endorsed the 1903 and

1904 Wright flights in letters to Ferber and others. Incidentally, it is still not known to what Archdeacon referred in the above lecture when he mentioned the report of the Wrights' "last tests"; but I believe this referred to the Wrights' 1904 flights, which Ferber had mentioned in his second article in *La Revue d'Artillerie*, published in August 1905, and undoubtedly read by Archdeacon.

But Archdeacon made himself outwardly believe strongly that the Wrights had never properly power-flown, and that the French would power-fly before them. Later, in the person of Gabriel Voisin, we are to see this attitude pursued *ad absurdum*.

Archdeacon's lecture on October 14th, which has been the subject of this section, was, of course, heard by a distinguished audience on this date; but it reached a world audience when it was published in full in *L'Aérophile* for January 1906.

❧ 58 ❧

The Wrights' new Impact on Europe: November, December 1905

When the Wright brothers were finally satisfied that they had a fully practical aeroplane in their powered *Flyer III* of 1905, they initiated a campaign of letter-writing: this had the dual purpose of informing the aeronautical world of their activities and achievements, and of finding a market for their machines.

We have seen that the European who first received the news of their 1905 successes was Ferber, when the Wrights wrote to him on October 9th 1905. The subsequent and abortive sequence of events—so far as Ferber was concerned—has been noted in the previous section. But the great impact of the Wrights on Europe came with the publication of a letter from the Wright brothers to Georges Besançon, in the Paris sporting daily *L'Auto*, in its issue of November 30th 1905.

The Wrights' letter from Dayton was dated November 17th 1905, and was received by Besançon during the last week of the month. Although it was perhaps the most important professional letter he had ever received, it is little known today. He was now in a quandary, on two counts: there was reason to believe that a German journal would soon receive and publish authoritative news of the Wrights, and his own journal, *L'Aérophile*, was running late for the December issue; it did not in fact appear until the first week of January 1906. So, in view of the vital importance of what he had received, he decided to let *L'Auto* have it. Here is the jointly-signed Wright letter:

WRIGHT CYCLE COMPANY
1127 West Third Street
Dayton, Ohio

Mr Georges Besançon,
84 Faubourg St-Honoré, Paris   November 17th 1905

Dear Sir:

Now that the season of experiment is closed, we are pleased to send

177

you the following account of our work this year, which may be of interest to the readers of your journal.

On account of a number of changes made in the machine since 1904, the first eight days this year, on which we had the machine out for trial, were not productive of any new records. It was not till the 6th day of September that we succeeded in passing our record of last year of four and one-half kilometres. The flooded condition of the ground, produced by the frequent rains of the summer, interfered greatly with the experiments, but progress was nevertheless rapid, and on the 26th of the month we passed for the first time the ten mile mark, marking 17,961 meters in 18 minutes and 9 seconds. The gasolene can would hold a supply of fuel sufficient for a flight of 20 minutes, but a little time is always lost in getting started after the engine is put in motion.

On the 29th of September a flight of 19,570 meters, in 19 minutes, 55 seconds, was stopped through the exhaustion of the fuel supply.

On the 30th one of the bearings in the transmission heated, bringing the flight to a close in 17 minutes and 15 seconds. We had no oil cups on any of the transmission bearings, but depended entirely upon a few drops of oil which were applied to the bearings just before starting, and which had been sufficient for the shorter flights.

October 3rd a larger gasolene reservoir, holding enough for an hour's flight, was placed on the machine. The flight on this day was limited to 24,535 metres in 25 minutes and 5 seconds by the heating of a transmission bearing.

October 4th an oil cup was fitted to the bearing which had given the most serious trouble, but after being in the air 33 minutes and 17 seconds the heating of another bearing in the transmission caused the operator to return the machine to the starting point and make a landing. A distance of 33,456 metres had been observed.

October 5th, after fitting an oil cup to the only remaining troublesome bearing, the machine was put in flight. Through an oversight the gasolene reservoir had not been refilled after a preliminary test of the engine, and but enough remained for a flight of 38 minutes and 3 seconds. In this flight a distance of 38,956 metres through the air was traversed.

All of these flights were made in circles, the machine returning and passing over the heads of the spectators at the starting point many times during the flight and finally making a landing without the slightest damage.

Although a wagon road, as well as the Dayton & Springfield Electric rail-road pass the full length of the field on one side, by selecting times when no cars were in sight we had been able to make flights of ten to fifteen minutes in almost perfect secrecy, except for the neighbouring farmers, who were witnesses to all of our flights. But after the flights became more prolonged we were unable to avoid the cars, and the news of what we were doing then spread so rapidly that, in order to prevent the construction of the machine from becoming public, we were compelled to suddenly discontinue experiments, just at a time when we felt that we were prepared to place the record above the hour.

Very respectfully,
(signed) *Wilbur and Orville Wright*

PS. If the editor of the Aérophile would like to personally investigate the truth of the reports that have been made concerning our flights of 1903, 1904 and 1905, we will be pleased to furnish you with the names of a number of well known citizens of Dayton who were present at these flights of 1904 and 1905, as well as the names of the men of the Kill Devil Life Saving Station, near Kitty Hawk, N.C., who were present at the first flights in 1903.

This was the first significant pronouncement to be published on the Wrights' flying, after a prolonged silence on their part, and a prolonged period of twilight in European aviation. La Vaulx (*op. cit.*) wrote later:

"One may imagine the stupefaction of most of the French aviators—who did not want to believe in the flight of December 17th 1903—when they read the letter ... When the first moment of stupor had passed, after the publication of this vital document, there was a general refusal in France to believe in the veracity of the Wright brothers."

Meanwhile, the Editor of *L'Auto* quickly despatched one of his best journalists, Robert Coquelle, across the Atlantic to Dayton in order to discover the truth about the whole astonishing story: he arrived in Dayton on December 12th. Others, too, were galvanised into action; telegrams and letters sped back and forth across the ocean, their sole object being to obtain proof, either that the Wrights had conquered the air, or that they were just plain liars. The chief targets now were naturally the witnesses who the Wrights offered to produce; they proved easy enough to find, being staid and prominent citizens of Dayton, and their straightforward testimony left no doubt at all in the minds of Coquelle and the others who questioned them. Coquelle immediately cabled his Editor:

"The Wright brothers refuse to show their machine; but I have interviewed the witnesses, and it is impossible to doubt the success of their experiments."

59

The Wrights; the Aéro-Club; and Archdeacon's Enmity: December 1905

Meanwhile, as may be imagined, there had been explosive scenes at the Aéro-Club de France. At a meeting of the Aviation Committee on December 23rd 1905, the members immediately showed themselves in their true colours. "The exchange of views," stated the official report, "started immediately on the subject of the news recently arrived from America about the Wright brothers". The members sorted themselves into two groups:

> (a) "the first [group] comprises those who believe that the flying-machine has been completely achieved [absolument réalisée], and who wish to see come to France the industry or sport which would result. In this group are MM. Drzewiecki, Besançon, Kapferer, and Captain Ferber".
>
> (b) "on the other side appears the second group—which are in the majority—who believe that the invention is little more than tentative (à peine au point), and who hope, in consequence, to have time to complete a better machine. This group therefore do not want to purchase the American machine, but prefer to see the liberality of French subscribers subsidise a consortium of inventors".

This second group was not named, and it would be unwise just to subtract the names of the first group from those listed as present, since some of them were sitting on the fence: but the leader of the second group was, of course, Ernest Archdeacon.

"A certain unanimity" was achieved at the end of the meeting, for the creation of an Aerodynamic Institute, "which would be the Pasteur Institute of Aviation".

However, Besançon insisted on their adjourning "la question Wright" until the next meeting, which was to be on December 29th, as he hoped to communicate new information which might modify the views of some members of the Committee.

The Wrights; the Aéro-Club; and Archdeacon's Enmity: December 1905

At the meeting on December 29th, the expatriate American member of the Aéro-Club, Frank Lahm, read a letter from a Mr Weaver, one of the Wright witnesses. This letter "confirmed and authenticated all the previous information". Other witnesses were mentioned, and the *L'Auto* correspondent, Robert Coquelle, also gave his opinion that the performances of the Wright machine were genuine. However, reported *L'Aérophile*:

> "In spite of these affirmative witnesses, the President of the Committee, M. Archdeacon, declares that he persists in believing that the machine is still only tentative (n'est pas au point), and that the Wrights only await the promise of a million francs in order to provide the capital which would allow them to construct a better machine."

At this distance in time, it is impossible to know for certain whether Archdeacon knew he would be quoted: since all the other meetings of the Aviation Committee were reported officially in *L'Aérophile*, there was no reason to suppose that this one would not be. But whether he realised this or not, Archdeacon's remarks can now only stamp him with disgrace in the eyes of history. His genuine patriotism had by now turned to chauvinism of the worst kind; his genuine desire to promote aviation had turned into a defensive, courage-sustaining whistle. The suggestion that the Wrights had not properly developed their machine, and were therefore angling for the purchase-money to enable them to build a better one, was a lamentable travesty of the facts, and he must have known it. Through Chanute, and through the Wrights' own communications, it was by this time (1905) obvious to all the Europeans who wished to know, what sort of men the brothers were; and such a suggestion from a man in Archdeacon's position was inexcusable by any standards.

Also, when all is said and done, the numerous names of local citizens which the Wrights gave as witnesses were certainly sufficient for the general purpose of verifying the mere fact of the flights. After all, they were not being called upon to judge times, or records, or whether the machine succeeded in just getting off the ground. All they had to bear witness to was an aeroplane going round and round and round the Huffman Prairie above their heads. For example, on Thursday October 5th 1905, over a dozen people were at the Prairie, and the machine made some *30 circuits of the field*. Any of those present who could not honestly testify that they had seen a long flight by an aeroplane would have needed their heads examining. Apart from the universally accepted honesty of the brothers—to which Chanute would always testify—it postulates a ludicrous state of affairs if one is asked to believe that all these solid men from a solid city like Dayton were bare-faced liars.

~ 60 ~

A pirated Sketch of the Wright Flyer III: December 1905

Robert Coquelle, who had been sent to Dayton by the Editor of *L'Auto*, also secured a most interesting 'scoop'. For he seems to have begged, borrowed or stolen an issue of the *Dayton Daily News* which the Wrights had apparently persuaded the Editor to suppress, or else reprint without an article on the Wrights. This article had included a pirated sketch—albeit a crude one—of the triumphant Wright *Flyer III* (Fig 29). This sketch was published in *L'Auto* in Paris in its issue of December 24th 1905. It was later to be followed by another pirated sketch which was also published by *L'Auto* (see Section 69).

This first pirated sketch managed to pack in quite a lot of correct information, despite the quality of the drawing. It shows clearly the following features of the Wright machine:

§ The front biplane elevator;

§ The hip-cradle in which the pilot lay;

§ The skid undercarriage lying on the launching yoke;

§ The launching yoke with its tandem wheels, running on the rail;

§ The two pusher propellers worked by chains from the engine lying on the lower wing;

§ The double rear rudder.

The appearance of this sketch in France had a strange and important impact of its own. Ferber, in his book *L'Aviation* (1909), wrote as follows:

"This drawing had great importance; it showed us the last details of which we were ignorant; and it was this drawing which caused the first aeroplanes of Delagrange and Farman—February and June of 1907—to have a forward cellular rudder [elevator]".

For some still unaccountable reason, both the French and the British

182

often came to refer to 'open' biplane structures as 'cellular'. despite the absence of the side-curtains which alone would qualify such structures for the term 'cellular': *vide*, for example, the name "Bristol Box-Kite" given to Bristol's first Henri Farman-type machine, which did not sport a single feature of a box-kite.

61

A Summary of Ideas and Achievements in 1905

The Europeans had continued their slow and confused progression. The only significant feature of Continental aviation—which in practice still meant French aviation—was in conception, not execution: this was the crystallising of their determination to pursue the inherently stable tailed aeroplane, rather than the inherently unstable machine espoused by the Wrights. In this they were maturing the 'stability concept' laid down by Cayley in 1809, published by Henson in 1843, popularised by Pénaud in the 1870's, and perpetuated by most of the model makers and full-size designers in Europe thereafter. In more detail, the European situation that matured in 1905 may be itemised as follows:

(a) the complete abandonment by the Europeans of the Wrights' doctrine of inherent instability;

(b) the consequent settled pursuit of inherent stability;

(c) the consequent abandonment of the pure Wright glider configuration of forward elevator, wings, rear rudder, and no rear horizontal surfaces;

(d) the development—following Ferber's inherent stability idea—of the Wright-glider-plus-tailplane configuration;

(e) the inauguration of the second type of European biplane configuration, ie the Wright-glider-cum-Hargrave-box-kite;

(f) the final, but still tentative, European attempts to fly gliders; then the abandonment of this basic flight philosophy of Lilienthal and the Wrights (which postulated that mastery of glider flight should precede attempts at powered flight);

(g) the tentative application of power to the first European biplane configuration.

The core of the European situation in 1905 was the final failure of her pioneers to comprehend the Wrights' philosophy and technique of aeroplane flight, which prevented the successful building and flying of

A Summary of Ideas and Achievements in 1905

Wright-type—or any other type—of gliders, despite the descriptions, photographs, and explanations published in 1903. It represented the strengthening of the 'chauffeur' attitude of winged automobilism, as opposed to the attitude of the true airman. This situation was worsened in 1905 by Robert Esnault-Pelterie lecturing to the Aéro-Club de France in January about his 1904 tests with Wright-type gliders, and the publication of his misleading lecture, with illustrations, in *L'Aérophile* for June 1905. This published lecture did much harm, both short and long term, and was one of the chief factors in the further retarding of aviation in Europe. Its one beneficial result was, by virtue of its large and clear illustrations, to put the visual idea of ailerons in the minds of the pioneers, ready for such time when they would come to realise the true nature of flight-control, which was not to be until late in 1908.

If the above remarks seem somewhat harsh, they are mild when compared with what the French said about themselves, and soon were to say in public. First of all, here is the Comte de La Vaulx summing up the state of European aviation in 1905:

> "In France on the other hand—after the trials over the Seine—by the end of 1904, and during 1905, the ardour of the aviators seemed to slacken. One did not think any more about the Wrights, who showed no sign of life. One could only fear that, after a passing effervescence, aviation began again to fall into neglect. On the other hand, the airships were progressing, and attention was being concentrated on them."

La Vaulx then paid a brief tribute to Ferber, who was not advancing at all fast; but he was rightly enthusiastic about Levavasseur's splendid progress in producing his two light engines (of 24 and 50 hp), and who was also working on aeroplane designs. Work was going on in various directions, and new machines were being considered; but, said La Vaulx:

> "This activity had not passed beyond the threshold of the workshops, and no new experiments had been witnessed, when—at the end of 1905—the Wrights decided to break their silence."

The European achievements during 1905 and indeed ever since 1902, had indeed been minuscule, and reflected little credit on anyone. Here is the lamentable list of gliders produced by the French pioneers from 1902:

1902	Ferber Wright-type glider
	Ferber Wright-type powered machine

1903	Ferber Wright-type glider (modified)

1904	Archdeacon Wright-type glider Esnault-Pelterie Wright-type glider Ferber tailed Wright-type glider
1905	Archdeacon glider No 2 Ferber powered glider Voisin-Archdeacon float-glider Voisin-Blériot float-glider

None of these machines was persevered with, and none of them was a success.

In addition, there was one interesting item in England, in 1905, when S. F. Cody, at Farnborough, built his glider-kite; this machine, as with the French, was not persevered with, and influenced no one; but it has the minor distinction of having been the second machine in history to be fitted with ailerons (Fig 30).

It was no wonder that Besançon, in his office of Secretary-General of the Aéro-Club, could sadly sum up—as mildly as he could—the past year's work in his speech to the General Assembly in the following March:

"I ask myself if the attentive and detailed examination of the period elapsed since our last General Assembly has not revealed, despite appearances, some sinking (tassement), some stagnation—extremely natural and explainable, without doubt, but none the less disagreeable to record, when one has been in the habit of recording each year incontestable improvements and striking progress."

Perhaps most symptomatic of all, was that the Aviation Committee of the Aéro-Club is not recorded as having met again during the year 1905, after its sitting of March 24th.

"It seems almost ridiculous," Wilbur Wright was—with justice—to say to Chanute in 1906, "that the French have never made any success at gliding in all these years." None of the pioneers of this vital period has satisfactorily explained why the French never persevered with gliding, despite the example, and the exhortations, of those who realised its importance. It was a thousand pities that these quotations from Lilienthal were not hung on the wall of every aviation pioneer in Europe:

"One can get a proper insight into the practice of flying only by actual flying experiments. . . . The manner in which we have to meet the irregularities of the wind, when soaring in the air, can only be learnt by being in the air itself. . . . The only way which leads us to a quick development in human flight is a systematic and energetic practice in actual flying experiments."

THE YEAR 1906

❦ 62 ❦

Discussion on the Wrights at the Aéro-Club: January 4th 1906

"L'affaire Wright" had swept the world of aviation, and was reaching a crescendo in early January 1906. The monthly dinner of the Aéro-Club on January 4th, was "particularly brilliant and animated". The official report was printed in *L'Aérophile*, in the January issue:

> "The great, the only, subject of conversation—l'affaire Wright; and naturally also the coming experiments of the Santos-Dumont helicopter. The celebrated aeronaut-aviator is assailed by questions about the details of the machine, his methods of testing, and his plans for public experiments.
> But the question of the genuineness of the Wright experiments provokes—at the 'delectable moment of the cigar'—a debate between the sceptics and the convinced so passionate that it goes on till midnight. M. Frank Lahm, the courageous doyen of the pilots of the Aéro-Club, defends the cause of his compatriots with a rare gift of argumentation and expression.
> In the heat of the discussion, Santos-Dumont declares that, in his view, the best means of arriving—'chez nous'—at a rapid solution, would be to create a huge prize (prix colossal), of 500,000 francs, for example, which would be awarded to the first aviator who covers 50 kilometres in one hour. Santos-Dumont adds that he is so convinced of this means, that he himself is ready to bet 100,000 francs that the prize would be won within six months of the competition opening."

No such competition, of course, was forthcoming, and the first man to attain this ambitious "goal" suggested by Santos—50 kilometres (*ie* 31 miles)—was Orville Wright, on September 9th 1908 at Fort Myer, when he flew 40 miles in 1 hour 23 minutes. Santos himself was never to remain airborne in an aeroplane for more than 10 minutes 27 seconds (in 1909).

❧ 63 ❧

The first *Aérophile* Article on the Wrights:
the December 1905 number, published in January 1906

In the first week of January 1906, the main Wright bombshell exploded among the French aviation community. For the December 1905 issue of *L'Aérophile*—which appeared early in January—ran an article which started by raiding the front page, for many months past the sacred preserve of the word-and-picture studies entitled "Portraits d'aéronautes contemporains". This article, presenting portraits of the brothers below its headlines (Fig 31), bore a long title:

> "*The Wright brothers and their powered aeroplane: the origin and sections of the debate; the facts advanced by the Wrights; objections and possibilities; first results of the enquiry*".

The article—signed simply "Aérophile", who was probably Georges Besançon, the editor—ran for $7\frac{1}{2}$ pages, and recapitulated the Wright story from the time of Chanute's Paris lecture of 1903, to date. It included the letters received by *L'Aérophile* from the Wrights; correspondence between the Wrights and Ferber; and between Ferber and Chanute; all of it attesting the plain fact that the Wrights had well and truly conquered the air with their powered *Flyer III* of 1905.

The Editor of *L'Aérophile* showed himself highly uncomfortable, sitting between the evidence before him and the vociferous chorus of those—mostly, one fears, activated by chauvinism—who could not, and would not, believe what the Wrights said. The Editor wrote:

> "From anyone else but the Wrights, the results announced would simply have been considered as 'bluff'. But their scientific past prevents us from holding this too summary opinion. Their previous experiments with gliders have never been seriously contested, and remain the finest to have been carried out . . ."

The Editor went on to ask why the brothers should try a useless bluff which was quite easy to unmask. However, he said, the great objection which had been made, from the outset of their powered flying, was their silence. A French editor could not understand any such silence from members of a nation who were "les premiers informateurs du monde", and about events which took place in the environs of a town of 85,000 inhabitants. (He did not realise it was not a case of "environs" as he understood the term). Then, see-sawing—as indeed he had to—the Editor tried to weigh the opinions of those in the Aéro-Club who kept in touch with Chanute and the Wrights, against the unbelievers. One of the greatest stumbling blocks to dismissing the whole business as 'bluff' was indeed Chanute, who was trusted by everyone in Europe—both pro- and anti-Wright—and who uncompromisingly supported the Wrights. Another aspect which bothered the Editor of *L'Aérophile* was the fact that the Wrights were demanding large sums of money for their invention; and that their own government in Washington did not appear to understand the importance of their invention, and support the Wrights as it had supported Langley. One wonders what the Editor would have thought about the US Government of that day if he had had access to the War Department's letters to Wilbur, which came very near the border-line of official insanity.

Then, said the harried editor of *L'Aérophile*, he had to contend with MM. Archdeacon, Ferber, Tatin, and many others, who maintained that the Wright machine did not incorporate any real discovery, but "simply a better construction, a more advantageous arrangement of the controls (organes) already known, without including the skill of the operator, which might count for much in the results obtained...". And so the arguments went on, for the $7\frac{1}{2}$ pages. Every logical and moral facet was examined. The Editor, not without justice, quoted a snide challenge issued publicly to the Wrights by Archdeacon in the journal *Les Sports*, which echoed his private letter of March 1905:

"I take the liberty of reminding you that there is, in France, a modest prize of 50,000 francs, which bears the name 'Prix Deutsch-Archdeacon', and which will go to the first experimenter who flies an aeroplane in a closed circle, not of 38 kilometres but only of 1 kilometre. It will assuredly not tire you very much to make a brief visit to France simply to 'collect' this 'little prize' . . . Thus, you will have obtained all, without anything being demanded of you, and you will have deserved—with the gratitude of your country—the esteem and admiration of the entire human race. If you don't do this, your invention will be quickly sneaked (chipé), and you will risk not only making no profit, but also not preserving for yourselves the glory—if it exists."

One can well understand the feelings of the French about this whole business—and even sympathise a little with Archdeacon—especially in view of their own collective disappointment, and even guilt, at having progressed so slowly over such a long time, a fact which they themselves recognised all too clearly, and deeply deplored.

The answers to all the questions the French were then asking are today well known to historians, but they were a matter of bitter controversy in 1905 and 1906.

The Editor then pieced together the statements from witnesses who had been contacted, and reprinted a cable from Robert Coquelle, who, as we have seen, had been specially sent over to the USA by the ournal *L'Auto* to investigate the Wrights. The Editor wrote:

> " 'The Wright brothers refuse to show their machine; but I have interviewed the witnesses, and it is impossible to doubt the success of their experiments.' . . . After this cable, it becomes more and more difficult, in our opinion, to challenge the magnificent powered aeroplane experiments announced by the Wright brothers."

In his sad peroration, the Editor drew up the balance sheet, and again—how often have we not met it before?—reminds his readers that France was the homeland of aviation, and issues both a warning and an exhortation:

> "If the results obtained by the Wrights are found to be definitely confirmed, let us not forget that aviation was born in France; that the theoretical bases, as also the first experimental results in mechanical flight, are due to a group of French aviators and researchers, such as Launoy, Bienvenu, Thibault, Duchemin, Ponton d'Amécourt, Pénaud, Marey, Tatin, Richet, etc. If by pure carelessness it has come about that we have allowed ourselves to be forestalled, we shall catch up when we really wish to, and even surpass. . . . If the news which we receive today from America is not true, it will be tomorrow. Let us remember the warnings we gave; and it is not long since we heard such authoritative voices as those of MM. Tatin, Archdeacon and Ferber; let us close with some words pronounced as early as 1903 by the distinguished President of the Aviation Committee of the Aéro-Club de France: 'Messieurs les savants, to your compasses! You, Messieurs the Maecenases, and you too, Messieurs of the Government, put your hands in your pockets—or else we are beaten!' "

❧ 64 ❧

The second *Aérophile* Article on the Wrights: January 1906

Close on the heels of the December 1905 issue of *L'Aérophile* (which appeared in January 1906)—with its news of the Wrights which caused such a commotion—came the second *Aérophile* article on the brothers, in its January 1906 issue (which appeared late in that month) confirming the reports of the Wrights' 1905 flights. It was entitled "The Wright aeroplane: continuation of our enquiry; detailed and decisive confirmations; the investigation of Robert Coquelle; letters from the Wright brothers, and from Mr Weaver; the Wright aeroplane bought by a French syndicate; the next experiments".

If the previous article had been a bombshell, this second article may be said to have provided the coup-de-grâce. There was now no dodging the truth; except by putting their heads in the sand; and that was what many of the French pioneers proceeded to do.

The article started by recapitulating what had been happening over the "affaire Wright", and about the leading part played in "the affaire" by *L'Aérophile*; and also reminding its readers that confirmation of the Wright claims had been sought in America. Now, said the Editor, the information received, "plainly corroborates the assertions of the Wrights". The Editor quoted the *L'Auto* correspondent, Robert Coquelle, who had visited Dayton. Coquelle, as we now know, acquired a somewhat romantic and inaccurate idea of the Wrights' performances; but he had interviewed a number of reliable witnesses, and had received undeniable evidence that the Wrights had achieved exactly what they said they had. In fact, Coquelle discovered—as we have discovered today—that the Wright brothers never knowingly told an untruth about their performances, or about anything else. The Editor of *L'Aérophile* thus summed up the situation:

"Following his enquiry, necessarily cut short by the lack of time, Robert Coquelle has returned to France fully convinced of the reality of the success obtained by the Wrights...."

After the detailed information gathered on the spot by our 'confrère', it became more and more difficult to doubt this cardinal point in the history of aerial locomotion—the complete success of the first powered aeroplane which, carrying a man, covered distances of several 'tens of miles'; was manoeuvred at the wish of the experimenter; and returned at will to its point of departure."

Then followed the text of a newly received letter to *L'Aérophile* from the Wrights, with details of their 1905 flights, accompanied by the names of seventeen reliable witnesses, fourteen of whom lived in Dayton.

Next was printed a long and detailed letter from Henry Weaver to his friend Frank Lahm in Paris—accompanied by a sketch map of the Huffman Prairie—in which Weaver reported on his visit to the Wrights at Dayton. Weaver, who had never before met the brothers, and only vaguely remembered hearing about them, was greatly impressed—as were most people—by the characters of the young men. Orville had taken Weaver to see some of the prominent citizens of Dayton who had seen the flights at the Huffman Prairie; and Weaver soon became completely convinced of what had taken place there.

After dealing with the French negotiations to purchase a Wright machine, the Editor of *L'Aérophile* closed his five-page article with the following historic statement:

"Thus we have the final proof that aerial navigation by purely mechanical means has just made the decisive step. We can only regret that it was not to be 'chez nous'. But at least let us have the satisfaction of seeing, on this particular occasion, the inventors to whom humanity will be indebted for this progress (which has been hoped-for from century to century) derive from their labours—with a glory before which the whole world will bow—the material gains which they have a right to expect. Here is one occasion, at least, when the legend of the inventor as an onlooker—unrecognised and ruined—at the triumph of his ideas, will be absent: and it will be a profound satisfaction for *L'Aérophile* to have been the prime cause of this occasion."

❧ 65 ❧

French Scepticism and British Belief: January 1906

There is no question, despite the overwhelming evidence available on the Wrights' flights—evidence now accepted by the unprejudiced French pioneers, including Victor Tatin, La Vaulx and the Editor of *L'Aérophile*—that the general opinion in France soon came to harden into a stubborn scepticism. The true position was admirably and calmly summed up by La Vaulx, who was one of the most responsible and mature-minded of the Europeans: it will be remembered that he had recently founded the FAI (Fédération Internationale Aéronautique) and was in the thick of the bitter discussions. He was later to say (*op. cit.*) of the Wrights' news:

> "The first moment of stupor having passed, after the publication of this cardinal document [the Wrights' first letter to *L'Aérophile*] there was a general refusal in France to believe in the veracity of the Wright brothers... The results obtained by the Wrights surpassed those achieved up to then in France to such a degree that—aided by amour-propre—it seemed quite simple to deny them. It was incomprehensible that the Wrights had succeeded so quickly where our aviators had failed, forgetting that the success of the aviators of Dayton was the logical result of a long series of efforts, starting with Lilienthal and continued by Chanute, which had cost the Wrights themselves five years of unremitting labour."

Considering the complete somnolence of Great Britain at this time—a somnolence shared by most of Europe—it was a great credit to the *Automotor Journal* that in the issue of January 6th 1906, it carried no less than three whole pages on the news from Dayton. Having weighed the evidence, the Editor came to the conclusion that the Wrights had well and truly conquered the air:

> "The whole subject is of extreme interest, and it will ultimately, no doubt, be regarded as epoch-making in the highest degree. We

are dealing with the first reports of absolutely the first successful attempts to accomplish mechanical flight, and it is not astonishing, therefore, that a large mass of material has already found its way into the Press. The discovery of exactly what the Wright Brothers have accomplished, and its publication to the world, under the circumstances, must certainly be regarded as a credit to the enterprise of French journalism."

Automotor made a thorough but curious job of reporting the whole "affaire Wright", dividing its long article into sections headed, "What the Wright brothers have actually accomplished"; "The evidence on which the statements are made"; "The means by which they have effected it"; and "The reasons for their reticence".

Some of the remarks and phrases which *Automotor* used on this important occasion are of great interest to the historian:

"As in their gliding experiments, the recent flights of the Wright Brothers, with the motor-driven aeroplane, have been extraordinarily conspicuous for a total absence of accident, due, no doubt, very largely to the remarkable mixture of bravery and caution which we have previously pointed out as a characteristic of all their proceedings. . . .

The witnesses, to whose evidence we shall next refer, and who were present at many of the actual flights executed, were most impressed, at any rate during the later flights, with the amazing manageability of the machine. It moved up and down, executed turns and figures exactly as the aeronaut managing it desired, and on one occasion performed an almost exact figure of eight inside a square of some 400 metres. . . . Wilbur and Orville Wright, as in their gliding experiments in North Carolina, managed, rode, or flew (whichever expression is preferred), the machine alternately."

Then the *Automotor Journal* writer got onto the all-important subject of control, and what we read here is as extraordinary as anything one could imagine:

"If what Mr. Chanute terms 'a whirling billow of air' tipped up one or the other side of the machine, this was corrected by movements of the vertical rudder, causing the aeroplane to swing round like a bird executing a curve, and so bring the depressed side up level again. . . .

They [the Wrights] made no secret at the time, and authorities on the subject like Mr. Chanute have been always fully convinced, that their success in gliding was due partially to the horizontal position of the aeronaut—a position, by the way, which it requires at first, at any rate, considerable pluck to assume—and the arrangement of the subsidiary planes, particularly the movable and controlling front plane. . . . The really important elements of their invention, therefore, consist of these three things, and we put them in order of increasing importance—the horizontal position of the aeronaut, the vertical tail or tails, and the forward approximately horizontal controlling plane, or planes. There may be other things which we do not know of as yet, but it is obviously those elements which have, so far as can be judged, made the difference between the extraordinary success which the Wrights have accomplished, and the ignominious failures which have been the record of practically everybody else."

The historian, even though he is conscious of the easy wisdom of hindsight, may be forgiven for feeling a certain amazement when, in this long and detailed account of the Wrights, there is no mention of any lateral control system, and—instead—the completely erroneous and misleading account of righting the machine laterally by rudder action alone. The Editor of *Automotor* need only have turned back to his own journal for February 1902 (page 197), where he would have read the very words of Wilbur which he (the Editor) had had the enterprise to print:

"The lateral equilibrium and the steering to right or left was to be attained by a peculiar torsion of the main surfaces, which was equivalent to presenting one end of the wings at a greater angle than the other."

Then, on page 243 of the March 1902 issue, where Wilbur's Chicago paper was concluded, there occurs among the points which Wilbur mentions which "stand out with clearness", the following:

"4. That with similar conditions, large surfaces may be controlled with not much greater difficulty than small ones, if the control is effected by manipulation of the surfaces themselves, rather than by a movement of the body of the operator."

Then, since we know that the Editor of *Automotor* kept a sharp eye on *L'Aérophile*, he surely must have seen the write-up of Chanute's lecture of April 2nd 1903, which mentioned the simultaneous use of warping and rudder in the 1902 Wright glider; and then Chanute's own article in the issue of August 1903, where he tabulated the improvements the Wrights had introduced as:

"1st. Placing the horizontal rudder or tail at the front . . .
2nd. Placing the operator prone . . .
3rd. Warping the wings to steer to right or left."

After all that, the Editor of the *Automotor Journal* then proceeds to offer his readers, now in January 1906, the remarks about the rudder action quoted above, to say nothing of his neglecting to tell his readers of the exhaustive work the Wrights had done on wing curvature, and the control of the movement of the centre of pressure. It is today quite impossible for us to understand the sort of blind spots which led writers and experimenters at this time—on both sides of the Channel—to fly right in the face of the basic written evidence of such clues and pointers, which, with only a slight extra effort of the mind, could so easily have led them to the whole true principle and practice of flight-control.

As we shall see, *Automotor Journal*—clinging to its idea of the rudder as the sole means of restoring lateral balance—attempted to explain Santos-Dumont's difficulties in terms of the lack of a vertical rudder to stop the rolling tendency of his machine.

❧ 66 ❧

Re-publication in France of the Wright Patent: January 1906

In *L'Aérophile* of January 1906, the basic Wright patent was exposed for all to see for the first time, under the misleading title "Derniers perfectionnements commus des machines volantes Wright".[1] The text here reprinted was not absolutely complete, because patents have, of course, to be repetitive in some respects. But the verbatim text is given for all the important matters, and this includes the whole of the control system, with the detailed exposition not only of the wing-warping and its aerodynamic raison-d'être, but of the warp-rudder linkage and *its* aerodynamic raison d'être. Here, in fact, was the key to all flight-control, which laid down the principles which were to be followed by everyone in Europe and elsewhere, for ever after, once the European designers came to appreciate what was necessary to control an aeroplane in the air.

The Editor, in his preamble, shows some confusion by saying that the only Wright machine of which details had been available up till now was the Wright 1902 glider described by Chanute in 1903: then, forgetting that a description of the warping—and a reference to the simultaneous application of warping and rudder—had been given in Chanute's lecture of April 1903, the Editor said that the machine now presented in the patent was a new machine of 1903, and incorporated "importantes différences": the modifications, he said, were of a nature to affect in great measure the "stabilité et la manoeuvre dans l'atmosphère".

In his comments, after giving the text of the patent, the Editor wrote:

"The Wright brothers have sought to combine, in this machine, lightness, strength, and ease of construction; to provide new means to maintain and re-establish its equilibrium; and to control it in the

---

[1] The actual patent had been granted and published in France in September 1904, but no one seems to have thought of reproducing it in a periodical for all to read.

vertical and horizontal planes. The construction and attachment of the rear rudder, the construction of the wings, the method of fixing the fabric to the ribs, and—above all—the mechanism by which the warping of them is brought into play voluntarily, as well as the improvements made to the forward rudder [*ie* elevator], are particularly to be observed. We believe that the French pioneers (chercheurs) will be able to find in the description of this machine—which dates, however, from 1903—some very useful information (de fort utiles indications). The machine described is minus a motor; but the mechanisms employed, and the principles of flight which have been applied, could be utilised in a powered machine. It is probable that all this will be found incorporated in the famous and still mysterious Wright powered machine of 1905."

Once again, the historian can only stand amazed at the immediate outcome of the revelations of this Wright patent; at the Editor drawing special attention to the improvements which would greatly affect stability and manoeuvrability; and his references to the wing-warping and the "principles of flight" involved in the whole design. It is true he did not single out the full and lucid description, and the raison d'être, of the warp-rudder linkage—which he certainly ought to have done—but it was, after all, placed firmly there in print in front of the reader's eyes in the body of the patent text.

One need scarcely say, after the pattern of the previous years in Europe, that the immediate outcome of even these revelations and 'pointing-out', was nil. No European pioneer seems to have taken the slightest notice, and I cannot find any contemporary reference to, let alone discussion of, the all-important subjects involved.

Later on, during this year and next, the half-hearted attempts to provide some sort of roll-control—and then only passively to restore lateral equilibrium and without the yaw component in lateral control—may have stemmed from the explicit descriptions of the patent, or from someone who had at some time read them.

It might be thought that the fact of the Wrights' flight-control system being thus made the subject of a patent, might have influenced the pioneers against adopting it. But such a supposition is quite ruled out, first by the fact that it was never a subject for discussion, and then by the fact that the pioneers later borrowed freely the parts of the Wright system which they came to understand, without bothering about the infringing of patents: indeed the pioneers were—later on—more than willing to challenge the whole idea of the Wright patents on various grounds, including prior disclosure, a disclosure which the Europeans took years to act upon!

☙ 67 ❧

Tatin's Admission
Censure of the French, and Exhortation: January 1906

It must have been a bitter blow to the body of French pioneers when they read, in *L'Aérophile* for January 1906, an article by the "grand old man" of French aviation, Victor Tatin, entitled "Progrès possibles de l'aviation en France"; for Tatin was the one man on the Continent whose prestige equalled—and in some ways exceeded—that of Archdeacon and Ferber; and Tatin now admitted that the Wrights had flown. He wrote for all to read:

> "The glory of having obtained the first results is therefore for ever lost to France ... (La gloire d'avoir obtenu les premiers résultats, échappe donc pour toujours à la France ...)."

Then, before going on to analyse the nature of the aeroplane as a whole, and lay down principles of construction, he castigated his fellow-countrymen for failing to take action after the earlier news from the Wrights; for remaining in a "regrettable state of expectancy"; and for being overcome by "indifference and inertia".
This article was one of the severest acts of censure with which the Europeans were ever to be confronted:

> "The recent news from America tells us that, if the problem of aviation has not seen an absolutely complete solution, at least the first steps along the road to success have certainly been achieved; since it is fairly undeniable that a machine has been able to fly long enough, and before a sufficient number of witnesses, for it to be impossible to refute the fact.
> "The glory of having obtained the first results is therefore for ever lost to France, which was nevertheless the cradle of aviation, and was for such a long time at the head of other nations in the matter of research. Unfortunately, for some years, in spite of all that could be stimulating in the news reaching us of the partial success and the well-founded hopes of the Americans, we have

remained in a regrettable state of expectancy (restés dans une regrettable expectative), when we had here in France (chez nous) all that was necessary to resolve—better and more rapidly—the problem of which the solution abroad has today aroused us; but aroused us a little too late, alas!

"Nevertheless, we cannot accuse all the French engineers of apathy, and we render justice to the all-too-small number of those who—like MM. Ferber, Archdeacon, and Henry Deutsch [de la Meurthe] —have made the most laudable efforts to shake up the indifférence and inertia of their compatriots (pour secouer l'indifférence et l'inertie de leur compatriotes), either by their own work or by founding prizes to stimulate emulation.

"Since we now apparently wish to pull ourselves together, and make up for lost time, what still seems to me to be possible—despite our evident backwardness—is that we study a little the ways and means which are capable of leading us to the desired result. This study will be able to show us whether there are grounds simply for resorting to the same features as the Americans (aux mêmes dispositifs que les Americains) and copying their machines, whilst at the same time seeking to perfect them—this, I believe, would be an admission of incapacity—or whether it would not be more worth while to rely on that originality of which French genius has so often given proof."

Tatin then says he will deal with the broad lines of this subject, not the details, and poses the question "of what does an aeroplane essentially consist?" He then goes over the familiar ground of wing area, weight, resistance, streamlining, power, thrust, etc.; but not a word about flight-control, or about experimenting first with gliders! Tatin was concerned with the powered aeroplane from first to last, without paying any attention to the business of flying it. This article—all too typical of the chauffeur's attitude—then enters on its peroration:

"Anybody could certainly, from today—with the means at our disposal and with the existing engines—achieve machines of which the features will appear in some years time, if not at once, as bizarre and crude, even grotesque (as, for example, the aviator lying prone in his machine), but of which the results, doubtless not yet very brilliant, would nonetheless be the proof that the machines could fly: it is this that the Americans have just made us see. But it seems to me essential that the existing types should be entirely recast if we wish them to make progress, rather than be a 'banale' copy of a machine which is in truth more than imperfect (un appareil vraiment trop imparfait); ... Let us then concern ourselves with research into possible improvements after each of our tests; the original and inventive genius of the French must bring to aviation the rapid progress which will again restore our Country to a position momentarily lost, of that I am quite certain. It is in France that there must be made flying-machines both speedy and unbeatable, as with the automobiles we make already: it is only a question of our getting down to it (il ne s'agit que de nous y mettre)."

Tatin had the mentality and outlook of an inveterate model-making, chauffeur-minded man, who always thought in terms of models; and then bigger models, in which men would sit and drive through the air. His writings give one the almost uncanny feeling that he looked upon the molecules of the atmosphere as being as well behaved, and as reliably still, as those which compose the earth beneath.

With its accent upon the theoretical structure of a powered aeroplane, its complete neglect of any flight-control consideration, and of the pros and cons of prior glider flying, this article conveys an acute sense of unreality, here at the beginning of 1906, and four long years after Ferber was sparked to emulate the Wrights.

But Tatin, despite his chauffeur's philosophy, did come to play an important part in European—and, indeed, world—aviation, by his advocacy of the monoplane. There seems little doubt, now that the persistent self-criticism and exhortations uttered by the French were slowly beginning to tell, that Tatin's enthusiasm for the monoplane was to bear fruit, although all too slowly; and that Trajan Vuia took up pioneering the monoplane at the suggestion, or inspiration, of Victor Tatin.

68

Santos-Dumont turns to Aviation: January 1906

The evidence is fairly considerable that Santos-Dumont first turned towards aviation, even if he did not actually make up his mind at the moment, when he was visiting the St Louis Exposition in 1904, and talked to Octave Chanute. It is certain that Chanute told him about the Wrights' success with gliding, and about their early powered flights. What also seems certain is that what finally persuaded Santos actively to take up aviation, and abandon his airships, was the furore over the Wrights in late 1905. This is confirmed by the following significant quotation from *L'Aerophile*, in its December 1905 issue, which, as already said, appeared in mid-January 1906. The article was headed "Santos-Dumont en Aéroplane", and the following appeared beneath it:

> "The Wright affaire (l'incident Wright) whatever may be the outcome, will at least have had the advantage of shaking our aviators out of their torpor (de secouer la torpeur de nos aviateurs). The Grand Prix d'Aviation, founded by MM. Henry Deutsch de la Meurthe and Ernest Archdeacon, has not given rise to a single competition since 1904. On January 2nd, the first competitor has been enrolled, and the most unexpected in the eyes of some people who have become accustomed to consider him as the fanatical champion of the dirigible balloon—Santos-Dumont himself. The celebrated 'aéronaute-chauffeur' requests of the Aviation Committee that they enrol him as a contestant for the Deutsch-Archdeacon prize, . . ."

This was indeed a bold gesture of confidence—or defiance—since the prize in question was for the first kilometre circle by an aeroplane, and Santos had not yet even built an aeroplane! What is more, Santos first announced he was going to build a helicopter; then, when Archdeacon expressed a lack of confidence in such a scheme—when compared with the advantages of the aeroplane—Santos decided to design an aeroplane.

The short notice in *L'Aérophile* was also remarkable for what the Editor said about the European pioneers, and the impact of the Wrights' news, which will bear repeating: this news, he said,

> "will at least have had the advantage of shaking our aviators out of their torpor."

The French had already been criticising themselves for their procrastination; but this was the first time these two emotive words had been used: "secouer" (to 'shake or shake off') and "torpeur". They were to become part of the clarion-call to action when the French authorities soon overflowed with exasperation with their countrymen; these words were used to greatest effect by the President of the Aéro-Club de France in August of this year (see Section 71).

The British *Automotor Journal* also supported the view that Santos had taken to aviation on account of the Wrights. In its issue of January 20th 1906, we find the following:

> "*Santos-Dumont adopts the Aeroplane.* Among the secondary results of the revelations which have been extorted from the Wright brothers, may be counted the decision M. Santos-Dumont has embraced with some suddenness, of, if not abandoning the navigable balloon type of airship, betaking himself to experiments with machines of the aeroplane type..."

The French "anti-Wright" daily, *Les Sports*, carried the news of Santos' new intentions, and this was followed by a detailed article in *L'Aérophile* for January 1906, entitled "Santos-Dumont, aviator: first enrolment for the Grand-Prix d'Aviation Deutsch-Archdeacon of 50,000 francs: the helicopter and the aeroplane of Santos-Dumont". The article described Santos' visionary scheme for a helicopter, with a sketch of it, which *Les Sports* had already reproduced: it was to have two rotors and a propeller for forward propulsion.

The Santos aeroplane design, also illustrated in this *Aérophile* article, was of the most sketchy nature, and consisted of a monoplane with an almost semi-circular wing-plan; tractor and pusher propellers; and a cruciform tail-unit (Fig 32).

It is difficult to tell what the other pioneers thought of these strange schemes at the time, beyond expressing great interest in what the famous Santos was doing. Both schemes were soon dropped by their author. Nothing much was heard about his aviation activities until July of this year 1906, when it became generally known that the construction work that had been under way in his workshop at Neuilly was not connected with either his helicopter or his monoplane, but with a completely new design he had conceived, this time for a cellular biplane. Santos was to allow the machine to be photographed on completion in July; it was, of course, the famous *14-bis* (see Section 70).

Ferber, in his London lecture of February 1908 (see Section 85) was quite definite about the reason for Santos entering aviation:

> "Santos Dumont at first was merely a balloonist, but after the considerable sensation caused in 1905 by the Wright brothers—and it may be remembered that this was solely due to the documents which I possessed and published at that time—he felt that the time had come for flying machines."

As might be expected, any suggested tie-up between the Wright brothers and Santos struck sparks from Gabriel Voisin; in his autobiography, he wrote:

> "On 30th July 1932, *L'Illustration* announced the death of Santos-Dumont in terms which merit the sharpest disapproval. Here is the text: 'Santos-Dumont has died at the age of fifty-nine. He was one of the great pioneers of aerial navigation. Executant rather than inventor, he did not create either the dirigible or the aeroplane, which were the works respectively of Renard and Krebs and of the 'brothers Wright'. Towards 1904, 'inspired by the Wrights', Santos turned to aviation.'
>
> If there is, in the whole range of pioneers, a man who owed absolutely nothing to the Wrights, it is Santos-Dumont. The machine which he piloted under official observation at Bagatelle in September 1906, and, later, his *Demoiselle*, are original works in conception and realisation. There is not the slightest connection, however distant, between Santos's aircraft and the Wright. To publish in a periodical as serious as *L'Illustration* a statement so diametrically opposed to the truth must require equal measures of falsehood and fatuity" (pages 212, 213)

Gabriel was unusually bitter in this case, as the piece concerned appeared in a respected French journal.

69

Miscellanea : 1906

In the magazine *L'Auto*, for February 7th, there appeared two more pirated drawings of the Wright *Flyer III* of 1905 (Fig 33), but how they were obtained it is now impossible to say. These drawings confirmed the biplane front elevator and the launching rail, and also incorporated the important feature of the machine's double rudder; but they made a new mistake in showing only one propeller. These drawings did nothing much except again remind the French of the now hated name of Wright. After the furore over the brothers in December and January, European aviation apparently settled back once again into its previous torpor; but various machines were now germinating; and, under the surface, the serious French inventors—having been goaded into action by their compatriots—were, so to say, making haste slowly. Perhaps the most significant long-term event occurred in March, when there appeared a prophetic and influential machine.

It is to the credit of Trajan Vuia, a Paris-domiciled Transylvanian—and a Doctor of Law of Budapest—that in 1906 he inaugurated the powered monoplane tradition of today with a machine of simple tractor configuration; but the machine itself was a failure. Vuia's immediate ancestors were Ader and Lilienthal; his immediate descendants, Blériot—who almost certainly was persuaded by Vuia to abandon the biplane and take to the monoplane—and Levavasseur's *Antoinettes*. The *Vuia I* (Fig 34) was a tractor monoplane, built by the designer, with a wing area of 20 square metres. The engine was a Serpollet steam-engine adapted to work on carbonic acid gas: it gave 25 hp driving a primitive propeller at 930 rpm. This machine, which had a rudder, but no elevator, ran on a four-wheel (cycle-type) undercarriage—incidentally the first pneumatic, and the second cycle-type, undercarriage of history—whose front wheels were steered in concert with the rudder. Control, apart from the rudder, was by a primitive type of wing-warping ("une sorte de gauchissement"), and a device to alter the angle of incidence of the wings in flight, in lieu of an elevator. The *Vuia I* made three take-offs at Montesson, on March 3rd, August 12th and 19th; its best, and last, hop covering 24 metres, and ending in a bad landing ("atterrisage brusque").

Vuia was unfortunately no airman, and had no conception of flight-control: he was a dogged 'chauffeur', bent on driving his machine into the air. It took his 'brusque' landing to show him that he had virtually no longitudinal stability or control. So he added a rear elevator of 3 square metres; reduced the wing camber; and increased the wing area to 23 square metres; the 'en route' variable incidence device was abandoned: engine and propeller were the same. This machine was known as the *I-bis*, and was also a failure. Between October 6th and March 30th 1906, it made a total of 8 take-offs at Issy and Bagatelle, its best hop covering only 10 metres. The *Vuia II* of 1907 was also a failure, and made only two brief hops. These machines are good examples of aircraft which, though failures in themselves, exerted strong influence on others.

It was virtually a misspent year for Blériot. In 1906, still employing the Voisin factory (as it had now become) at Billancourt on the Seine, he ordered what may be termed the *Blériot III/IV*, which never succeeded in flying in any of its transformations (the Voisin-Blériot float-glider of 1905 was named by Blériot the *Blériot II*). The *III* was a float biplane with ellipsoidal main wings and ellipsoidal tail-unit of the same size, and a combined lifting surface of 60 square metres. A 24 hp Antoinette engine drove two tractor propellers through flexible shafts. There was a narrow biplane elevator within the front wings, and a rudder within the rear wings. The machine was first tested in May 1906, on the Lake d'Enghien, but never rose. This was the first aeroplane (after Levavasseur's own primitive machine of 1903) to be powered by one of the excellent Antoinette engines which, in versions of 24 and 50 hp soon became the mainstay of European flying.

The *III* was then modified to become the *Blériot IV*, still on floats, in which the ellipsoidal tail-unit was retained, but in which the front part of the machine was totally transformed: there was now a Wright biplane wing structure (with two side-curtains widely enclosing the pilot), and a biplane elevator out front. A remarkable added feature were the two small and narrow ailerons fitted to the rear struts at mid-wing position, this being the third application of ailerons in history, and clearly inspired by Esnault-Pelterie's, which had been illustrated during the previous year. There were also two 24 hp Antoinettes driving two pusher propellers. This machine was tested on Lake d'Enghien (in October 1906), but also failed to rise.

The *IV* was then modified to become a landplane, with wheeled undercarriage, but now with a single 50 hp Antoinette driving the two pusher propellers: it was tested at Bagatelle in November 1906, but never flew (Fig 35). So ended Blériot's current concern with biplanes.

Although anxious to get up into the air and fly, Blériot—like most of the Europeans—was still too much of a 'chauffeur' to realise that mastery of gliding flight was the best road to powered flight: he was also, one feels, somewhat lacking in original thinking, but he became a talented adapter and courageous pilot later on.

The name of J. C. H. Ellehammer, the Dane, has often come into prominence as a putative rival to Santos-Dumont for the honour of being first to fly in Europe; but, tentative as Santos' flights were, Ellehammer—in his second machine—did not even achieve the free flight which his admirers have so often claimed for him. His aircraft was a tractor semi-biplane—a monoplane with a loose sail stretched above—with a 3-cylinder 18 hp motor. On September 12th 1906, tethered to a central post on the small island of Lindholm, the No *II* lifted Ellehammer over a circular track for some 42 metres; but the pilot was only a passive passenger, the machine having a fixed rudder and automatic (pendulum) longitudinal control (Fig 36). If Ellehammer had concentrated on his excellent engines, he might have played a major role in history; but his work in aviation was not historically important, and influenced no one.

Next, it is the historian's duty—and a melancholy duty indeed if he is an Englishman—to record the continuing neglect of practical aviation in Britain, a neglect as deplorable and inexcusable as it is still inexplicable, in view of the wealth of technological talent then present in the country. Here is a serious statement from a well-known authority of those days, Patrick Alexander (quoted in the *Daily Mail* of November 24th 1906):

> "Great Britain and the British Empire stand easily in the van of progress. We know more about the science of aeronautics than any other country in the world. As yet, we have not attempted to apply our knowledge, but silently and quietly we have been studying the subject, exhausting every possible theory and fact, until today our scientists may lay claim to have conquered the air on paper. To achieve the victory in practice will not be a difficult matter."

Comment on such an incredible *obiter dictum* would be superfluous!

When the statutory 'assemblée générale' of the Aéro-Club was held on March 1st, the President (L. R. Cailletet) made yet another of those pathetic whistle-to-keep-courage-up statements, when he said:

> "The science and the zeal of our colleagues, Messrs. Archdeacon, Captain Ferber and Tatin have given such a stimulus to this branch of aeronautics, that one may consider imminent the solution of the problem of aerial navigation by means heavier-than-air."

The Rebirth of European Aviation 1902-1908

The great American pioneer, S. P. Langley, his spirit broken by the failure of his full-size Aerodrome—which twice crashed into the River Potomac in 1903—and by the foul-mouthed and ignorant jeers of his detractors, had died on February 27th. *L'Aérophile* carried (in its March issue) a four-page survey of his aeronautical work by Ferber, signed with his pseudonym 'F. de Rue'; and a beautifully written tribute by Chanute appeared in the April issue: in the latter, Chanute attributes some of the misery suffered by Langley to his 'chagrin' at learning that the Wright brothers "had made four flights (trajets) through the air on December 17th 1903 with a powered machine much more simple, and much less costly than his."

Also in March, the Comte de La Vaulx had been created a Chevalier de la Légion d'Honneur, a fitting reward for his work in aeronautics.

* * * * *

From a long-term point of view, the most important achievement of the year was the perfecting, by Leon Lavavasseur, of his two Antoinette aero-engines (named after Antoinette Gastambide, daughter of the firm's chief), of 24 hp and of 50 hp. His first engine was fitted to his abortive bird-form monoplane of 1903; then, with outstanding success, his motors powered motor-boats from 1904 onwards. Now, from this year 1906, Antoinette motors were to become the chief power-standby of European aviation until 1909: Antoinette motors might even be said to have made early European flying practicable. There was no comparable advance in European propellers, which remained primitive devices until Chauvière's sophisticated airscrew of 1909, first used successfully in 1909.

* * * * *

The reader will have noticed how, from time to time, the French pioneers say proudly, plaintively, or aggressively, that aviation was a French science. Here, for example, is Victor Tatin in 1904:

"Must we one day read in history that aviation, born in France, only became successful thanks to the Americans?"

But this is simply not true, and the best witnesses are the French themselves; for Sir George Cayley has for a very long time been held by the French to be the true father of aviation. Even Tatin himself, three years later (in 1907) wrote these words:

"In following the chronological order, one finds, at the head of the inventors of the aeroplane (en tête des inventeurs de l'aéroplane) Sir George Cayley; this man of genius... the masterly work of Cayley (l'oeuvre magistrale de Cayley)..."

Here is the journal *L'Aéronaute*, in 1877:

"It is Cayley who has truly (véritablement) founded, in Great Britain, the school of aviators which are today very flourishing."

Here is Alphonse Berget, in 1909:

"This inventor, the incontestible precursor of aviation, was an Englishman, Sir George Cayley, ... the name of Sir George Cayley should be inscribed in letters of gold at the beginning of the history of the aeroplane."

And here is Charles Dollfus, creator of the Paris Musée de l'Air, and Europe's leading historian, in 1923 and 1932:

"The aeroplane is a British 'invention'; it was conceived in its entirety by George Cayley, ..."

"Sir George Cayley, the true (véritable) inventor of the aeroplane."

70

Santos-Dumont and his No 14-bis : July 1906

On July 18th Alberto Santos-Dumont formally registered his name to compete for two prizes; first, the 'Coupe d'Aviation Ernest Archdeacon' for the first aeroplane to be airborne for 25 metres (85 feet); second, the Aéro-Club's prize of 1,500 francs for the first aeroplane to be airborne for 100 metres (328 feet), the machine in each case to rise from level ground.

The *14-bis*, the fruit of Santos-Dumont's advertence to heavier-than-air flying, was so-called because it was flight-tested slung beneath his airship No 14. After toying with the idea of a helicopter and with the idea of a monoplane, Santos designed and built at his 'aerodrome' at Neuilly-Saint-James (near Paris) a cellular biplane of the canard-type, *ie* with fuselage and 'tail-unit' forward of the main planes (Fig 37). A covered-in fuselage extended forward from the wings to support the control surfaces in the form of a single box-cell, pivoted to turn up and down as an elevator, and left and right as a rudder. The wings, set at a pronounced dihedral angle, had an area of 52 square metres, a span of 11.50 metres, a chord of 2.50 metres, and a wing gap of 1.50 metres. There were six side-curtains. A 24 hp Antoinette engine (later exchanged for a 50 hp Antoinette) was placed in the centre section, driving direct a 2.50 metre diameter metal pusher propeller at 900 rpm. The pilot stood in a wickerwork balloon basket in front of the engine. The elevator-cum-rudder was 1.50 metres high, with a 2 metre span, and a 2 metre chord. The machine was built of fabric-covered pine and bamboo (round the basket); had two main wheels (with rubber shock absorbers), and a skid under the forward box. The weight was 300 kg. Its ancestry is evident enough; a basic Wright configuration of biplane wings and forward elevator, combined with Hargrave box-kite modification to form a cumbersome ensemble.

The *14-bis* was first tested suspended from a pulley running on a wire; this was at Neuilly. Then, in July, the machine was tested suspended beneath Santos' airship No 14, also at Neuilly.

❧ 71 ❧

The Great French Manifesto: August 1906

Whatever the generality of French pioneers thought about the chances of Santos-Dumont and Blériot succeeding with their experiments, the leaders of French aviation seemed to have little faith that what was going on in their country meant more than the continual muddling and confusion they all knew so well, which had been bedevilling their colleagues since 1902, when the news of the Wright gliders had sparked Ferber into reviving aviation on the Continent. For, now, a most remarkable and unprecedented step was taken to 'blast' their confrères out of their torpor and into some sensible and sustained action, whereby they might produce a proper and practical aeroplane.

In the "Bulletin officiel de l'Aéro-Club de France" which was always published in *L'Aérophile*—in this case the August number—there appeared two long documents. The first was entitled simply "Towards Aviation (Vers l'Aviation)", and was signed by the President of the Club, L. P. Cailletet. The second was the most famous manifesto ever issued on the subject of aviation in any country: it was entitled "For the Success of French Aviation (Pour le succés de l'aviation française)", and was signed by no less than three men, each representing a respected body of scientists and pioneers, the first being the Aeronautical Committee of the much revered Academy of Sciences.

Here are the two documents in translation:

TOWARDS AVIATION

"To shake us out of our inexcusable torpor (pour secouer notre inexcusable torpeur), to attract in France that universal interest—and the active sympathy which its capital importance merits—in that ultimate problem of modern locomotion, that is to say, aerial navigation by purely mechanical means, M. Ernest Archdeacon, the active and devoted President of the Aviation Committee of the Aéro-Club de France, has had the happy idea of appending to a manifesto he has written—and which is reproduced below—the authorised signatures of those savants, engineers, and pioneers best qualified by their world-wide reputation or their special work.

This happy initiative has won, in the first place, the support of M. Poincaré, the eminent President of the Académie des Sciences, and of all the Aeronautical Committee of that illustrious body.

Having won this high approbation, M. Ernest Archdeacon read his vibrant appeal first to the Aviation Committee, and then to the Council, of the Aéro-Club de France, which decided unanimously to countersign it, and give it the widest possible publicity.

This affirmation of what will be—if we really desire it—the brilliant reality of tomorrow, is not just a profession of faith by one individual. It takes on a decisive importance, given to it by the leading scientists of our country, and by men such as MM. le commandant Renard, Captain Ferber, Victor Tatin, Rodolphe Soreau, Drzewiecki, Louis Blériot, and Georges Besançon, who carry authority with us in the matter of aviation, and who are grouped for the most part in the Aviation Committee of the Aéro-Club de France.

Under such patronage, this manifesto on French aviation should make the great impact it deserves. It will stimulate the ardour and emulation of those inventors and 'savants' concerned with seeking the solution. The nobility and scientific interest of the goal to be pursued, and the very important applications which one can already foresee, will provoke—let us hope—the essential gesture, in such a case, of those Maecenases anxious to preserve for their country the glory of this vital progress.

The Grand Prix d'Aviation mentioned in the following manifesto is, of course, only one of the innumerable ways in which aviation can be given the necessary encouragement. Among the thousand other ways advocated, the authorities of the Aéro-Club de France believe that one of the most worthy would be the creation of an institute of aerodynamics, which would, as it were, become to aviation what the Pasteur Institute is to microbiology. Such establishments already exist in various countries, notably at Koutchino in Russia.

To hasten this supreme triumph of the human intelligence, hoped-for from century to century, and now so near, the only difficulty for both private individuals and the Government itself, is to decide what rewards are to be offered to the pioneers.

May we be listened to!

(signed) L.-P. CAILLETET
Member of the Institute, President of the Aéro-Club de France, for the Council of the Aéro-Club de France

* * * * *

FOR THE SUCCESS OF FRENCH AVIATION

"The final discovery of heavier-than-air aerial navigation is imminent. If France wishes to do what is necessary, we can still arrive before the others, and present the first demonstration (expérience) in public of a flying-machine. But we must hurry up.

All the specialists in aviation, in America as in France are—except for some points of detail—absolutely in accord on the general conception of the flying-machine, which is none other than a kite of large dimensions, in which the pull of the cord is replaced by that of a pair of propellers.

Till recently, we were up against the lack of lightness in the engines; but with the marvellously light engines now produced by modern automobilism, the question of the motive power has been victoriously resolved.

Unfortunately, the second—and the greater—difficulty, that of the equilibrium and the actual flight-control of the machines (la conduite matérielle des appareils) is far from being solved.

The solution depends, for the most part, on the control system, as well as on the training of the pilot (or, la solution dépend pour la plus large part des organes de conduite, ainsi que de l'apprentissage du pilote).

Much has recently been talked about the experiments (expériences) which the Wright brothers are *supposed* to have made in America.

Although they have always worked in the greatest secrecy, there is certainly some truth in what has been reported.

On the other hand, it seems more and more to have been confirmed that their great secret is in the handling (maniement) of their machines, at which they have been working for six years.

It is thus that they have arrived at learning, little by little, 'the difficult craft of the bird'.

Nevertheless, everything leads us to believe that their machine is still only tentative (pas définitivement au point), and cannot be sent to France for many months (et ne pourra débarquer en France avant de longs moins).

There is therefore still time for us to catch up on the slight advance which the Americans have been able to make (de rattraper la légère avance que les Américains ont pu prendre).

We have, indeed, better and lighter engines than theirs, and a galaxy of learned specialists who are only waiting to be encouraged to get going (une pléiade de savants specialistes qui ne demandent qu'à marcher, pour peu qu'on les encourage).

There must be found for aviation one or more Maecenases, more eager for glory than for pecuniary gain, and doing it for what M. Deutsch de la Meurthe and the Lebaudy brothers have done for the airship.

Could there not be founded, to help aviation, a very large cash prize, a prize large enough to make certain that the pioneer who wins it will be generously indemnified for his expenditure?

There exists, at this moment, the Prix Deutsch-Archdeacon of 50,000 francs, for the first experimenter who covers, with a heavier-than-air machine, a closed kilometre circle.

M. Archdeacon, deeply regretting that the prize is not greater, is the first to recognise that the sum is entirely insufficient in view of the immense effort demanded of the French inventors, and the necessity to succeed quickly.

More especially, as there will probably be nothing patentable in the whole thing—the Wright brothers themselves agree on this—which will ensure that the inventor will be properly remunerated for the expenditure of his time and money.

If one succeeds in founding, by subscription, a prize of 500,000 francs, for example, for the same performance (avec le même programme) as the Deutsch-Archdeacon prize, the efforts of the pioneers will find in it a very real encouragement.

If the total amount of the Grand Prix d'Aviation is put up by one individual, the whole affair will obviously proceed much faster; and such a man will, into the bargain, win the well-deserved glory of having his name given to the prize.

In conclusion, we can only repeat the following: '*the discovery of aerial navigation is imminent, but there is still just time for you to carry out in France the first public demonstration (expérience).*"

For the Council of the Aéro-Club de France
The President: L.-P. CAILLETET

The Aeronautical Committee of the Académie des Sciences
BOUQUET DE LA GRYE, POINCARÉ, MAURICE LÉVY, MASCART, J. VIOLLE, L.-P. CAILLETET, Members of the Institute

For the Aviation Committee of the Aéro-Club de France
The President: ERNEST ARCHDEACON

These are remarkable documents, which show the intensity of feeling on the part of the authorities, and their exasperation at the inexcusable delays, and the 'torpor' which lay at the root of the situation. It is highly significant that in Cailletet's piece, he speaks of "this affirmation of what will be—*if we really desire it*" (my italics).

In the triple-signed manifesto, it was inevitable that the Wrights should be played down as much as possible, even in a document written by those who knew the facts. It would certainly have been Archdeacon who set the tone for the whole composition. To speak of the "final discovery of heavier-than-air aerial navigation" as being "imminent", after what they knew the Wrights had done, was not only dishonest, but showed to what lengths they would go to keep up the courage of their colleagues, and drive them on to some kind of progress.

The Great Manifesto: August 1906

They quickly say, skirting the real question of the "final discovery", that France can—*if she wishes*—stage the first "demonstration in public". But, they add, "we must hurry up." Then comes the damning-with-faint-praise of the Wrights, when we read (their italics) that

"Much has recently been talked about the experiments which the Wright brothers are *supposed* to have made in America. Although they have always worked in great secrecy, there is certainly some truth in what has been reported."

They rightly speak of the 'handling', *ie* flight-control, as being the Wrights' great achievement; yet they call their machine 'tentative', which was Archdeacon's original word for it.

The whole manifesto was, in reality, only one more urgent exhortation to their colleagues to get up and get going, with the heartfelt cry at the end that "there is still just time for us to carry out in France the first public demonstration." It was therefore natural enough for Archdeacon to lose his head the moment Santos-Dumont managed to leave the ground, and hail it as a great triumph.

Note. In this manifesto, I have translated 'expérience(s)' as 'demonstration(s)' instead of 'experiment(s)', as it seems to meet the case better.

72

Santos-Dumont makes his first Tests
and first free Take-off: August, September 1906

Not only was Santos-Dumont deservedly popular with the Parisians, and with his colleagues in the Aéro-Club, but there was intense public expectation over the coming tests with his *14-bis*. The general public knew next to nothing about aviation, and nothing at all about its history, and about the Wright brothers. All they knew was that their popular Santos was about to try and fly in a heavier-than-air machine, and they fervently hoped he would succeed.

The first free test with the *14-bis* took place at the Polo Ground in the Bois de Boulogne, on August 21st; but the propeller was broken at the outset, and the tests were abandoned. Then, next day at the same place, there was another test; but although the wheels just left the ground a few times, the *14-bis* could not take-off. Whereupon Santos removed the 24 hp Antoinette engine and ordered a 50 hp replacement from Levavasseur.[1]

With the new engine, Santos made his next test on September 4th at Bagatelle; but although he could now taxi much faster, a slight accident to the front elevator caused him to abandon tests for the day. That same day, he informed the Aéro-Club that he would attempt to fly next day, September 5th. This proved impossible because of a high wind. He then fixed September 7th for the trial; but after a brief ground-run in the morning, the trial was postponed until the afternoon as rain threatened later in the morning: so, at about 5 o'clock that day, Santos tried again: he made three runs, but could not manage to take-off. Next day, September 8th, again at Bagatelle, the propeller was again broken, and again the trial was abandoned. Santos then ordered a new aluminium propeller, larger in diameter, to be fitted.

Then, on September 13th, still at Bagatelle, Santos tried again. He made one ground run, but went to the end of the field; so he turned

[1] A number of writers, including myself, have in the past erroneously stated that Santos changed his engine after his take-off on September 13th.

round, and started again. This time he finally got the *14-bis* off the ground amidst a "tonnerre d'acclamation" from the crowd. After being airborne—as the official judges said—for some 4 to 7 metres, he sank back onto the ground, but then again broke his propeller when it struck the ground.

After describing this trial in *L'Aérophile* for September 1906, A. de Masfrand wrote as follows:

> "But it was established—with profound satisfaction—that on September 13th 1906, for the first time, in public trials officially controlled, a powered aeroplane carrying its pilot, had succeeded in leaving the ground entirely under its own power; had sustained itself, and had proceeded for some moments (instants) through the air. This first success, this first take-off (essor) will deservedly impress public opinion: it will, above all, rejoice that minority of isolated men, so long unappreciated and laughed at, who fought vainly to convince their contemporaries of the possibility of purely mechanical flight. This is the first step, the most difficult, on the newly-opened path to aerial locomotion."

The word 'essor' in French is generally used of birds, and can mean, in certain contexts, 'flight' or 'soaring'; but it is generally used for a bird which 'takes wing', in other words which takes off. This is, without doubt the sense in which Masfrand used it in *L'Aérophile*. But, at this time of desperate non-achievement, some Frenchmen, including Archdeacon looked upon Santos' 4 to 7 metres—under a third of the length of a tennis court—as a proper flight. Archdeacon believed that this performance should count as the first powered flight in Europe, and became rather unpopular, as he would not accept that Ader had been airborne for 50 metres in 1890. Yet many of those who support the French in all their claims, are quick to belittle the Wrights' flights on December 17th 1903, when the first covered 120 feet and the last 852 feet: and the 852 feet, which lasted for 59 seconds, was still well ahead of Santos' best which was only to measure 220 metres (722 feet), in $21\frac{1}{5}$ seconds.

73

Santos-Dumont wins his first Prize: October 23rd 1906

In its issue of October 1906, *L'Aérophile* carried an article by Ferber, headed "Santos-Dumont's second Flight (La deuxieme Envolée de Santos-Dumont)", which led off as follows:

"The 23rd of October 1906 will remain a memorable date. On this day, at 4.45 p.m. Santos-Dumont, in his aeroplane No *14-bis*, and propelled by a 50 hp Antoinette engine, left the ground and traversed in full flight, a distance greater than 50 metres, and under 100 metres. The average height was about 3 metres, but I personally believe I observed a maximum of 5 metres. The members of the Aviation Committee present were able, without dispute, to award the Coupe Archdeacon (Cup), which was for the first who traverses at least 25 metres. This event is of the highest importance, for it fixes, under official supervision, a quite definite result.

To be sure, we well knew that the solution was near. Since Lilienthal and Chanute, who had shown the possibility of supporting a man, great progress has been made. M. Archdeacon has stirred up public opinion; I repeated Lilienthal's experiments; Voisin has experimented over water; and the French invention of extra-light engines has increased the hopes of all; but there unfortunately remained too great a number of sceptics.

Henceforth there is a precise fact; leaving the ground, a man in a flying-machine has traversed more than 50 metres; this is not one of those apocryphal or simply affirmed results, like those of the Wrights.

Apropos of this, I have always believed in the success of the latter, because I realised, from my own experiments, what was actually possible; but today, I believe that if the Wright brothers will not make a public experiment, they will not only lose the profits they anticipate, but even the glory of being the first inventors. . . ."

After describing the actual flight, and the wild enthusiasm of the crowd, which surged over to where Santos sat in his aircraft, Ferber wrote:

"One salutes the triumphant one, and the new era which opens up for us. For, there is no doubt about it, it is really a new world which opens before mankind. . . . A new era commences . . ."

The length of this winning flight, which won Santos the Archdeacon Cup, was not accurately measured, but was taken to be about 60 metres.

Although Ferber pays high tribute to Santos for this effort, it is nothing compared with what Archdeacon is to say at the banquet which was thereupon fixed for November 10th. For the public, this was indeed the first time they had ever seen a heavier-than-air machine leave the ground, and they were justifiably excited to a high pitch. But there was no excuse for a man like Ferber to voice the wicked nonsense reported above. He could have described the wild enthusiasm, and said that it was fully justified from the general public, without disgracing himself as he did. His remarks about Lilienthal and Chanute, with no mention of the Wrights, his statement that he repeated Lilienthal's experiments, along with his remarks about "apocryphal or simply affirmed results" were outrageous by what they omitted. He was later—when he had to appear before the bar of history in his own book—to head one of his chapters "Ferber in pursuit of the Wrights . . .", and to write (in 1907); "Just think of it, that without this man [Wilbur Wright] I would be nothing" (see Section 79). And, back in 1904, Ferber had written of the 17th December 1903:

"For the first time a piloted flying-machine had *really flown*, and the honour of this memorable experiment falls to the name of Wright."

It was also Ferber, it will be remembered, to whom the Wright brothers wrote in October 1905, announcing the news of their triumphant *Flyer III*. And in his book (1909) Ferber was to say:

"I was the first in the entire world to know—long before the others—this sensational news."

The explanation of these turn-abouts might be that, at this particular time, Ferber thought it prudent to play down the Wrights. After all, there was a mood of violent patriotism about, which he could not allow to find him praising the Wrights. Later, as we shall see, he was to play fair where they were concerned.

74

The Banquet
in honour of Santos-Dumont:
November 10th 1906

An indication of the collective feelings of guilt at the pitiful progress made by Continental powered aviation, and a true measure of the absurdity of the situation—seen historically—was the banquet given in honour of Santos-Dumont on November 10th 1906. What many people do not realise is that this banquet was given, not after Santos won the Aéro-Club's prize on November 12th for a 'flight' of 100 metres, but after he had won the Archdeacon award for a 'flight' of only 25 metres, in which he had actually covered 60 metres. Two days after the banquet, on November 12th, he was to win the second prize.

The authorities were so possessed by their emotions that they could not even wait for Santos to try for the second prize, which he had already entered for. They were bent on holding a grand banquet for a 'flight' of 60 metres, less than three times the length of a tennis court.

The hysteria already evident at the previous trials—understandable in the public—now became rampant at the banquet. It was held in the evening of November 10th at the Café de Paris, and was attended by just about everyone who was anyone in the aeronautical world, along with many other well-wishers who had come to pay their tribute to the popular little Brazilian. I am a great admirer of Santos, and what I write here is in no way in denigration of him. But the situation simply got out of hand in the frenzy of adulation that was poured out for what were scarcely more than tentative tests with a wholly unpractical aeroplane.

The President of the Aéro-Club (L.-P. Cailletet) made a graceful speech in which he paid tribute to Santos and his success, a success which was, he said "due entirely to your energy, your courage, and your indomitable tenacity". So far, so good. This was fair enough.

But when Ernest Archdeacon took the floor, his hatred for the Wrights, his hysterical joy in Santos' minuscule achievement, knew no bounds:

"Gentlemen, as I take the floor, I am experiencing today one of the purest joys of my existence. First, because I am celebrating the first truly decisive experiment in this science of aviation, of which I have made myself the apostle; next, because I am awarding to my friend Santos-Dumont this Cup, offered by me more than two years ago, which has not seen a competitor. . . . If I had ever been capable of the sin of envy, I would envy friend Santos-Dumont today, who has just assuredly gained one of the greatest glories (une des plus belles gloires) to which a man can aspire in this world. He has just achieved, not in secret, or before hypothetical and obliging witnesses, but in broad daylight and before a thousand people, a superb flight of more than 60 metres at three metres above the ground, which constitutes a decisive step in the history of aviation. . . . For me, after this experiment at Bagatelle on October 23rd, having seen what I have seen, I predict that the question will now march with a giant's stride, faster even that I have hoped for in my most optimistic dreams. . . . It will be partly thanks to me that my country will have been the first officially to have given birth to aviation, perhaps the greatest discovery made by Man since the beginning of the world. . . ."

It is hard to believe that these ludicrous words could be spoken by an honest pioneer within a few months of the Wright furore, and of Tatin writing:

"The glory of having obtained the first results is therefore for ever lost to France . . ."

"Thus we have the final proof that aerial navigation by purely mechanical means has just made the decisive step. We can only regret that it was not to be 'chez nous'."

But Archdeacon, in his blank refusal to listen to reason, and accept the obvious fact that the Wrights had not only well and truly flown, but flown for long distances, and for over half an hour, thrust his head even deeper into the sand. Such behaviour was completely unworthy of a man of his mentality and attainments. So far as he was concerned, it was only patriotism which counted. But, despite his ecstatic tributes to Santos-Dumont, he must have been deeply chagrined that—after all the exhortations that had been flowing forth since 1903—he was not presenting his Cup to a Frenchman. But anyone would do, provided he was not American; and Santos' experiments had at least taken place on French soil, and with a French engine, a point he made well in one part of his speech.

Particularly reprehensible were Archdeacon's remarks, aimed directly at the Wrights, about Santos' tests having been carried out "not in secret, or before hypothetical and obliging witnesses"; this was a straight imputation that the Wrights were dishonest.

Archdeacon had been making such a fuss over Santos, that he unwittingly became the instigator of one of the biggest non-events in aviation history. For he was foolish enough to claim that Santos' first free take-off on September 13th—in which he was only airborne from 4 to 7 metres—was the first powered flight in Europe. To make matters worse, Archdeacon also said that he had investigated Clément Ader's claim to have flown in 1890, and had decided his *Éole* had not flown. As a matter of fact it had achieved a very respectable free take-off, and although the hop it made was not sustained or controlled, it did at least cover about 50 metres. Archdeacon made these statements in the magazine *Les Sports*, in its issue of September 14th. The effect on the ageing Ader was to touch off a veritable explosion in the old pioneer; and not only an explosion. For, soon after, Ader wrote his celebrated series of articles in *Les Sports*, alleging that in 1897 he had flown for 300 metres, a story which was completely fabricated.[1] Archdeacon immediately wrote to the Ministry of War and asked them to release General Mensier's report, which had been written immediately after the 1897 trials. But the Ministry refused: it was not until 1910 that this report was issued at last, in which it was made clear that Ader had not flown at all. But the delay allowed his mendacious story to penetrate book after book, and become accepted as fact among many of the less critical members of the aeronautical fraternity.

---

[1] See my Science Museum monograph on Ader, already mentioned.

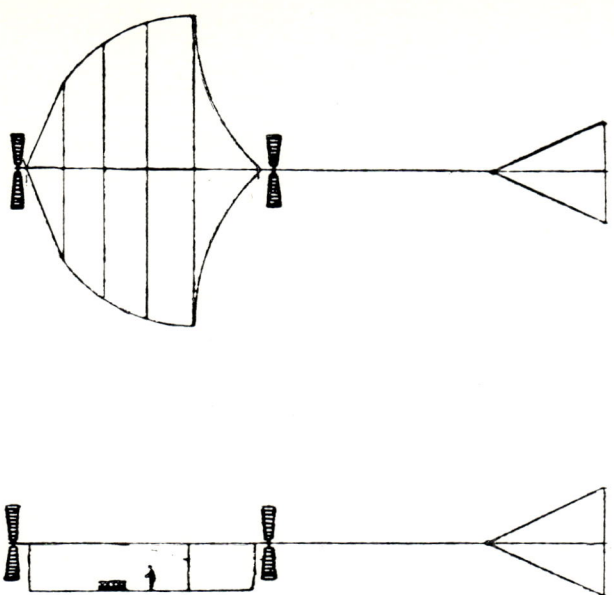

Santos-Dumont design for a Monoplane: 1906 Alberto Santos-Dumont, famous for his little pressure airships, turned to aviation for the first time in 1906 with this first design for an aeroplane; but it was never built.

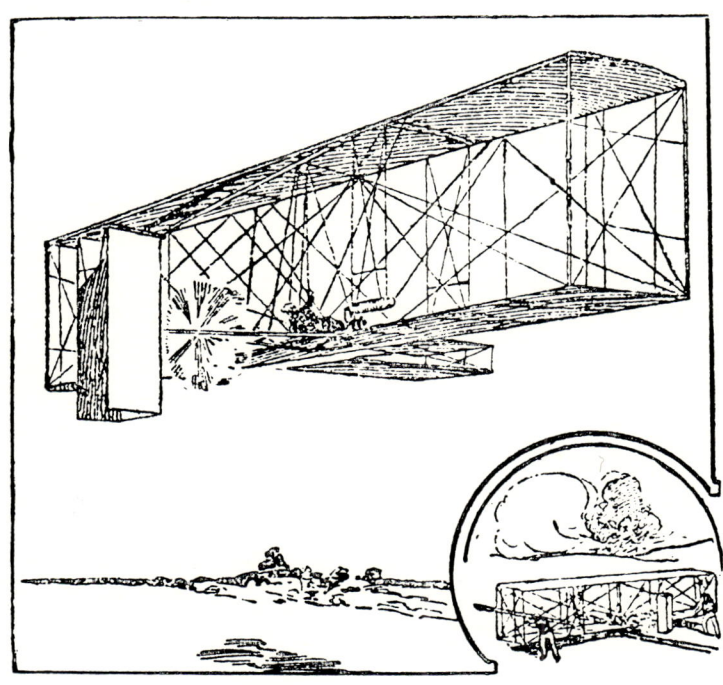

Wright Flyer (published 1906) Yet another pirated sketch of the Wright *Flyer III*, which this time showed the double rear rudder. It appeared in *L'Auto* for February 7th 1906.

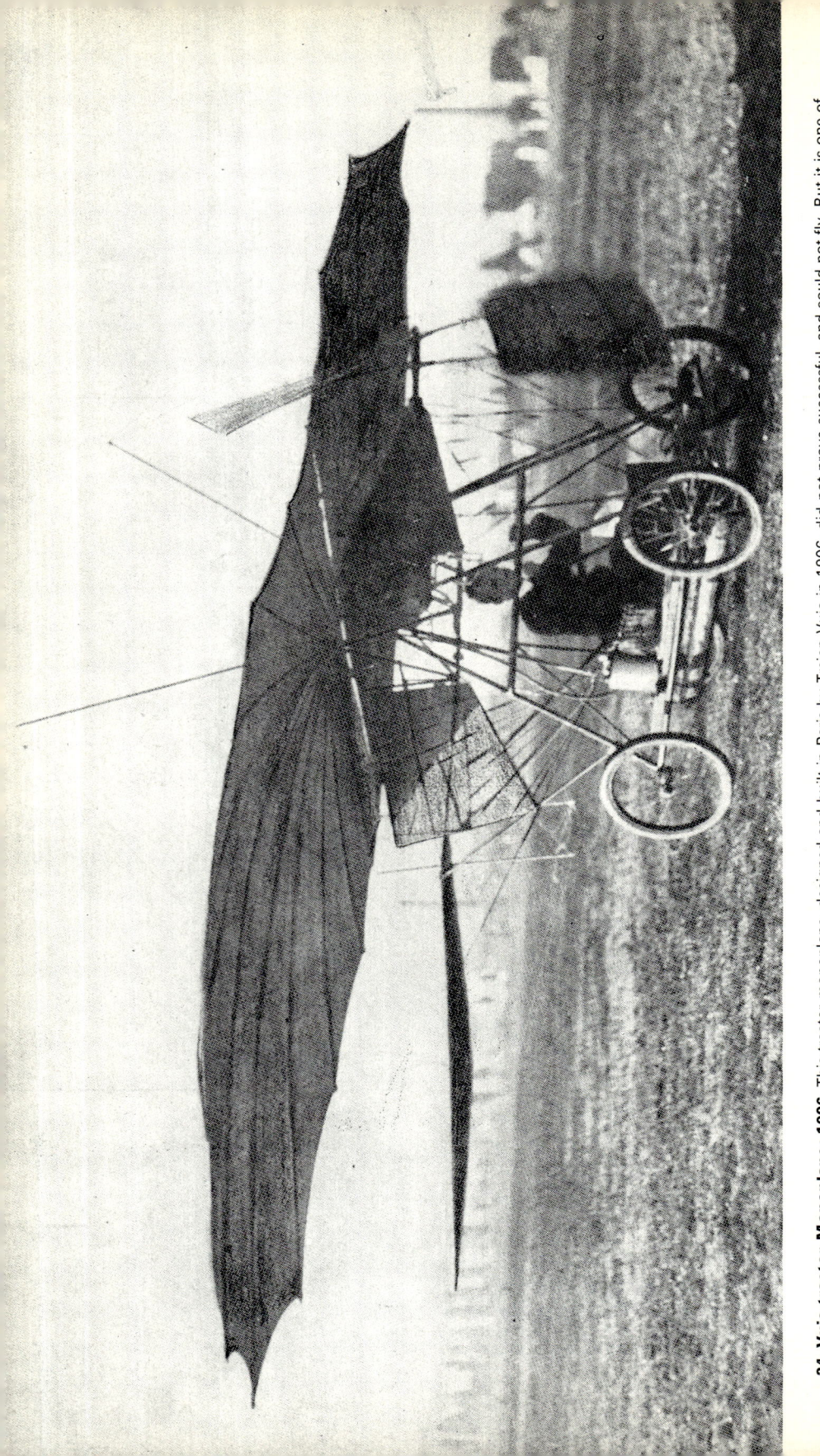

34 Vuia tractor Monoplane: 1906 This tractor monoplane, designed and built in Paris by Trajan Vuia in 1906, did not prove successful, and could not fly. But it is one of those machines in the history of aviation which, despite its lack of success, was an inspiration to others which set them off on the pursuit of the successful monoplane. The Blériot

5 Blériot IV: 1906 The year 1906 was a largely wasted one for Blériot. He was a brave pioneer, but relied heavily on others for his design ideas. In 1906 he produced this unpractical and unsuccessful biplane, with its huge cellular tail. The front half betrays its Wright ancestry, but interesting features are the small ailerons between the wings, which would have been too diminutive to control the machine in roll if it had ever been able to fly.

6 Ellehammer II: 1906 It is often claimed that the Dane J. C. H. Ellehammer was the first to fly an aeroplane in Europe; this machine is cited as the successful candidate. But although it took off before the famous Santos-Dumont *14-bis,* its achievement cannot be accounted a flight as it was tethered to a central pole when it became airborne, and merely whirled around it: it never flew in free flight.

37 Santos-Dumont 14-bis: 1906 The first full-scale aeroplane designed and built by Santos-Dumont was this box-kite canard type of machine called *14-bis,* which made the first heavier-than-air flights Europe in 1906. Santos had attended the St Louis Exposition in 1904 and had there talked with Chanu who told him of the great progress being made by the Wrights: Santos thereupon determined to aband airships and take to the aeroplane.

38 Santos-Dumont 14-bis (modified): 1906 Santos won two major prizes with the *14-bis;* one flying for 25 metres or more, and the other for flying for 100 metres or more. For this latter success, mounted ailerons in the outer panels of the box-kite wings, and made his record hop of 220 metres (so 722 feet) in $21\frac{1}{5}$ seconds.

9 Blériot VI: 1907

Blériot's second machine of 1907, his No VI *Libellule* (Dragonfly), which was a tandem-wing monoplane — inspired by Langley — with a tractor propeller. It made a few hops, and was then abandoned. Interesting features were the forward wing-tip elevons, which were intended only as lateral stabilisers.

10 Blériot VII: 1907

Blériot's third machine of 1907, his No VII. This was an unsuccessful, but prophetic, machine in that it was to set the style for the Blériot monoplanes of the future, with mainplanes, covered-in fuselage, elevators, and rudder.

11 REP No 1: 1907

The REP No 1 of 1907. Designed and constructed by Robert Esnault-Pelterie — hence REP — this was his first powered machine with his own engine. It was not successful. Having abandoned the Wrights' wing-warping as dangerous in 1904, he quietly re-adopted it in his powered machines.

42 Santos-Dumont No 19: 1907 This diminutive machine, which was to progress and become t popular *Demoiselle* in 1909, represented a complete change of outlook by Santos. After his clumsy a ineffectual *14-bis*, the *No 19* was to become the world's first light-plane. In its successful form, as t *Demoiselle*, it was popular but difficult to fly.

43 Voisin Glider: 1907 This glider, with its Chanute-Wright wings, and box-kite tail-unit, was built a tested briefly in 1907 by Gabriel Voisin. In later years, in order to further his own claims, and to spite t Wright brothers, Gabriel said this glider was tested in 1899, which is a complete fabrication. It is a we documented production of 1907.

Voisin-Delagrange 1 : 1907 This machine represented the début of Gabriel Voisin as a designer and ilder of powered aeroplanes. It was made for Leon Delagrange, but only succeeded in making a few ort hops. But it already has the 'look' of the famous Voisin biplanes to come, which were to have box-kite ngs as well as the box-kite tail-unit seen here.

De Pischoff Biplane : 1907 Although this machine was unsuccessful at its trials in Paris in 1907, it of great importance in aviation history because it was the first full-scale biplane with a tractor propeller, d its curiously prophetic all-round appearance. There is little doubt that Breguet, in 1909, was inspired De Pischoff into taking the tractor biplane into the sphere of practicality.

46 Voisin-Farman (modified): 1908 This is the second modification of the *Voisin-Farman I*, originally flown by Farman in late 1907. It was on this machine that Farman, on January 13th 1908, made the first official circular flight in Europe and won the Archdeacon prize.

47 Roe Biplane (Roe I): 1908 In 1908 A. V. Roe built this Wright-derived biplane and tested it at Brooklands in England. It was given an accelerated take-off downhill; was towed behind an automobile and it may have made brief hops on its own; but it could not fly properly as a powered aircraft.

48 Cody No 1(b): 1908 On October 16, 1908, the American S. F. Cody made the first powered flight in Britain, at Farnborough, on this Wright-derived biplane. Cody was greatly helped in the design by the Farnborough staff; but much credit goes to Cody himself, who (in 1909) became a British citizen and went on to ever greater success in later years.

❧ 75 ❧

Santos-Dumont wins his second Prize: November 12th 1906

After his first success, Santos now fitted two octagonal (not hexagonal) ailerons to the *14-bis*, one in each outer cell of the wings; they were operated by a body harness to obtain a measure of roll control (Fig 38). Then, on November 12th, he went out to win the Aéro-Club prize of 1,500 francs offered for the first flight of 100 metres. The venue was, as before, Bagatelle.

On this day, Santos made four trials, two in the morning and two in the afternoon. The first attempt resulted in a hop of about 40 metres. The second saw two hops, of some 40 and 60 metres; the latter ended when Santos tried to do a turn, and came too near the trees; one of the wheels was slightly damaged, and it was repaired during lunch-time. In the third trial, Santos again made two hops, one of 50, and the other of 82.6 metres, in which he started again to turn, and was halted by the proximity of the fence of the polo ground.

The fourth and last trial won him the prize. He covered—"exactement mesuré"—220 metres (722 feet) in $21\frac{1}{5}$ seconds, at 37.3 kph (23-24 mph). This was over double the distance necessary to win (100 metres), and was the sensation of the day.

After the enthusiasm and the plaudits over the last effort by Santos, there was not much more to be said when he thus finally won the Aéro-Club's prize of 1,500 francs. "An immense clamour of enthusiasm" greeted him when he came to a halt. Again one must point out that, to the crowd, such excitement was understandable since they had never seen any aeroplanes in the air; and here was one actually taking off the ground before their eyes.

The *Aérophile* account was written by A. Cléry, who, after detailing the events of the day, ended his piece by saying he hoped, in view of Santos' efforts to turn, which he made during the trials on this day, that he would go ahead and win the Grand Prix of 50,000 francs for a closed circuit flight of a kilometre; but this had to wait until January 13th 1908, when Henri Farman won the prize.

The *14-bis* was to make only one more appearance; this was on April 4th 1907 at Saint-Cyr, when it covered 50 metres. It was, alas, a sterile and unpractical vehicle in every way, and influenced no one; but it had served its creator well, even though its performance, when compared with the Wright machines, was pitiful.

❧ 76 ❧

The European Attitude to Flight-Control: 1906

Only three European powered aeroplanes made free take-offs in 1906, the *Vuia I* and *I-bis*, and the Santos-Dumont *14-bis*; but, in the midst of the general "torpeur", there was much discussion of aeroplanes on both sides of the Channel. If this discussion had been more enlightened, the "torpeur" might even have been dispelled sooner; for the most astonishing aspect of aeronautical discussion in 1906 was, once again, the neglect of flight-control problems, especially lateral control; but this time there was a difference. For in the January 1906 issue of *L'Aérophile*, the most important points in the basic Wright patent were published for all to see, and these included the full and explicit description of the technique, and the aerodynamic raison-d'être of wing-warping, and its co-ordination with the rear rudder, with explanatory illustrations. Yet I cannot find any contemporary discussion or assessment of this all-important subject of lateral control. It is still one more symptom of the astonishing lack in Europe of what I have called 'aero-talent'.

To study the columns of the best-informed British and French journals of the time—*The Automotor Journal* and *L'Aérophile*—is to experience an almost mesmeric fascination for the sheer obtuseness of the undoubtedly intelligent men active in European aviation. The editorial staff of *The Automotor Journal*, as is obvious from their columns, kept a sharp eye on their specialised contemporary *L'Aérophile*; and, by their own showing, were acquainted with Wilbur's Chicago lectures; yet they could include lengthy accounts of the Wrights' gliding achievements, and speculations on their powered *Flyers*, without having absorbed even the simple principle of the Wright system of lateral control, let alone the co-ordination of warp and rudder. They could even write the following about the Wright gliders (*Automotor*, January 6th 1906):

"If what Mr Chanute terms 'a whirling billow of air' tipped up one or the other side of the machine, this was corrected for by movements of the vertical rudder, causing the aeroplane to swing round like a bird executing a curve, and so bring the depressed side up level again."

We here see described the actual flight-control method of Farman *et al* in 1907-08, which was brought about by the prolonged inability to understand lateral control, despite its clear exposition, which was laid before the European pioneers in the Wright patent.

The Editors were still completely at sea on the subject nine months later, when speaking of Santos-Dumont's *14-bis*:

> "There certainly also appear to be grounds for supposing that something in the way of a rear tail will have to be adopted, for it is only a vertical rear tail that has hitherto, so far as can be judged, been successful in bringing an aeroplane into a horizontal position when, from gusts of wind or other causes, one side is lifted more than the other. In fact, horizontality is obtained by manipulating the rear tail or rudder so as to sail round into the wind and bring the machine once more on a level keel. That a machine unprovided with means for accomplishing this end, can ultimately maintain itself for any length of time in the air, would appear extremely doubtful." (October 20th 1906.)

Then, when Santos said the *14-bis* had a bad tendency to roll on its October 23rd tests, *The Automotor Journal* wrote:

> "Now it is just on this point of lateral control that we have always been sceptical, as we explained at some length in an article published in our issue of October 20th. Without rearward vertical rudders we did not then, and do not now, see how this lateral stability can permanently be maintained. But it is a very great point that even without them, the machine should have been able to travel practically straight the distance that it did." (November 3rd 1906.)

On November 12th 1906, Santos made his best hop-flight of 220 metres in $21\frac{1}{5}$ seconds, which was, as we have seen, acclaimed as epoch-making, and won him the Aéro-Club's prize for the first flight of 100 metres. *The Automotor Journal* gave an excellent account of this flight, and even went to the trouble of printing—in its issue of November 24th—a photograph of the *14-bis* during this prize-winning performance, rightly pointing out that many of the illustrations in the press, purporting to be of the November 12th flight, were in fact of the machine in its earlier tests.

Now it so happened that, in order to correct the machine's rolling tendency, Santos had fitted octagonal ailerons in the outer wing-cells (Fig 38), and the starboard aileron was shown large and clear in the photograph published in *The Automotor Journal*: yet there is not a word in the text about possible roll-control by these ailerons, or in the long caption to the photograph, despite the importance with which the journal's writers had previously been treating the subject; or in any follow-up references in 1907. This can only mean that their French correspondent made no mention of it, and no one on the editorial staff in London questioned what the newly-added and highly conspicuous object was in the photograph they published.[1]

The situation in France had been just as bad throughout the year, and I can find no mention by anyone of lateral control until the end of the year, despite the Wright patent published in January. This is all the more remarkable, in view of the weighty articles on aeroplanes published in *L'Aérophile* and elsewhere. The most significant of such articles was one in the October issue of *L'Aérophile* by the almost sacrosant Victor Tatin: it was a detailed survey of the construction of an aeroplane, illustrated with 12 diagrams, entitled "Étude et Construction d'un Aéroplane". It occupied no less than $17\frac{3}{4}$ pages of the issue, one of the longest articles ever to appear in this journal: yet there was not one word in it about lateral control, and only the most naive remarks about control of any kind. It is quite clear that this eminent old 'savant aviateur'—who had never attempted to fly a glider—simply, as they say, "did not have a clue" about practical flying, and had never read a word the Wrights had written. He spent more than two pages on the subject of launching the machine, and—in his penultimate paragraph—devoted 13 lines to landing it. But he seems to have believed it would fly itself, once it was airborne, by means of the now obsessive European panacea of inherent stability. This article is a classic example of the 'chauffeur' philosophy. Here are two revealing quotations, the first taking it for granted that his experience with models was all that was necessary when it came to building full-scale piloted aeroplanes:

> "The small experimental models (petits appareils d'étude), constructed on these principles, enjoy such stability that, if well made, it is absolutely impossible for them to be capsized, no matter in what position one places them when committing them to the air (absolument impossible de les faire chavirer, quelle que soit la position qu'on leur fasse prendre en les abandonnant dans l'air)".

Then Tatin came to the control system: it is hard to believe, but the following quotation, which takes up a bare $6\frac{1}{2}$ lines of text, is his entire contribution to the subject of flight-control; and this is October of 1906, four years after the Wrights' modified No 3 glider had been perfected, and a year after they had achieved the first practical powered aeroplane.

---

[1] *L'Aérophile* even included special mention of the ailerons.

Tatin had dealt with the small surface of the fixed rear tail-plane; then he came to

"Another small surface, which is movable about a horizontal axis, and forms the elevator (gouvernail de profondeur), either to ascend, by modifying the angle of attack of the main planes, or to descend towards the ground; or, finally, to suppress the vertical oscillations which appear to be important enough to cause a disturbance in the progress of the machine.
On top of the fixed tailplane, there is a vertical rudder for direction in the horizontal plane, which presents nothing special, its function being that of all rudders, which is understood without description (et n'offrant rien de particulier; sa fonction, étant celle de tous les gouvernails, se conçoit sans description)."

We are faced with an insoluble problem in this stubborn refusal to consider what was one of the key factors in the construction and operation of flying-machines. No one has ever come up with a plausible explanation, or even a plausible excuse, for this wholesale neglect of three-axis control.

77

Summing-up for 1906

Although the Wrights had not flown in the year 1906, it was their unseen presence which pervaded the whole European scene, and underlay every move the Continental pioneers made during the year. This was primarily due to the shattering impact of the news of the Wrights' successful 1905 season.

Despite the shock of the Wrights' news; despite self-criticism; despite the efforts of Archdeacon, Ferber, Voisin, and their friends; the Aéro-Club de France had, at the end of 1906, to take a long and melancholy look back at another full year with little to show on the credit side. Only three powered machines, all built by men outside the main stream of endeavour—and none of them Frenchmen—had succeeded in making free take-offs; but none had been able to remain airborne for even half the 59 seconds which the Wrights achieved, or travel as far over the ground, as they did on their first day three years before, on December 17th 1903.

But perhaps the most important 'news' of all in 1906 was the negative news that the Wright brothers did not fly at all during the year. Their last flight in 1905 had been on October 16th, and they were not once to leave the ground until May 6th 1908, a period of $2\frac{1}{2}$ years. Realising the full epoch-making significance of what they had achieved, the brothers had become understandably secretive in face of the efforts they knew were being made to discover their secrets. They felt they could only preserve their patent rights, and their rightful priority, if they came to a clear-cut business agreement with a client—preferably their own, or a foreign, Government—that such a client must guarantee to purchase their machine provided there was mutual agreement about the desired performance, and provided the machine performed as agreed. But the clients insisted on viewing the machine (or drawings of it) before signing a contract; this seemed to the Wrights an unreasonable condition, and was totally unacceptable, especially in view of the many would-be spies who were out to learn all they could. So the Wrights refused to fly.

The ethics of this strange situation are somewhat confused, and it is all too easy for us today to pass judgment on the Wrights, and assert that they should have presented their invention to the world, regardless of reward. Nevertheless, from the historian's viewpoint, their $2\frac{1}{2}$ years interregnum was a major tragedy. It severely retarded the whole development of aviation, and the history of the world might well have followed a radically different course—possibly even a better one—had the Wrights made public flights at the end of 1905 with their perfected *Flyer III*. They would undoubtedly have been copied rapidly and widely, especially in Europe, where there was a talented and receptive audience of men interested in flying, and an excellent reservoir of automobile engineers and mechanics to back up their efforts. The Wrights would also, of course, have received at once the full and universal honour for their achievements which will always be their due, an honour which historians have since had to waste much time in defending against those men of dishonesty and ill-will who even today seek to denigrate the American brothers.

The roster of European powered aeroplanes, after all the exhortations of the past four years, was indeed a pathetic one:

§ *Vuia I* and *I-bis:* Best distance was 24 metres (say 80 feet);

§ *Blériot III, IV:* Never left the water, or the ground;

§ *Ellehammer II:* No free flights; but best of tethered 'hops' was 42 metres (say 140 feet);

§ Santos-Dumont *14-bis:* Best distance and time was 220 metres in $21\frac{1}{5}$ seconds (say 722 feet):

THE YEAR 1907

78

General Survey of 1907

The year 1907 witnessed the true, but still tentative, beginnings of practical powered flight in Europe, following the "false dawn" of 1906. Three important configurations, all confirming the European inherent stability idiom, became well defined during the year; but only one machine was to achieve significant flights. These configurations were as follows:

(a) The pusher biplane, Wright-cum-Hargrave-derived, with forward elevator, 'open' biplane wings, and box-kite tail-unit, but with no control in roll. This type was represented by the first two successful Voisin machines, the *Voisin-Delagrange I* and the *Voisin-Farman I*; the latter, after important modifications by Henri Farman, was the first European aeroplane—and the only one in 1907—to achieve productive flying (October 1907 onwards), and the first to stay in the air for a single minute (November 9th).

(b) The tractor biplane, derived from Ferber, but with no surfaces forward of the propeller: this was first seen in December in the De Pischoff machine which, although itself not successful, was to exert much influence later.

(c) The tractor monoplane with main wings, fuselage, and tail-unit, developing from the Vuia (which itself made some hops in 1907), seen in the promising *Blériot VII* and the diminutive *Santos-Dumont 19*, both of November, but both still only tentative.

Starting with the monoplane, we find Blériot—having abandoned the biplane—attacking the problem of practical aviation with extraordinary energy and intrepidity: but he dissipated his energy by changing from one configuration to another. He first built his *No V* 'tail-first' canard monoplane, in which he only managed some 4 take-offs in April before it crashed and was abandoned. Then came his *No VI*, a Langley-type tandem monoplane named *Libellule* (Dragonfly), fitted with wing-tip elevons and a 24-hp Antoinette (Fig 39): he made a

235

number of short hop-flights in this during July and August at Issy before it, too, crashed and he narrowly escaped death. He had modified it en route, and in September exchanged the 24-hp for a 50-hp 16 cylinder Antoinette;[1] his best hop was 184 metres.[2] Then "avec infatigable ténacité", he built a third monoplane which was of vital importance (Fig 40). For, at one stroke, Blériot established the 'modern' tractor monoplane configuration with long enclosed fuselage (also enclosing a 50-hp Antoinette engine) with the main wings forward and a rear tail-unit, the machine resting on a two-wheeled undercarriage with a third wheel near the tail. Long to remain indecisive about his control systems, Blériot curiously decided to make this *No VII* inherently unstable, with only the rudder and elevons forming the tail-unit. He made some six flights in all at Issy during November and December; on the last hop-flight he crashed the machine and abandoned it: the best flights were two each of about 500 metres (say 1,640 feet), one including some tentative turns.[3]

Also new in the monoplane field was Esnault-Pelterie, who built the first of his tapered-wing bird-like tractor monoplanes, powered by his own excellently designed 25-hp 7-cylinder fan-type engine (Fig 41). This *REP No I* (REP were his initials) was particularly interesting for its designer's return to a primitive version of the Wrights' wing-warping which he had done so much to discourage. The machine made short flights at Buc in November and December, the best being 600 metres, but it had little influence on contemporary design, its very short fuselage providing poor longitudinal stability. But his later machines were to fly well in 1908 and 1909.

Vuia made further short hops with his *No I* monoplane in March, and tested his *No II* briefly and unsuccessfully in June-July at Bagatelle, when he abandoned it, and ceased to play an important part in aviation.

Santos-Dumont, after his new biplane had failed to fly in March (see below), also abandoned the biplane and next built a combined aeroplane and airship (his *No 16*); but this curiosity was destroyed on the ground during its first test in June. Then, in the winter, he turned his attention seriously to the monoplane and built the first of his remarkable miniature machines, which in 1909 were to become famous as the

---

[1] This was the first 16-cylinder Antoinette engine to become airborne.
[2] Say 604 feet.
[3] There has always been doubt about the amount of credit due to Blériot for the actual design and construction of his aircraft. Dollfus, who knew Blériot well, has made a study of this vexing subject. He has established that Blériot himself decided upon the small 'canard' No V of 1907 the basic idea of which came from Santos-Dumont's *14-bis*. The tandem monoplane No VI—derived from Langley—was designed and built by Louis Peyret, probably as a result of a general suggestion by Blériot. The prophetic No VII was due entirely to Blériot; and the VIII and IX also seem to have been his ideas, along with the abortive biplane (No X). The famous cross-Channel No XI was, however, designed by another friend, Raymond Saulnier. The No XII (and XIII) were probably Blériot's own design.

Demoiselle (Dragonfly). The new machine, the *No 19*, was tested in only three 'hops' (November) before being abandoned: it was a tractor monoplane built of bamboo poles, with a wing span of only 5 metres; a supporting area of 9 square metres, and a 20-hp two-cylinder Dutheil-Chalmers engine lying on the jointure of the wings: the pilot sat under the wings and controlled the machine by means of a combined rudder and elevator at the rear, two rudders at the side, and a forward elevator: it had a pronounced dihedral angle on the wings, but no control system for control in roll except leaning his body from side to side (Fig 42). It was damaged on November 21st at Buc, and not revived as a type until the end of 1908. It had made only three take-offs, the best covering 200 metres.

Turning now to the biplane, one must again mention Santos-Dumont, who completed a small biplane based on the *14-bis*, his *No 15*, but working the other way round, as a tractor biplane: its most interesting feature should have been the pair of small rectangular ailerons mounted ahead of the wings on short outriggers.[1] It was intended to fit a 16 cylinder 100-hp Antoinette, but that was not yet available, and a 50-hp was installed. This machine was damaged when taxying on March 27th. After repairing it, and fitting the 16 cylinder engine, he still could not make it fly. [P.S. Its wings were of plywood.] Santos now turned to the monoplane, as we have seen.

Meanwhile the most important practical steps in aviation were being taken by the Voisin brothers at their Billancourt factory, As a result of the experience with their two 1905 float-gliders, and having briefly flown (in May this year) a modified type of Chanute hang-glider with box-kite tail (Fig 43), they came to standardise a type of machine which was to persist until the year 1910—the stable pusher biplane with forward elevator, and tail-unit containing the rudder, but with no form of control in roll.[2] In its first form, it showed its origins clearly in the Wright biplane wings and forward biplane elevator[3] (the latter soon reduced to monoplane form), and the Hargrave box-kite tail-unit. The Voisins built two similar machines, one of which was for Henry Kapferer: this never flew. The other—with a 50-hp Antoinette motor—was for Delagrange: it made six hops in March (Fig 44): it was then tried on floats on Lake d'Enghien, without success; and finally fitted

---

[1] Although the ailerons were included in the designs, they were not fitted to the machine when built. The idea came from the Blériot *IV* of 1906.
[2] The form of these early Voisin biplanes which is most familiar had the so-called 'side-curtains' added both to the ends of the main wings and also inboard, thus making the wings into giant box-kites, with a box-kite tail-unit. The tail-unit was in this form from the start, but the 'side-curtains' on the wings did not start appearing until about May (one curtain each side of the pilot), and did not appear fully-fledged (four in all) until July-August 1908. These standard Voisins disappeared in 1910 after ailerons or warping had become universal.
[3] This adoption of the forward biplane elevator is discussed in Section 60.

with wheels again, it made two take-offs at Issy in November 1907, and crash-landed on the second.

Neither of these first two Voisins was a success; but in the summer a third machine was built for a newcomer to aviation, the English-born—but not English-speaking—Henri Farman. Farman, who was the son of a prominent English journalist domiciled in France, was an artist turned motor-racer, and came to exert a far-reaching influence on the whole sphere of aviation. He first flew his new Voisin—the *Voisin-Farman I*—also with a 50-hp Antoinette, in September and thereafter pursued his new profession with a vigour and pertinacity only matched by Blériot. He proceeded to modify and improve his Voisin; he went on altering it right to the end of 1908, and by the end of that year had become the most successful and famous of the European pilots. On October 26th (1907) he had won the Archdeacon Cup with a flight of 771 metres in $52\frac{2}{3}$ seconds. In November he made turns, and on the 9th an unofficial circle of some 1,030 metres in 1 minute 14 seconds. This was the first European flight longer than the Wrights' 59 seconds in 1903. It was obvious now that he would soon win the Deutsch-Archdeacon prize of 50,000 francs offered for the first man to fly a kilometre circuit, *ie* to round a post placed 500 metres away from the start, and return (see later). By the end of the year he had already made important alterations to his Voisin, especially the exchange of the huge box-kite tail-unit for a similar, but much smaller, unit, the fitting of a monoplane elevator, and the rigging of his wings at a slight dihedral angle. (See Postscript on page 240.)

A. de Pischoff built a biplane which he tested without much success at Issy (1907-08) (Fig 45) its chief role being further to consolidate the tractor biplane tradition started by Ferber in 1905; it was, in fact, de Pischoff who was the first to dispense with all forward control surfaces, and so crystallise the modern tradition, although in a somewhat primitive form, with a cruciform tail-unit 'sprouting' from the rear of the centre section: de Pischoff used a 3-cylinder 25-hp Anzani fan-type engine, driving the first sophisticated propeller in Europe, a Chauvière. In England, where interest in aviation was beginning to grow, an important exhibition of model aeroplanes was held at Alexandra Palace in April: no first prize was awarded, but the second was given to A. V. Roe for his 8-foot span Wright-type model.

It was also in this year that the ageing Horatio Phillips made the first tentative hop-flight in Britain in a strange and quite impractical machine incorporating a large number of his high aspect-ratio aerofoils in four tandem frames, and powered by a 20-22-hp engine driving one tractor propeller. Phillips succeeded in becoming airborne for about 500 feet. It was his last appearance in aviation.

At Farnborough, S. F. Cody took the intermediate step between kites and aeroplanes by fitting a 12-hp Buchet motor to one of his modified man-lifting kites, and 'flying' it suspended from a wire at Farnborough, without anyone on board. Cody then commenced construction of his *British Army Aeroplane No 1* which was completed in 1908. Meanwhile, J. W. Dunne, working in great secrecy for the British Government, had built his first swept-wing biplane (the *D.1*) in search of inherent stability, inspired by the form of the winged seed of the Zanonia plant: it was tested only once—when it was damaged—this year as a glider at Blair Atholl (Scotland). It was next fitted with two 12-hp Buchet motors, but was damaged on the ground at its first take-off attempt, and abandoned.

One of the important events of the year in aeronautics took place with the publication of an erudite book, *Aerodynamics*, by Frederick W. Lanchester, well-known as an automobile engineer, who next year issued his second book, *Aerodonetics*. Lanchester is recognised as one of the great figures of aeronautical history. Today, although the profundity of Lanchester's theory of sustentation is universally recognised, his books were very difficult to understand, even by specialists; consequently his theory was not applied by anybody, and his books made no impact on practical aviation. The German engineer Ludwig Prandtl (1875-1953) had independently evolved an essentially similar theory, expressed in a much clearer mathematical form, and, largely through translations and expositions of Prandtl's work, the theory of lifting wings was made known to the scientific world, and later made its impact on aeroplane design and construction. It was not until 1915, in a paper which he read to the Institution of Automobile Engineers, that Lanchester re-cast his theory in a form more capable of application to practical aircraft design. The circulation theory of wings is today referred to as the "Lanchester-Prandtl Theory"; but as communication is the key to technological influence, one must pay primary tribute to Ludwig Prandtl and to the school of aeronautical research workers which he established at Göttingen.

The Wright brothers had been continuing their complex negotiations for the sale of their aeroplanes on both sides of the Atlantic; and although Wilbur paid a long visit to Europe this year, a satisfactory conclusion of their affairs was not to be reached until early in 1908. But during 1906 and 1907 they had been building some half dozen improved engines, and two or three improved and 'finalised' aircraft, one of which was to make aviation history. This finalised type of Wright biplane, which we will call the type A, was a two-seater, basically similar to the *Flyer III*, and with a Wright 30-hp engine. In expectation of a French manufacturing agreement, a new Wright was shipped to France in July 1907, and remained in store at Le Havre until the summer of 1908. The brothers had also taken stock of European aviation and found it so backward that they felt they had nothing to fear from Continental progress; which indeed proved correct.

Also in 1907, there appeared the first tentative man-carrying helicopters. After some six centuries of experiments with models, this type of machine arrived at the stage of lifting a man for the first time in history. There are two claimants for this 'first', Paul Cornu and the Breguet brothers; but the French authorities have established that the Breguet-Richet machine—Richet collaborated—was the first to take off with a man (September 29th at Douai); but it was not a free flight, in that four men steadied it, and although not helping to lift the machine, they certainly aided its stability in the air. The Cornu helicopter, on the other hand—a much more compact two-rotor machine—was the first to fly freely with a man without any connnection with, or help from, the ground on November 13th, near Lisieux. So, although the matter is somewhat academic, the proper 'first' should go to Paul Cornu. Both machines had Antoinette engines, the Breguet a 50-hp and the Cornu a 24-hp, but the flights—although genuine enough—can only be regarded as extremely tentative, the practical helicopter not appearing until the 1930's.

Machines off the main lines of influence which merit inclusion here were the following: 1. The Tatin-de La Vaulx monoplane built by Mallet with a 50-hp Antoinette driving two pusher propellers: it took off twice at St Cyr in November 1907, but was abandoned when it crash-landed after the second take-off; 2. The tailless monoplane by I. Etrich and F. Wels, with swept-back wings, intended to be powered by a 24-hp Antoinette, but tested at Oberalstadt (Bohemia) only as a glider: this machine later led to the famous *Taube* monoplanes of 1909-13. 3. The first man-carrying triplane of history, which Ellehammer introduced—his *No III*—this year, with an excellent 5-cylinder 30-hp Ellehammer radial: the machine made many hops, but was not successful.

Postscript. Henri Farman used both 'i' and 'y' indiscriminately in his christian name, with 'i' appearing in the inscriptions on his early Voisin machine and its modifications, and 'y' on those aircraft he himself designed from 1909 onwards.

❧ 79 ❧

Ferber's
Tribute to Wilbur Wright:
June 1907

Wilbur Wright arrived in Paris on May 27th 1907, in the course of his European tour to investigate the markets for the Wright machines. The Editor of *L'Aérophile* got to hear of his presence in Paris, and wrote to Ferber asking him why he (Ferber) had not notified him of Wilbur's arrival. The resulting letter from Ferber to Besançon, the Editor, is one of the most interesting letters to be found in the history of aviation, and includes an extraordinary tribute to Wilbur. It is not known on what day Ferber wrote this letter to Besançon, but it was published in *L'Aérophile* in the June 1907 issue. As a brief foreword, Besançon introduced the letter as follows:

"WILBUR WRIGHT IN PARIS. Wilbur Wright, the elder of the famous aviators of Dayton in Paris! This news, at first kept secret, ended by leaking out, and the Press has sought by every means to gain accurate information on this unexpected voyage; and, above all, about the real reasons for it. Captain Ferber has given us the interesting details on the subject which follow:"

Then comes Ferber's letter, which was printed in full:

"My dear Besançon,
You ask me why I did not notify you of the arrival of Wilbur Wright as soon as I knew of it two weeks ago, and as punishment you want to condemn me to writing you the impressions I had on seeing him walk into my office to ask me for information on the only light motor there is in the world?
Well, I am about to carry out the sentence. First, he asked me not to give away his arrival just yet—which is reason enough—and then I was overjoyed to note that, in spite of the information power of the modern press, there is still room for people who used to be called 'newsmongers' and knew the news long before others. On the subject of aviation, I am a 'newsmonger' and proud of it.

As to my impression, it was profound, and I grasped his hand and looked upon him with great emotion. Just think, that without this man I would be nothing, for I should not have dared, in 1902, to trust myself on a flimsy fabric if I had not known from his accounts and photographs that it would carry me! Think, that without him, my experiments would not have taken place, and I should not have had had Voisin as a pupil. Capitalists like Archdeacon and Deutsch de la Meurthe would not in 1904 have established the prizes you know of. The press would not have spread the good seed on all sides. Your magazine would not have quadrupled its circulation, and other special journals would not have been born!

Without our press campaign of 1905, in which you played the important part, the most accurate news from America would not have come to France, and our country would not have become 'the only market' for aeroplanes, so much so that Wright is obliged to come here to sell his invention. And still people confuse the aeroplane with the balloon! But ask the children in the Tuileries, and they will tell you the difference.

Without this press campaign, Santos-Dumont, the great balloonist, would not have realized the moment had come, would not have lent his speed of execution and daring courage to the cause, and the public would not have been struck by the evidence. Delagrange would have kept on sculpting delightful statues and not have ordered an aeroplane from Voisin.

Through an appropriate turn of affairs, the noise made over these last two pioneers has brought the wolf out of the woods. I mean that M. Wright has placed himself in the hands of a backer and comes to us ready at last to do business.

It is still the same proposition as I tried to bring to a head in 1905: 'The Wright brothers agree to do 50 kilometres in the air, after which they are to be paid a million and half francs.' (The passage of time has raised the indemnity.) With the question put this way, no one risks anything, and I have never understood why I was not taken up in 1905. Today, after the experiments of Santos, Voisin, and myself, I think that 50,000 francs is the maximum that should be paid for an aeroplane. That is what I said to our colleague, M. Hart O. Berg, who is the financier to whom O. and W. Wright have entrusted themselves. But M. Berg told me, with the great experience of financiers: 'Captain, perhaps you are right. Absolutely, the thing is worth less than in 1905. But relatively, it is worth much more today because, with the publicity you have made, and the experiments of Santos-Dumont, people believe the thing is possible and will give their money. You mark my words.'

In conclusion, my dear friend, I think so too, and am glad of it, because we are about to enter the active period that I foresaw so long ago.

'Truly yours', as one says in America.
(signed) FERBER"

It would be impossible to find a more thoroughgoing tribute to the Wrights than this, or a more explicit public admission of Wright influence, by the man whom Dollfus called "the great French precursor of practical aviation: he was the first in our country to build, and test methodically, full-size gliders". Ferber was a curious and complex character, and when it suited his purposes, he could forget what he owed to the brothers; these latter occasions were generally when he was in the company of his Aéro-Club colleagues, and patriotism in claiming aviation as a wholly French invention was the order of the day. Here, on the other hand, we are seeing Ferber as an honest man paying just tribute where tribute was due, as he does in his writing. I think it would be in order to repeat those particular lines of tribute:

"Just think, that without this man I would be nothing, for I should not have dared, in 1902, to trust myself on a flimsy fabric if I had not known from his accounts and his photographs that it would carry me! Think, that without him, my experiments would not have taken place and I should not have had Voisin as a pupil. Capitalists like Archdeacon and Deutsch de la Meurthe would not in 1904 have established the prizes you know of. The press would not have spread the good seed on all sides. Your magazine would not have quadrupled its circulation, and other special journals would not have been born!"

80

Henri Farman's first Flights: October, November 1907

To the French, there was only one Man of the Year in 1907, and that was Henri Farman. Having been airborne for about 30 metres on September 30th, he went on to make regular flights through October and November of 1907. On October 26th, he flew for 771 metres (say 2,531 feet) in 52⅖ seconds, at a height of about 6 metres. This was greeted in *L'Aérophile* for October by the following title and sub-titles to its report:

> "The Great Stages in Aviation
> HENRY FARMAN FLIES FOR NEARLY A KILOMETRE
> He beats the World's Record for mechanical flight and wins the Archdeacon Aviation Cup."

With a flight of 186 metres in 15⅖ seconds, on October 23rd, he won a prize of 200 francs and a plaquette for the first man to fly for 150 metres. Then, on October 25th, he flew for 200 metres.

On the 26th, as noted above, he flew for 771 metres and won the Archdeacon Cup. The phrases used about this achievement, were, as with Santos, flowery in the extreme. We read of "the courageous aviator", "absolutely sure of himself", and of the ovation which greeted him being expressed "frenetically (frénétiquement)"; Farman was said to have "brilliantly carried off the Archdeacon Cup"; and so on, and so on.

At the monthly dinner of the Aéro-Club de France on November 7th, there were given two rapturous speeches, one by the President (Cailletet) and the other by Archdeacon. The President went quickly through French aviation history from the eighteenth century, starting with the words:

> "It seems that to the genius of France is reserved the glorious mission of initiating the world into the conquest of the air."

In the course of his speech, the President also pointed out that

"In these researches [*ie* in aviation] our dear country can still claim the monopoly, and the Aéro-Club de France should be proud to count among its members the principal promoters of aviation."

When Archdeacon took the floor, he started off in the same vein as at the banquet to Santos-Dumont, by saying:

"You can believe me, without difficulty, when I tell you that this evening is one of the most marvellous (une des plus belles) of my existence. . . . It is incontestible that the superb flight achieved by Henri Farman on October 26th marks an absolutely decisive stage in the history of aviation."

After referring to other glories, and paying various tributes, he had this to say about the Wrights:

"The famous Wright brothers may today claim all they wish. If it is true—and I doubt it more and more—that they were the first to fly through the air, they will not have that glory before History. They would only have had to eschew these incomprehensible affectations of mystery, and to carry out their experiments in broad daylight, like Santos-Dumont and Farman, and before official judges (controleurs), surrounded by thousands of spectators. The first *authentic* experiments in powered aviation have taken place in France; they will progress in France; and the famous 50 kilometres announced by the Wrights will, I am certain, be beaten by us well before they will have decided to show their phantom machine (l'engin fantôme)."

Near the end of his speech, Archdeacon referred to the well-known journalist François Peyrey. In view of Peyrey's rapturous accounts of the Wrights in 1908, it is as well to note what Archdeacon said:

"I cannot here cite all the names [*ie* of the many members of the press who have helped to propagand aviation] but there is, however, one I must mention; that is our friend François Peyrey, who incontestably was the first journalist to become well versed (adepte) in aviation. In 1903, he started a campaign for aviation: he accompanied me to Berck in 1904 . . . Today, Peyrey still continues—with the talent of which you know—his excellent work of popularisation . . ."

And, one must remind oneself, this "absolutely decisive stage in the history of aviation" did not even equal the duration achieved by Wilbur Wright when he flew for 59 seconds on December 17th 1903!

On November 9th 1907 Farman went on to win the Archdeacon Cup with a flight of about 1,030 metres (3,380 feet) in 1 minute 14 seconds, during which he made an official circle, or an almost closed U-turn.

This was to be the best of Farman's efforts during 1907, and far away better than anyone else in France.

Incredible to relate, since Ferber was sparked by the Wright glider configuration in 1902, *it had taken the Europeans five and a half years before one of them could stay airborne in a powered aeroplane for a single minute.* Can one wonder that they were accused by their own confrères of suffering from an "inexcusable torpor"?

… 81 …

European Stagnation and Procrastination: 1907

Looking back on 1907, which should have been a year of achievement and rapid progress—especially after the crescendo of self-criticism and exhortation in 1906—we find the same old story, the *mystifying* old story. European aviation had recruited a number of good men, some of them attracted away from automobilism; but it all added up to a massive muddle of making haste slowly, and a refusal to see and follow the obvious paths to success. The 'chauffeur' philosophy was still dominant, and still no one realised that mastery of glider construction and control —which they could easily have accomplished within a year of Chanute's revelations in 1903—would have led rapidly to mastery of powered flight. The main reason for this neglect was their basic mistake of regarding flight-control as only passive and corrective—as an extension of inherent stability—and not as both a corrective and as a positive dynamic means of manoeuvre: both Lilienthal and Chanute had been explicit about this; but both were now forgotten. The Europeans also failed to realise that even the corrective function of flight-control needed a control system which would be instantly effective about all three axes: hence their neglect of proper control in roll. In view of Chanute's partial revelations of 1903, and the explicit revelations in the Wright patent published in 1906, it is almost incredible that this dual function of flight-control was not properly comprehended, let alone appreciated and utilised, by the Europeans until they saw Wilbur Wright's display of mastery in France, from August to December of 1908.

Nor had the Europeans progressed beyond the most primitive of ungeared propellers, revolving at engine speed, despite the modest presence of the first Chauvière propeller on the de Pischoff biplane. No one came forward, in this important but neglected sphere, to initiate proper research and development. Yet again we meet this incomprehensible stagnation and procrastination.

Continental pioneers were understandably unwilling to risk any effective catapult, or other launching methods, with their embryo machines: they would certainly not have known how to control them if they found themselves suddenly committed to the air, and would assuredly have suffered many a casualty. Not having had to face the problems of flight-control in gliders, they underestimated or neglected flight-control in general—especially control in roll—in their powered machines; they therefore concentrated on forcing them into the air in small straight-line hops. Henri Farman alone had begun to acquire the 'airman's' attitude by the end of 1907; Blériot was to embrace this attitude next year; but neither really espoused it until they saw Wilbur Wright fly in 1908.

The only strong suit of the Europeans was their traditional pursuit of inherent stability, and their correct insistence on the adoption of a tailplane to bring it about longitudinally; and wing dihedral for lateral stability. But this persistent pursuit of stability was largely nullified by their equally persistent refusal to appreciate the vital role of flight-control. This neglect of control—one of the two chronic European diseases—was worse confounded by the second disease, the refusal by the European pioneers to think out and adopt any definite configuration and pursue it through test and modification, until success was reached. Blériot's production of no less than three basic configurations in this one year, 1907, with no proper testing and development of any of them, was typical of the Continental attitude. It was again a question of time-wasting 'slapdashery'. Blériot had been working at the problem of flight since mid-1905; but only at the end of 1907 had he even arrived at a promising configuration: he was to labour until mid-1909—two more years—before he could achieve a practical aeroplane.

In Blériot we have a typical, and incomprehensible, case-history of European pioneering. Equipped with all the information and clues that anyone could wish for, and with ample funds, a talented and fearless man *took no less than four years—forty-eight months, two hundred and eight weeks—to progress from an unsuccessful (but feasible) float-glider to a practical powered machine; and, even then, he had others to aid him in the designing of his aircraft;* and he used other people's aero-engines and propellers!

The case-history of the Voisin machines is equally extraordinary. Gabriel Voisin was tentatively piloting Archdeacon's Wright-type No 1 glider in the spring of 1904; but it was not until November of this year, 1907, that the powered machine he built for Henri Farman (and modified by him), could remain in the air for sixty consecutive seconds; and not until November 1907, and January 1908, that this same machine could make its first two wavering circles, yawing round on rudder alone. The standard Voisins were indeed never to have any control in roll, right to the end of their days in 1910.

European Stagnation and Procrastination: 1907

As the year 1907 came to an end, there was only one European aeroplane—the *Voisin-Farman I*—which could remain airborne for a single minute!
Here is the melancholy record for the year 1907:

Date	Location	Duration	Distance	Aircraft and pilot	Total take offs	Minutes 0	½	1
1907								
March 30	Bagatelle	6 seconds	60 metres	Voisin-Delagrange I (G. Voisin)	6	O		
April 5	Bagatelle		6 metres	Blériot V	4	O		
Spring/Summer	Streatham		500 feet	Phillips II	(?)	O		
July 5	Bagatelle		20 metres	Vuia II	2	O		
July 25	Issy	10 seconds	150 metres	Blériot VI Libellule	11	O		
September 17	Issy	17 seconds	184 metres	Blériot VI (modified)	6		●	
November 5	Issy	40 seconds	500 metres	Voisin-Delagrange I (modified) (Delagrange)	2		●	
November 9	Issy	1 minute 14 seconds	1,030 metres	Voisin-Farman I (H. Farman)	20			■
November 16	Buc	55 seconds	600 metres	REP 1 (Esnault-Pelterie)	5			■
November 16	Issy	45 seconds	500 metres	Blériot VII	6			■
November 17	Issy	18 seconds	200 metres	Santos-Dumont 19	3		●	
November 18	Saint-Cyr	7 seconds	70 metres	De La Vaulx	2	O		
December 6	Issy		7 metres	De Pischoff I	7	O		
1907–08	Lindholm			Ellehammer III	(?)	O		

Note. The durations are shown to the nearest half-minute

THE YEAR 1908

✵ 82 ✵

European Biplanes: January to July 1908

The year 1908 opened with what seemed to contemporary Europeans as an outstanding achievement—the first official kilometre circular flight in Europe. This consisted of taking off, flying between two posts, then round a single post 500 metres away, and back to pass airborne between the twin posts again. This 'kilometre bouclé' was performed by Henri Farman in his *Voisin-Farman I* (*mod*) at Issy on January 13th 1908, and won him the Deutsch-Archdeacon prize of 50,000 francs, with medals going to the Voisins for construction—which, in spite of Farman's modifications, they well deserved—and to Levavasseur for the Antoinette engine (Fig 46). This event is described in detail in Section 84.

Farman had to achieve his circle on rudder alone, with a wide yawing turn, since he had no warping or ailerons; with the outer wings thus moving faster, and the inner moving slower, the machine was in a banked attitude during the whole manoeuvre, but had to be turned in as large a circle as possible to avoid stalling, side-slipping or rolling into the ground.

After his kilometre circle, during which he actually covered about 1,500 metres in 1 minute 28 seconds, Farman made one more flight that day of 500 metres; then, also at Issy, he made three flights on January 15th, the best being about 1,500 metres. Next, he carried out further modifications to his machine, in which he was to make significant flights. In the course of its long career, this machine was to survive longer, and undergo more transformations, than any other early European aircraft, from its first hop-flight on September 30th, 1907 to its last take-off on May 23rd 1909.

The *I-bis* had wings of 40 square metres area. There were only minor differences in dimensions and weight from the original Voisin which Farman had purchased in September 1907. Farman's own important modifications of incorporating dihedral and a small tail-unit had already been made before the end of the year 1907. Now the fabric had been renewed, "Continental" rubberised linen (as used for airships) being substituted for the previous varnished silk. A new engine was installed, an 8-cylinder 50-hp air-cooled Renault, which drove direct the same primitive Voisin propeller at 1,100 rpm: but this engine was only used in one take-off (the first, on March 14th), and was then replaced by the original 50-hp Antoinette. It is not known why the Renault proved unsatisfactory. In May, two inner side-curtains were added between the centre section struts, closely enclosing the pilot.

The *I-bis* made some 38 take-offs from March 14th to August 9th 1908, at Issy, at Ghent in Belgium, and in the USA; but only two significant flights resulted:

March 21st, at Issy; 2 kilometres 4.80 metres in 3 minutes 31 seconds.
July 6th, at Issy; duration flight of 20 minutes 20 seconds.

Farman also made the first two brief flights in Europe carrying a passenger—Archdeacon in both—one of 131 metres in 11 seconds on May 29th; the other of 1,242 metres on May 30th; both were made at Ghent.

The flight of July 6th was a new European record, *the first time a European had exceeded a quarter of an hour's flight duration*, for which Farman received the Armengaud prize of 10,000 francs for the first quarter-hour's flight.

The Wrights had made *their* first quarter-hour's flight on September 26th 1905, when they flew for 18 minutes 8 seconds.

After a brief and unsuccessful visit to the United States in July and August, Farman was back in Europe in August. His *I-bis* was now due for its next modification.

Farman and Delagrange carried on a friendly rivalry throughout much of the year in their respective modified Voisins. They also paralleled one another till the Autumn in the various alterations made; but in the Autumn, Farman—who was more talented aeronautically—set about incorporating important features in an effort to produce the results which he saw the Wright machine achieve.

* * * * *

Delagrange, after his crash-landing in November last, had the Voisins build him a second machine—probably using much of the first—which became the *Voisin-Delagrange II* (called 'Leon Delagrange No 2'). It closely resembled the Farman, with the latter's modifications of monoplane elevator, dihedral on the wings, and small tail-unit. In May—after a crash-landing on May 3rd—two side-curtains were fitted to promote lateral stability, and the machine was given a new designation, 'Leon Delagrange No 3' which we will call *Voisin-Delagrange III*. In late July the machine had its two inner side-curtains removed, and four fitted, which enclosed the outer bay on either side.

The machine, in its two first versions, *ie* as the *Voisin-Delagrange II* and *III*, made some 40 take-offs from January 20th to July 10th 1908 at Issy, Rome, Milan and Turin, which included many good flights, and some circles, the best being as follows:

March 20th at Issy: 1 kilometre 500 metres in 2 minutes 30 seconds
April 10th at Issy: 2 kilometres 500 metres in 3 minutes
April 11th at Issy: 3 kilometres 925 metres in 6 minutes 30 seconds
May 30th at Issy: 12 kilometres 750 metres in 15 minutes 25 seconds
June 23rd at Milan: 14 kilometres 270 metres in 18 minutes 30 seconds

These Voisin-Delagrange and Voisin-Farman biplanes were the only European machines of any kind to make consistently good flights in 1908.

No monoplane could be rated as practical in 1908.

* * * * *

Although, in England, Cody's biplane was ready in the early summer of 1908, it was not properly tested until September-October, and it was A. V. Roe who in England was next after Phillips to test (at Brooklands) a full-size powered machine. After winning a *Daily Mail* prize for his Wright-type biplane model in March 1907, Roe built a full-size Wright-derived machine, minus rudder, with a span of 30 feet, a weight of 600 lbs, and a 9-hp JAP engine: this was far too weak, and was exchanged for a 24-hp Antoinette (Fig 47). In June 1908, the machine was 'air-towed' behind a car, and also made a few self-propelled take-offs (probably after down-hill runs) covering up to about 150 feet. It never flew in any real sense. Although Roe was later to do outstanding work, this first machine is of little significance historically.

Back in France, Ferber came briefly into prominence again—and for the last time—with a tractor biplane which kept his would-be inherently stable 'tailed' concept before his contemporaries' eyes, and continued to influence them in that direction. This machine was the *Ferber IX* which, owing to Ferber's having joined the Antoinette company, confusingly bore also the alternative designation of *Antoinette III*: it had a wing area of 30 square metres, a 50-hp Antoinette engine and a weight of 400 kilograms. The forward elevator was retained and there was a fixed tailplane and fin. At first, two flimsy triangular rudders were attached to the outer rear struts; then these were exchanged for a rear rudder. There was a two-wheel tandem undercarriage. This machine made some 8 take-offs from July 22nd to September 19th 1908, at Issy, the best result being 256 metres covered on August 19th, and 500 metres on September 19th (with a crash-landing), with Legagneux piloting it on both occasions. Ferber then abandoned this machine and did no more original work. He was to be killed when his standard Voisin ran into a ditch at speed after landing, on September 22nd 1909, at Boulogne (see Postscript on next page).

A biplane designed and built by the brothers Ernest and Paul Zens, at Gonesse, has a certain interest in its extended lower wings, which seem to have consisted of smaller surfaces laid on the main wings and overlapping them, not—as it sometimes appears—raised above them; in the different dihedral angles of the wings; and in the rocking elevator out front, which was also intended for steering. There was a fixed tailplane well out behind. It had a 50-hp Antoinette. The machine was tested at Gonesse on August 4th; after making a few hops, it was damaged and then abandoned.

The pusher biplane designed and produced by René Gasnier, also powered by a 50-hp Antoinette, was a somewhat similar machine, but had only minimal success. It was an interesting variation on the Voisin theme, with main-plane dihedral, a fixed monoplane tail and, as with the Zens, a forward rocking elevator to act also as a rudder. The machine made some 45 take-offs from August 17th to September 17th 1908, at Rochefort-sur-Loire, its best performance being 500 metres on September 17th, when it crash-landed and was abandoned.

In Denmark, Ellehammer persevered—but did not advance aviation technically—with his *No IV*, which was a tractor biplane with the same engine as the triplane, and the same pendulum control system. It started making take-offs on January 14th 1908, and at Kiel on June 28th, made the first official hop-flights in Germany: these were two of 11 seconds each, which could have been exceeded had the extent of the ground allowed. Then the machine, which was not practical, was abandoned.

Although it stands outside the exact date-span of this section, I will include here some comment on S. F. Cody's ingenious 'British Army Aeroplane No 1', as it was officially called, because the machine was completed by the Summer of 1908, although it was not properly tested until September-October. Cody was at this time an expatriate American employed at Farnborough by the British Government, and working on airships and aeroplanes for the Commanding Officer, Colonel J. E. Capper. It was on October 16th 1908 that Cody most creditably made the first powered flight in Britain at Farnborough, and covered 1,390 feet in 27 seconds. It has always been noted that it was Wright-derived, with a forward elevator, rear rudder, and an auxiliary rudder on the upper wing for roll control (Fig 48): a 50-hp Antoinette drove the two pusher propellers. It originally had between-wing ailerons, but these were removed before the first flight. Dr P. B. Walker, in Vol II of his official history of the Farnborough Establishment (see Bibliography), writes as follows of the Wright brothers' influence:

> "Fortunately, it is now possible to form an opinion based on a detailed knowledge of the design as revealed by the various surviving photographs. These leave no room for doubt: the Wright influence was paramount. The only thing to explain is how the necessary information reached Cody, and here the answer is simple: through Colonel Capper. Not only did Capper have the knowledge acquired during his visit to [the Wrights at] Dayton in 1904, but he was a conscientious reader of all the published technical literature in the aeronautical field. There are good reasons for believing that he kept Cody supplied with all the published illustrations of the work of the Wrights; ..."

Cody was an ingenious and indefatigable pioneer, but he only influenced aviation history indirectly, by helping to make Britain airminded, and thus produce her own designers. Cody was naturalised in this country in October 1909.

Postscript. Ferber had touched down and was about to take off again when his machine hit a ditch at speed; it threw him out; overturned; and then trapped him beneath the engine.

≈ 83 ≈

European Monoplanes:
January to July 1908

The lone figure of Esnault-Pelterie produced the second version of his monoplane in 1908, the *REP. No 2*. It was now 'cleaned up' all round, and a great improvement on the first, with a wing area of 17 square metres, a span of 8.6 metres, and a weight of 350 kilograms. There was the same 7-cylinder air-cooled REP engine of 30-hp driving direct a four-bladed propeller. The wing anhedral, with downward warping control, had been retained; but a large expanse of fin area was added to improve directional and lateral stability, along with a larger tailplane, a rear rudder, and twin auxiliary elevators on the fuselage just behind and below the engine, which were later removed. The tandem undercarriage and wing-wheels were as before, but efficient hydraulic brakes were fitted—history's first. This machine made an unknown number of tests at Buc, from June 8th 1908 to the end of November, which included a flight on June 8th of 1,200 metres at an unconfirmed speed of some 50-55 mph ("une vitesse vertigineuse"), an extraordinary achievement at that time: it also attained an unofficial height record of 30 metres on that day. Esnault-Pelterie's machines, although increasingly successful, did not appeal to his contemporaries as worthy of imitation, probably owing to their low degree of inherent stability.

Meanwhile, Blériot had come near to his finalised monoplane configuration in his *No VIII*, except in lateral control, which first flew on June 17th 1908; between then and October 31st it made—in its three versions—over 30 take-offs at Issy and Toury. It was a tractor monoplane with a long 'trellis' fuselage, a wing area of 22 square metres, an initial span of 11 metres later reduced to 8.5 metres, and a 50-hp Antoinette engine driving direct a four-blade metal propeller. When first completed, it was an enclosed fuselage type (reminiscent of the *VII*), but did not fly in this form. In its first tested form, two inadequate triangular ailerons were fitted at the wing-tips, a small tailplane was flanked by twin elevators, and there was a rudder. It was tested from

June 17th to June 29th at Issy; and of its half-dozen take-offs, the best achievement was 700 metres on June 29th.

Either during these tests, or immediately after, rectangular flap-type ailerons (down-moving only) were fitted—the first of their type in history—and the machine was redesignated the *Blériot VIII-bis* (Fig 49): in this form it made some fair flights, including several turns, among its more than 20 take-offs at Issy during July and September; the best was of 8 minutes 24 seconds duration, on July 6th.

This is the proper place to introduce formally the figure of Leon Levavasseur, artist turned engineer, who always remained an artist in whatever he did. He was the mainspring of the Antoinette firm ('Société Antoinette'), Antoinette being the name of the daughter of Jules Gastambide, the head of the firm. As already noted, Levavasseur's first successes were with his remarkable series of Antoinette engines, used first in his abortive aeroplane of 1903; then successfully in motorboats from 1904 onwards. It was Antoinette engines that virtually made early European aviation possible. But Levavasseur was to become just as great an aircraft designer as he was an engine designer.

In 1907, Levavasseur designed a monoplane for two members of the firm—MM. Gastambide and Mengin—which emerged in February 1908 as the *Gastambide-Mengin I* (Fig 50). This machine and its successors provide for the monoplane the same instructive evolutionary series as do Henri Farman's machines for the biplane. It had tapering wings of 24 square feet area, with a pronounced curvature, and set at a dihedral angle. The engine was a 50-hp Antoinette driving direct a two-bladed metal propeller at 1,250 rpm: there was a four-wheel undercarriage. This tentative machine had no control surfaces; but, following Phillips, Levavasseur built wings with the upper surface more cambered than the lower; but unfortunately adopted, not the thick leading-edge, but a thin Henson-like one. Up till now, attention had been paid to the curvature and lifting qualities of the thin double-surface wing, with similar curvatures on top and bottom, especially by the Wrights; but here we have the tentative beginnings of the efficient modern aerofoil with a high lift-drag ratio. The *Gastambide-Mengin I* made some 4 take-offs at Bagatelle (piloted by the mechanic Boyer) from February 8th to 14th, 1908, its best hop being 150 metres on the 13th. It is uncertain whether there were any tests after the 14th, when it crash-landed; probably not.

The machine was rebuilt by July 1908, when it was sometimes referred to as the *Gastambide-Mengin II*, but also—and more correctly as we now know—as the *Antoinette II*. It made its first take-off on July 22nd, and was extensively tested during July and August at Issy, piloted by Welferinger (of the Antoinette firm); it also appears to have been slightly modified in its stabilising and control surfaces along the way: it has now taken on something of the true Antoinette 'look' of the later developed machines: it had triangular ailerons, twin rudders, and an elevator. On August 20th, Welferinger carried Robert Gastambide for two short hops (the best was 100 metres); and on August 21st the machine made a creditable flight of 1 minute 36 seconds, during which it performed a complete circle—the first by a monoplane. This seems to have been its last test.

Before leaving normal monoplane development in the first half of 1908, a word should be said of the machine, given the English title of the *Flying Fish*—with the designation 'Henri Farman No 2'—which the Voisin brothers were building for Farman during February and March 1908; this machine was abandoned by Farman before ever being tested. It was later sold to a German, Lieutenant Fritsche, but never flew.

❧ 84 ❧

Farman flies the first European Circle: January 13th 1908

By January, the *Voisin-Farman I* had been slightly modified by having its cellular tail-unit reduced, and on January 13th 1908—before the necessary official witnesses—it made the first circular flight in Europe covering a kilometre, at Issy-les-Moulineaux, thus winning the Grand-Prix d'Aviation Deutsch-Archdeacon of 50,000 francs. The flight duration was 1 minute 28 seconds (Fig 46).

European aviation was still in such a parlous state that the first achievement of this essential manoeuvre for any flying machine—carried out in the aerodynamically worst manner—was heralded as a literally epoch-making event, as were Santos-Dumont's efforts last year. One must go back to that time in spirit, stand there with the witnesses, read their words, and look at the photographs they looked at next morning, to appreciate the extraordinary effect it had on its contemporaries, and to assess its historical significance.

First of all, here is the account from the London *Times* for January 14th 1908:

> "*THE CONQUEST OF THE AIR*
> *SUCCESS OF AN ENGLISHMAN*[1]
> (from our own Correspondent)
>
> Paris, Jan. 13.
>
> In the domain of the air, at all events, the number 13 would seem to bring good luck. In any case, today has been an epoch-making date, that of the victory before official witnesses of human intelligence in its efforts to solve the problem which brought Icarus to grief, which tormented the brain of Leonardo da Vinci, and which during the last few years has become possible of solution only through the invention of a light and stable motor whereby to animate the gigantic wings of a machine heavier than air. This morning, a little

[1] Farman remained a British citizen until 1937, when he became a naturalised Frenchman.

after 10 o'clock, Mr. Henry Farman succeeded in rising on such a machine of his own invention, in flying over a kilomètre towards a goal previously fixed, which he rounded in perfect conditions of stability, and in returning to his starting point, where he alighted without a hitch. Nothing of the kind has ever before been accomplished. Mr. Farman thus wins the 50,000 f (£2,000) prize of aviation offered by MM. Henry Deutsch and Archdeacon. But he wins as well a unique fame.

The details of an event of such historic importance are worth chronicling. Mr. Farman's aeroplane, which was brought out upon the Plain of Issy towards 10 o'clock this morning, consists of an apparatus built on the principle of the Chanute two-surface gliding machine. The total surface of the apparatus is 52 square mètres and the total length is 10 mètres. The screw is worked by an Antoinette 50-horse power motor of eight cylinders. It is extraordinarily light, being one of the few motors possible in aviation owing to this peculiarity. While the members of the Aviation Committee of the Aéro-Club measured and marked out the distance on the Issy Plain Mr. Farman calmly took his place on his aeroplane. The conditions of the contest were that a machine heavier than air should travel at least one kilomètre, making the circuit of a goal fixed at 500 mètres from the starting-point, and in setting out, as well as on its return, should pass between two posts separated by a space of 50 mètres.

When the starting signal was given the aeroplane rolled for about 50 mètres on the ground, then rose to about 100 mètres (*sic!*) and headed towards the turning-post. There was then a slight but steady descent, but after this Mr. Farman kept himself steadily at the same level, turning at the point indicated, and returning to the point that he had just left, with the ease of a gigantic bird coming to earth. The entire flight had taken only 1 minute 28 seconds, which would make 40 kilomètres 909 mètres an hour.

Mr Farman, who is only thirty-three years old, and who began life as a pupil of the painter Cormon, is the son of the well-known Paris Correspondent of the *Tribune*, and had long been known to sportsmen, having taken part in many automobile races. He intends to go to England to compete for various aviation prizes. On Thursday he will be presented with the grand gold medal of the French Aéro Club, and a banquet will be organised in his honour."

The Aéro-Club gave Farman a sumptuous banquet, with speeches couched in terms similar to those which commemorated Lindbergh's atlantic flight of 1927. The President of the Aéro-Club, L.-P. Cailletet, declared:

"The date January 13th 1908, will henceforth be a historic date: it will recall to the generations to come the brilliant victory (la victoire éclantante) achieved this day by Henry Farman.

It is he, indeed, who—the first among men—has succeeded in

rising and steering in the air, thus resolving one of the most extraordinary problems posed since the beginning of the world . . . . The victory of Henry Farman . . . shows us that men are capable of resolving the most difficult of questions—one is tempted to say the most impossible—when they are actuated by a burning conviction and by a never-failing energy."

Ferber, in a lecture he gave in London on February 9th, paid this extraordinary tribute:

"On Monday January 13th, Henri Farman has astonished the entire world by victoriously demonstrating that a man can make a mechanical flight with a machine heavier than air; that he can depart from a fixed point, manoeuvre, and return to the point of departure. He has thus realised the dream which everybody has harboured for such a long time."

Gabriel Voisin was later to write of this flight in his autobiography:

"It had thus demonstrated the possibility of voyaging by a heavier-than-air craft, and of manoeuvering in altitude and in azimuth, a thing which nobody, until that day, had been able to do."

Archdeacon, recording his thoughts in an article entitled "Après le succès" (*L'Aérophile*, May 1st 1908) wrote:

"One must indeed, for history, *fix the actual and indubitable advent, in France, of aerial locomotion by aviation*. Eh bien! I do not hesitate to say that date ought to read January 13th 1908. . . . It is *Farman who is the first incontestibly to win the mastery of the air by aeroplane* (qui le premier, à incontestablement conquis la maîtrise de l'air, en aéroplane)" (the italics are Archdeacons')

To the British, all but a score or so of whom had never seen an aeroplane, the event was just as memorable. Here, again, is *The Times* of January 14th 1908, which we have already seen:

"Today has been an epoch-making date, that of the victory before official witnesses of human intelligence in its efforts to solve the problem which brought Icarus to grief. . . "

The equally sober *Illustrated London News* headed its full-page photograph "MAN FLIES AT LAST . . ."; and its text described the flight as "the most extraordinary feat yet performed in the navigation of the air".

Similar, and even more laudatory, headlines appeared all over the world, and curiously echoed the reception of Santos-Dumont's hop-flights of 1906. Such statements, and the feelings they expressed, offer a revealing reflection of the primitive standards of achievement which Europe had been forced to accept in aviation over the years. In that November of 1906 Santos had become airborne for a maximum duration of $21\tfrac{1}{5}$ seconds; now, in January 1908, Farman had made a circle, and remained airborne for 1 minute 28 seconds.

To bring so many wild thoughts of praise—and equally wild words—into historic proportion, I feel one must repeat that the occasion of such acclaim was simply the flying of a circle by an aeroplane, and flying it precariously on rudder alone. As *L'Illustration* put it, "the turning constituted the difficult part of the programme". When the first circle in history was flown on September 20th 1904, the pilot (Wilbur Wright) started his simple diary entry thus:

> "Rain N.E. wind P.M.
> Completed circle. Distance . . ."

Finally, as a two-piece tail to the story of Farman's circle on January 13th 1908, here are two quotations, one from 1904, the other from 1912. First, here is Amos Root, the bee-keeper, who witnessed Wilbur flying the first circle and wrote:

> "In making this last trip of rounding the circle, the machine was kept near the ground, except in making the turns. . . . When it first turned that circle, and came near the starting-point, I was right in front of it; and I said then, and I believe still, it was one of the grandest sights, if not the grandest sight, of my life."

And here is the Editor of *Flight*, writing—the italics are his—of the death of Wilbur in 1912:

> "When Henry Farman won the first Grand Prix *three years later* [ie after the Wrights' successful 1905 season] by flying a circular kilometre, the world perforce hailed it as the greatest thing that had yet been done, because so far as the world at large was concerned, that was the first circular flight that mankind had 'officially' observed. Three years previously, Wilbur Wright had already flown distances exceeding *twenty miles* in length, and when in 1908, he also went to France and showed what he could do with the machine that had for so long been idle in its crate, the world at last no longer doubted his attainments in the past."

85

Ferber lectures in London: February 1908

A comparatively unknown event occurred in London on February 7th, 1908, when The Junior Institution of Engineers staged a symposium on "Aerial Navigation" at the Royal United Service Institution in Whitehall. This symposium—which should of course have been put on by the Aeronautical Society—included papers by Captain Ferber and the Comte de La Vaulx; but the former's, on the "Development of the Aeroplane", is the only contribution which concerns us.[1] Ferber's paper, given at this time (February 1908) exhibits an almost monumental naïvete. When one remembers that he took to using a Wright-type glider in 1902, and should have been power-flying by at least 1905, if not earlier—or rather *somebody* in Europe should have been power-flying by that time—the whole tone and tenor of Ferber's paper is astonishing. It provides, in its way, a fascinating microcosm of the floundering Continental failure to conquer the air, as they should have been able to do, with ease and expedition.

Ferber started by dealing with Farman's first circle on January 13th:

"On Monday, the 13th of last January, Mr Henry Farman astonished the world by a triumphant demonstration that man was capable of flight—that with a machine heavier than air, he could set out from a given point, perform evolutions in the air, and return again to the starting place. Thus he realised a dream which has occupied the minds of men in all ages.

How, it may be asked, was such a wonderful practical result obtained, and almost without funds? It was not brought about by scientific calculations, but was due to three causes, moral, sporting and commercial, which I shall point out, and which have permitted the application of a fairly safe method of experiment, first employed by Lilienthal. This method consists of taking one's position in the

---

[1] The translation used for the quotations is the official one, given in the March (1908) issue of the Institution's *Journal ond Record of Transactions*.

machine, trying it as near the ground as possible, correcting the defects observed, and repeating the process until satisfaction is obtained.

The three causes to which I have referred are as follows. Firstly,—The aviator received the moral support of the sympathetic atmosphere created by the presence of men, all of whom were familiar with the subject, and who for over two years had been convinced of the possibility of success. Secondly,—Flight was achieved by an apparatus designed by Voisin, who was my first pupil, and who, confident that the problem of aerial flight would be solved, had had the hardihood to establish two years previously, the first flying machine factory in the world. Thirdly, and lastly, flight has been achieved, because Levavasseur, also knowing that the flying machine problem must be solved, had by 1903 sought and found the light Antoinette motor.

It was on this account that mechanical flight was realised first in France, and had Farman failed or vanished on the 13th of January, ten aviators would have replaced him in a short time, working on the same principles and by the same methods."

His delightful remark about if "Farman failed or vanished on the 13th of January, ten aviators would have replaced him in a short time" would not have been properly appreciated by his British audience; for the only man remotely likely to have replaced Farman was Delagrange, who flew his first circle on May 30th, at Juvisy. Then we come to this:

"It is not possible for a great discovery such as the one in question, to be made entirely by one man. Quite a number of favourable circumstances are necessary. There was Lilienthal, in Germany, who by 1891 had shown the way to learn to fly; he was made fun of because hardly anybody was capable of understanding his work. In America, there were men like Chanute and the Wrights who have guided us, but being too isolated they have been ill-advised, and wishing to retain everything have lost all."

This sort of double-talk was unfortunately to become rather familiar where Ferber's utterances are concerned. We have seen him in more detached mood, and the reader will recall his considered statement in 1904 about the Wrights' first flights in 1903, when he wrote:

"for the first time a piloted flying machine had *really flown*, and the honour of this memorable experiment falls to the name of Wright".

Ferber also failed to point out that the realisation of this "dream which has occupied the minds of men in all ages"—already realised long since by the Wright brothers—had taken Europe *nearly six years to come about!* I doubt if even Ferber himself knew exactly what he meant by his remark about the Wrights having "lost all". We next have a remarkable admission:

> "In France, the man who rendered public opinion favourable, without which fruitful work is impossible, is unquestionably M. Archdeacon, thanks to his enthusiastic speeches, numerous writings and handsome prizes. It must be confessed also that he was assisted by the French characteristic of ardently attempting to realise the dreams of the future, such as are so admirably treated by your famous author, Mr H. G. Wells."

This is the first time we have seen it stated that the pioneers needed a favourable public climate to work in! The Wrights could not have cared less; in fact they did not particularly like publicity. But one receives from Ferber's words an interesting hint that the French inventors needed a "climate of opinion" favourable to their work, and wonders if this was in any way true. But they certainly never put any 'ardour' into their work, unless it was of a very slow-burning kind.

Ferber goes on to say that:

> "In this sympathetic atmosphere have been working and are still at work, always on the same lines, firstly the Comte de La Vaulx, who is with us tonight. He has made a most interesting aeroplane according to the designs of the well-known engineer Tatin, who would have been able to drive an aeroplane a long while ago, had be been able to procure the necessary funds. But alas! as we aviators often remark, 'It is much easier to make a machine that will fly than to "raise the wind" for it'."

This piece about Tatin is, of course, absurd. Tatin was a venerable pioneer, very model-minded, who neither knew nor cared about proper flight-control. If he had really been capable of designing and building an aeroplane, he would have done so. The Wrights earned only a modest income from their cycle business, and never received a penny from anyone; yet they not only carried on their business, but built aeroplanes as well.

We get some interesting side-lights on Ferber's contemporaries. Here is what he said about Esnault-Pelterie and Blériot:

> "After the Comte de La Vaulx, comes Esnault Pelterie, who has done something original, having himself invented the motor and devised a very practical method of starting. At the end of each wing he places a wheel, and beneath the centre line two supporting wheels. At starting it leans on one side-wheel, then as the speed increases, it rights itself, and runs on the two central wheels until it rises from the ground. This aeroplane is very graceful, but it seems somewhat lacking in power and longitudinal stability.

The Rebirth of European Aviation 1902-1908

We next come to M. Blériot, who has been an enthusiast for many years, and who has constructed no less than seven machines. One, in 1900, was of the aviplane [*ie* ornithopter] type; the second, in 1905, was built by Voisin, like that of Farman. The third, similar to the previous one, was tried on the Lake d'Enghien, without success, with two Antoinette motors, of twenty-four hp, driven by Voisin. The fourth, similar to Farman's model, was broken at Bagatelle in 1906. It was driven by Peyret, an old soldier, formerly under my command.

Real progress commenced in 1907, because then, instead of employing hired drivers, the inventor commenced to drive the machines himself. I often say, in fact, that the drawings and calculations of a machine are nothing; the difficulty begins with the construction, which necessitates the devising of a multitude of details; but this difficulty even is nothing in comparison with that of learning to manage the machine.

Blériot succeeded in raising himself with No 5 machine, but it lacked stability. With No 6, he first employed 24 hp Antoinette motors, then 50 hp motors, and also raised himself, but the instability remained. One day he travelled 18 kilometres (11.2 miles), but fell from a height of 25 metres (82 feet). However, with No 7 machine, he flies at the present time at the rate of nearly 80 kilometres (50 miles) per hour. [He has the aircraft numbers confused—CHGS.]

Santos Dumont at first was merely a balloonist, but after the considerable sensation caused in 1905 by the Wright brothers—and it may be remembered that this was solely due to the documents which I possessed and published at that time—he felt that the time had come for flying machines. Like all balloonists he started with helicopters, or lifting screws, but it was not long before he saw the great difficulty of the problem and wrote to Levavasseur, who advised him to take up the aeroplane. Nevertheless, he made the mistake of putting in front of the aeroplane that which should have been behind, and it really required all his skill to cover 220 metres with this machine. He did not repeat this performance, and last February his aeroplane got broken on a windy day. Then he abandoned the aeroplane for some time and took up the question of the hydroplane, in the endeavour to realise a speed of 100 kilometres (62.1 miles) per hour on the water. He thus lost time, and when, the other day, he wished to catch up Farman—with a rational aeroplane this time, but too small—he was only able to accomplish very short flights."

Ferber next goes on to describe the history of the Farman and Delagrange machines, and to outline the story of how Gabriel Voisin came into aviation through him (Ferber), which is described in Section 38.

Ferber then tells of how he imitated Lilienthal, and blithely omits any tribute to the Wrights as those who completely changed his path of development:

"You have seen that Voisin trusted at the outset on the ideas and the method which I professed. Whence came these ideas? I had noted the experiences of Lilienthal, and contrary to the majority of people, who believed that he made experiments with a parachute, I realised that they were really flights, and that this man had found a way to learn flying which ought to lead to complete success, for it allowed of comparatively safe experiment and of continual repetitions.

I considered it my duty to repeat these experiments and to popularise them, in order that this great invention might first be born in my own country. I am glad that this has happened, as I hoped."

Only seven months ago, he had gone on record in *L'Aérophile* (for June 1907) in these words about Wilbur Wright:

"Just think, that without this man I would be nothing, for I should not have dared, in 1902, to trust myself on a flimsy fabric if I had not known from his accounts and photographs that it would carry me! Think, that without him, my experiments would not have taken place and I should not have had Voisin as a pupil . . ."

Ferber went on, paying a well-deserved tribute to Pilcher *en route*, and half takes back his implied neglect of the Wrights. His mood was changeable, and he does not quite seem sure of himself, as he brings his survey to a close:

"You have also had in England a man who understood the matter in the way which I did, and who was capable of obtaining the final result. Unfortunately for you, this man met his death in 1899 in an experiment. His name was Pilcher, and it is only right to render him justice.

In America, another man understood the question in the same manner. I refer to Octavius (*cis*) Chanute, and we must also do him justice, for with entire disinterestedness he told all he knew to those who wished to work at the subject. He has trained two remarkable pupils, the brothers Wright, whose works, published up to 1903, have assisted my own progress. You have heard how I borrowed the idea of the lifting-rudder and taught Voisin the manipulation of it. We must also tender them their due, although pitying them for not having understood that there is no secret in a flying machine, and that the skill required to drive it is not worth a million. But we should perhaps excuse them, because, blinded by a great and legitimate pride, they have believed, as they have often told me in letters, that they were ten years ahead of other workers.

However this may be, in 1901 I repeated Lilienthal's experiments, and from the commencement of 1902 the experiments of Chanute and Wright, which have led me to the knowledge of very stable designs, and of rules for management, which I have made public since, in writings and discussions.

It is thus that I have demonstrated by 260 aerial glidings that landing presents no difficulty provided that one can arrive tangentially to the ground, the machine then running along the ground at full speed."

It is particularly interesting to find Ferber here implying that he was a master of gliding flight, whereas he had never built a successful glider, and therefore had never really mastered glider flight in any serious sense.

Finally, we find Ferber saying this:

"I ought to point out, finally, that the cellular form of the heavier-than-air machine, which is used at present, and which was introduced into France by my influence, is a bad solution, because it presents great resistance to forward motion. It is not the French solution—the French solution being that which Pénaud indicated in 1868, and which he exemplified in model form with an indiarubber driving spring. His model is monoplanar (sic), and approaches very nearly to the form of a bird, there being a long tail for steadiness, and rudders behind.

Now that the possibility of solving the problem is demonstrated, we are all going to apply ourselves to the solution which is theoretically the best; but I believe, however, that I did not do wrong in trying first to realise that which already worked well elsewhere, for one must always take what exists in order to improve it afterwards."

Coming from Ferber, this is an all but meaningless statement; for he never in his life took to the monoplane, despite the fact that for a time he was even a member of the Antoinette firm. When he finally abandoned designing aeroplanes, he bought the very last type of machine one would have thought possible, a standard Voisin!

❧ 86 ❧

The Wright Brothers in 1908 (General Survey)

The Wrights did not once leave the ground between October 16th 1905 and May 6th 1908—a period of $2\frac{1}{2}$ years—nor did they allow anyone to view their machine. This astonishing interregnum, which severely retarded the whole development of flying, was due basically to the continued thwarting of the Wrights' demand that any client must guarantee to purchase their machine provided they (the clients) agreed the desired performance, and provided that the machine performed as agreed. But the clients, particularly the US Government, insisted on viewing the machine (or drawings of it) before signing a contract, totally unacceptable conditions to the Wrights, especially in view of the many would-be spies who had heard reports of the 1904 and 1905 flying, and were out to learn all they could.

But this $2\frac{1}{2}$ years' interregnum was not wasted technically, because the brothers built some half-dozen improved engines, and two or three new *Flyers*, pending a satisfactory agreement with their own or a foreign Government; or—failing that—some commercial firm. These machines (Fig 51) may now be called collectively the Wright type A; these include the machines built in 1907 (first used in 1908) and later, along with those built under licence in France, Britain, and Germany in 1909: the Wrights themselves built about seven of these standard machines during the period 1907-09, as well as the somewhat modified one built for the US military trials in 1908: six were to be built under licence by Shorts in England, but the numbers built on the Continent are not yet known. All the machines were similar, with only minor differences—such as the Wilbur and Orville 'sub-systems' of warp and rudder control—and represent not only the culmination of the Wrights' achievement, but the type of Wright machine which was first seen in public, and which directly inspired the last and triumphant phase of world aviation in which the powered aeroplane was to be established as a new and practical vehicle in the year 1909.

The Rebirth of European Aviation 1902-1908

The approximate specification of the type A was:

Wing area: 510 square feet
Chord: 6 feet 6 inches
Elevator area: 70 square feet
Length: 31 feet
Weight (empty): 800 lb

Span: 41 feet
Gap between wings: 6 feet
Rudder area: 23 square feet
Engine: 4 cylinder, 30-40-hp
Speed: 35-40 mph.

This type, similar in general configuration to the 1905 *Flyer III*, was however a two-seater, with upright seating: it still retained the skid undercarriage and derrick-and-rail launching, although it could (and occasionally did) take off from the rail on engine power alone. Also, as before, it was inherently unstable, and it was this feature particularly which struck the Europeans as undesirable in an ordinary 'workaday' aeroplane (see Section 95).

At last, after interminable negotiations, the Wrights signed a contract with the US Army in February 1908, and with a French company the following month. It was decided that Orville would conduct the Army tests, and Wilbur demonstrate in Europe, where (at Le Havre) a *Flyer* already lay crated, having been sent over to France in July of 1907.

To regain their skill, the brothers took the 1905 *Flyer III*, now adapted to take two—both seated upright—to the Kill Devil Hills, and practised there from May 6th to 14th 1908: they made some 20 flights, among which were the world's first two passenger flights, when each brother took up C. W. Furnas on 14th May, the best of these latter flights lasting for 3 minutes 40 seconds.

Wilbur came to France later in the same month, and—after various delays—completed the assembly of the 1907-built *Flyer* in his friend Leon Bollée's factory at Le Mans. Intense and widespread interest, suspicion, and scepticism, was focused on Wilbur by both the French aviators and the Press, following the reports of the Wrights' power-flying in 1904 and 1905. Therefore, when Wilbur announced his first flight in public—he had so much confidence in the machine that he made no secret flight tests—the most critical of technical audiences assembled on August 8th, 1908 at the small local racecourse at Hunaudières (5 miles south of Le Mans).

On that memorable day, Wilbur took off, made two graceful circles, and landed smoothly: he was in the air for only 1 minute 45 seconds; but this was only to test that all was well with the machine.

Between August 8th and 13th Wilbur made nine flights, the longest lasting just over 8 minutes. Then he received permission to use the great military ground, the Camp d'Auvours, 7 miles east of Le Mans. Here, from August 21st to the last day of December 1908, he made over 100 take-offs, and was airborne for about $25\frac{1}{2}$ hours, thus making some 26 hours for the combined French locations that year (see next section).

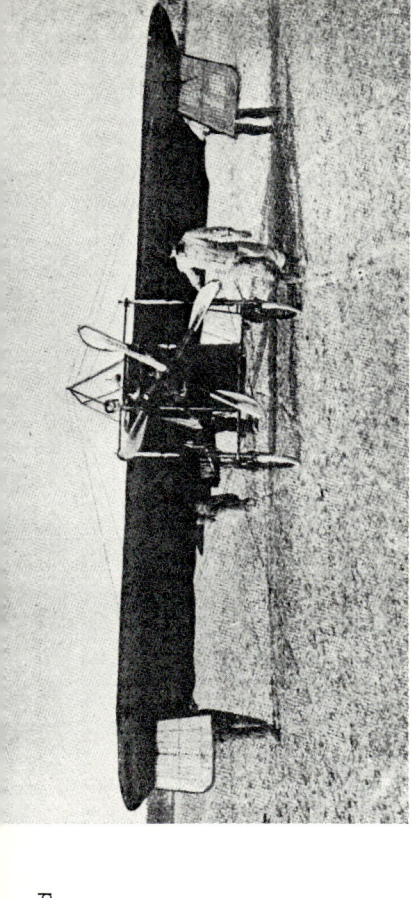

This aircraft had only limited success, but it was the first machine to incorporate large flap-type ailerons, although even now they were only intended to preserve lateral balance. Blériot, in 1909, was to adopt the Wrights' wing-warping on his next, and most successful machine, the XI.

50 Gastambide-Mengin I: 1908 This was the first tentative ancestor of a famous line of *Antoinette* monoplanes, designed and built — along with their engines — by one of the great geniuses of early aviation, Léon Levavasseur.

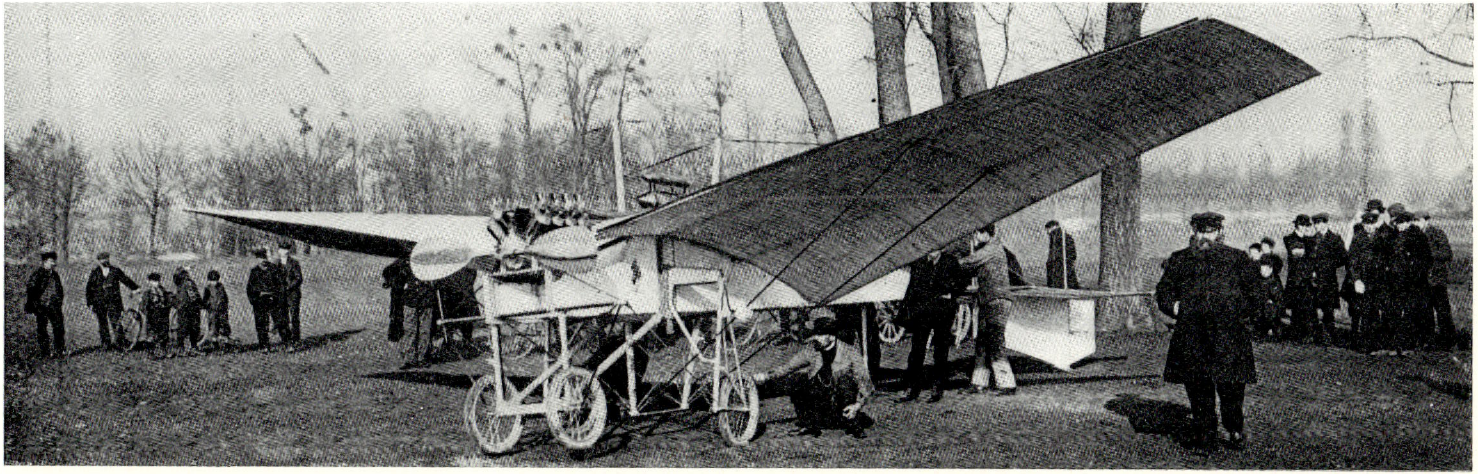

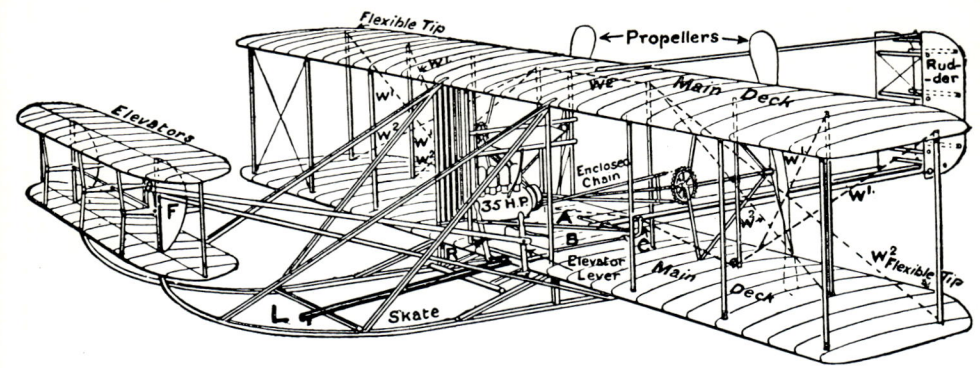

51a Wright Type A: 1908 On August 8th 1908, a Wright machine first flew in public, and, in the words of the Count de La Vaulx, "revolutionised the aviators' world". This was the standard *Wright A (France)*, which was flown by Wilbur first at Hunaudières, and then at Auvours, near Le Mans in France until the end of 1908; then at Pau and at Centocelle (near Rome) in 1909. Note the launching tow-rod 'L' in this contemporary diagrammatic drawing.

51b Wright A (France): 1908 Wilbur is at the controls, flying at the Hunaudières race-course. After his first flight here on August 8th, the critics and sceptics had to admit they had never seen anything to compare with the remarkable way the machine was controlled. They also agreed that the Wrights' claims to having first flown in 1903 were correct, and a chorus of appreciation and praise went up all over Europe. Wilbur's success led to him being offered the nearby military ground at Auvours for further flights.

51c/d Wright A (France): 1908 Two fine photographic studies of Wilbur flying the *Wright A* (*France*) in France. The skids allowed landing on even rough ground; but it also meant that take-offs had to be made along the Wrights' rail system, a point much criticised at the time. But, later on, small wheels were added to the skids, and normal take-offs were made just as easily.

52 Wright A (Fort Myer): 1908 It is here being flown at Fort Myer (near Washington, D.C.) in September 1908, during the U.S. Signal Corps trials. These trials — as successful as Wilbur's in France — were to be abruptly terminated by the crash on September 17th 1908, in which Orville's passenger, Lieutenant T. E. Selfridge was killed, and Orville severely injured. The following year the replacement Wright machine was again flown brilliantly by Orville, and became the first military aeroplane of history.

53 Voisin-Farman I-bis: 1908 After Wilbur Wright's spectacular flying in France in 1908, the shrewder Europeans soon learnt the vital lessons of his consummate flight-control, especially control in roll, and Farman was the first to benefit. Late in 1908 he fitted the large flap-type ailerons, seen here, to his already much modified *Voisin-Farman I-bis* biplane. He was soon to dispose of this machine and order a new aircraft from Gabriel Voisin.

4 Henry Farman III: 1909 This *Henry Farman III* was the first machine wholly designed and built by arman, in 1909. The machine which he had ordered from Voisin was sold behind his back by Gabriel oisin. As a result of the ensuing quarrel, Farman went off to work on his own, and produced this *No III*, hich became one of the great classic aircraft of history.

5 Standard Voisin Biplane: 1909 This was evolved from the Voisin machines of 1907 and 1908. It as an inherently stable box-kite biplane, with side-curtains on the wings to aid lateral stability. But though safe to learn on, and to fly, in fine weather, it was an essentially backward-looking aircraft. Its orst characteristic, which Voisin stubbornly refused to rectify, was the lack of any control in roll, despite e example set by the Wrights. The Voisin had to be yawed round, when turning, and was always in anger of side-slipping if the resulting bank could not be corrected by opposite rudder. But there were any men — and a few women — who experienced their first flights on these near-obsolete biplanes, and ey played an important part in early European aviation, but a part soon to be totally eclipsed by the new *enry Farman III* illustrated above.

56 Antoinette IV: 1909 The most graceful of all early aeroplanes were the *Antoinette* monoplanes 1908-09 named after Mlle Antoinette Gastambide, daughter of the head of the firm. The *Antoinettes* we designed by Léon Levavasseur, who also created the equally famous *Antoinette* engines. This is t *Antoinette IV,* fitted with ailerons.

57 Antoinette VI: 1909 Seen here is the *Antoinette VI,* with wing-warping for control in roll. Levavasse built all his later machines with warping, which he copied from the Wrights' system. The most famo *Antoinette* pilot was the half-English Frenchman Hubert Latham, who twice attempted to cross the Engli Channel in 1909, and twice failed because of engine failure.

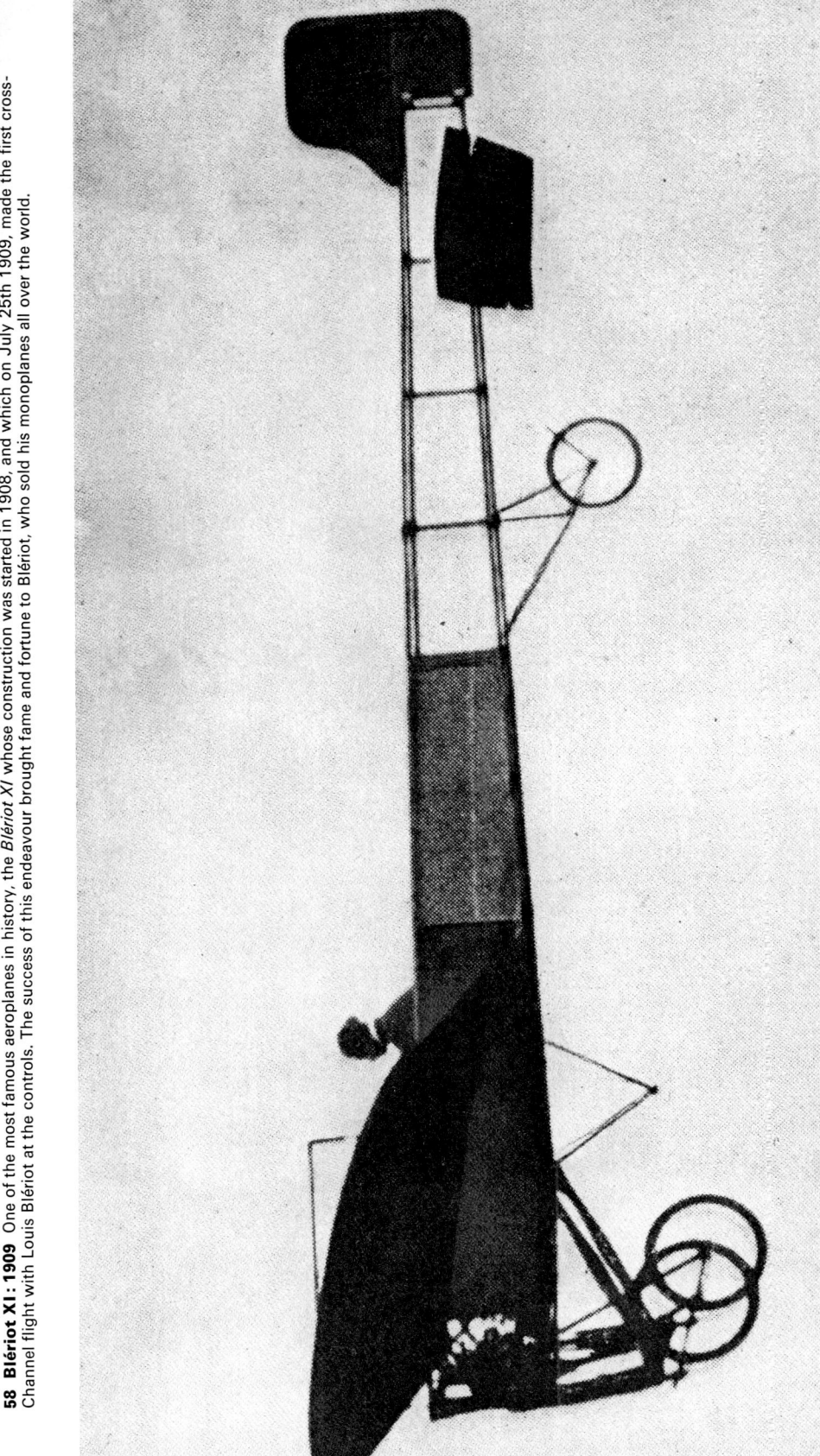

58 Blériot XI: 1909 One of the most famous aeroplanes in history, the *Blériot XI* whose construction was started in 1908, and which on July 25th 1909, made the first cross-Channel flight with Louis Blériot at the controls. The success of this endeavour brought fame and fortune to Blériot, who sold his monoplanes all over the world.

59 Wright A (Lefebvre) : 1909
Seen at the Reims meeting, in August 1909. This was the standard type of Wright biplane which the brothers themselves first flew in 1908, and which was also built under licence in France, Germany and Britain. Lefebvre taught himself to fly his Wright machine, and was one of the most successful pilots of these famous aircraft. The ability to manoeuvre is well shown by this photograph of the biplane in a steep bank around one of the pylons at Reims. All Wright machines at this time were built to be inherently unstable, and had to be flown by the pilot the whole time; they were exceptionally sensitive to their controls.

87

Wilbur Wright's Flights in France in 1908

Here is an abbreviated record of what Wilbur Wright accomplished in France in 1908:

Total number of take-offs: over 100

Total number of hours airborne: over 26 hours

Number of times when passengers were taken: about 60

Total time airborne with passengers: over 8 hours

Number of flights of between $\frac{1}{4}$ and $\frac{1}{2}$ hour duration: 14

Number of flights of between $\frac{1}{2}$ and 1 hour duration: 6

Number of flights of over 1 hour (and below 2 hours) duration: 6

Number of flights of over 2 hours duration: 1

A table of Wilbur's flights lasting for $\frac{1}{4}$ hour and over is given below: the letter 'P' stands for a passenger flight.

WILBUR WRIGHT'S BEST FLIGHTS AT AUVOURS IN 1908[1]

1908	Hours	Minutes	Seconds	Km	Miles		Mph
September							
5		19	48				
10		21	43				
16		39	18				
17		32	47				
21	1	31	25	66.6	41.4		27
24		54	3	55	34.1		38
25		36	14				
28	1	7	24	48.12	29.9		27
October							
3		18	23				
3		55	37	56	34.7	P	37
6	1	4	26	70	43.4	P	40
10	1	9	45	75	46	P	40
28		15	2			P	
29		17	34			P	
29		19	25			P	
30		15				P	
November							
10		15				P	
11		15				P	
11		20				P	
16		19				P	
16		21				P	
18		19				P	
December							
4		25				P	
18	1	54	53	99.8	62		32.5
30	1	52	40	96.8	60.1		32
31		42					
31	2	20	23	124.7	77.5		33

[1] These figures—or lack of them—are taken from the French records.

⚔ 88 ⚔

Orville Wright's Flights at Fort Myer: September 1908

When Orville Wright embarked on the acceptance tests for the US Army in September 1908, he flew—on another Type A—as spectacularly as his brother in France (Fig 52): the tragedy that ended these trials did nothing to diminish his triumph. Wilbur wrote to their sister: "Tell Bubbo (Orville) that his flights have revolutionised the world's beliefs regarding the practicability of flight. Even such conservative papers as the London *Times* devote leading editorials to his work and accept human flight as a thing to be regarded as a normal feature of the world's future life."

Supervised by officers of the US Signal Corps, Orville started the tests at Fort Myer (near Washington, DC) on September 3rd 1908, and crashed—killing his passenger Lieutenant T. E. Selfridge—on September 17th.

The fatal crash—the first in powered aviation—happened at 5 pm on September 17th. The machine was of the standard Wright type A, with 'Orville' controls. Orville had taken up Lieutenant Selfridge as an official passenger, and was making the fourth round of the field when the trouble struck. A blade of the starboard propeller developed a longitudinal crack, which caused it to flatten and lose its thrust, thereby setting up an imbalance with the good blade: the consequent violent vibration loosened the supports of the propeller's long shaft, causing the latter to 'wave' to and fro, and thus enlarge the propeller disc: the good blade then hit, and tore loose, one of four wires bracing the rudder outrigger to the wings, the wire winding itself round the blade and breaking it (the blade) off. Orville cut the motor, and tried to land; but the rudder canted over and sent the machine out of control: even then, in the ensuing nosedive, Orville was succeeding in bringing up the nose with the elevator; but it was too slow, and the *Flyer* crashed, killing Selfridge. Orville was seriously injured, but made a good recovery.

The death of Selfridge did nothing to deter the United States Army's interest: having seen the outstanding performances Orville put up, they were determined to go ahead as soon as Orville had recovered.

Total number of take-offs: 14

Total number of hours airborne: over 6 hours

Number of flights of one hour's duration: 4

The following is a table of all the flights made by Orville at Fort Myer: the letter 'P' denotes a passenger flight.

Flight	1908	Hours	Minutes	Seconds	Km	Miles		Circuits
	September							
1	3		1	11				
2	4		4	15	4.8	3		
3	7			55	1.4	⅞		
4	8		11	10				13
5	8		7	34				8
6	9		57	31				57
7	9	1	2	15				55
8	9		6	24			P	6½
9	10	1	5	52				58
10	11		10	50	11.44	7		9
11	11	1	10	24				57½
12	12		9	6			P	
13	12	1	14	20				71
14	17		?	?	ACCIDENT		P	4½

❧ 89 ☙

The Wright Brothers and aerial Records

In view of the many uncomplimentary remarks made by the Europeans about Wright aeroplanes and their practicality, and so on, it might be of interest to look at the not unimportant matter of flight records.

FLIGHT DURATION

On October 5th 1905, Wilbur Wright flew for 38 minutes 4 seconds at the Huffman Prairie. This duration was not exceeded by a non-Wright machine for three years, *ie* until October 2nd 1908, when Henri Farman flew for 44 minutes 32 seconds at Châlons.

But on September 9th 1908, Orville Wright had already been the first in history to fly for an hour: this he did at Fort Myer (USA), his time being 1 hour 2 minutes 15 seconds. Then, a fortnight later, on September 21st 1908, Wilbur flew for 1 hour 31 minutes 25 seconds at Auvours.

Then, on December 31st 1908, Wilbur performed the spectacular feat of flying at Auvours for 2 hours 20 minutes 23 seconds.

The first Frenchman to stay up for an hour was one of Wilbur's pupils, Paul Tissandier, who, on May 20th 1909—at Pont-Lond (near Pau)—flew a French-built *Wright A* for 1 hour 2 minutes, and covered 57.5 kilometres. This was beaten by Roger Sommer when he flew a Farman biplane for 1 hour 50 minutes 30 seconds at Châlons, on August 1st 1909.

The Wrights' records were not beaten until August 7th, 1909, when Sommer flew on a Sommer, at Mourmelon, for 2 hours 27 minutes 15 seconds.

ALTITUDE

On September 12th 1908 Orville Wright flew to 200 feet at Fort Myer. Then, on December 18th 1908, Wilbur flew to 360 feet at Auvours. Neither of these heights was reached by a non-Wright machine until

July 15th 1909, when Paulhan reached 357 feet at Douai in a Voisin. On October 4th 1909, Orville then reached 1,600 feet at Berlin, which was beaten on January 7th 1910 by Latham, when he flew to 3,444 feet at Châlons.

PASSENGER-CARRYING

On May 14th 1908, Orville Wright carried C. W. Furnas as passenger for 3 minutes 40 seconds at the Kill Devil Hills, covering 2½ miles. This modest duration was not exceeded by a non-Wright machine until February 14th 1909, when Henri Farman carried a passenger for just over 3 miles at Mourmelon. But long before that, on October 6th 1908, Wilbur had carried M. A. Fordyce for 1 hour 4 minutes 26 seconds, covering 70 kilometres (43.4 miles); then, next week, on October 10th 1908, he had taken M. Painlevé for 1 hour 9 minutes 45 seconds, covering about 75 kilometres (say 46 miles). These records were first improved upon by Orville Wright, who, on September 18th 1909, took Captain Engelhardt for 1 hour 35 minutes 47 seconds at Berlin. These Wright passenger flights were not exceeded until January 31st 1910, when Van den Born, flying a Farman biplane took a passenger for 1 hour 48 minutes 50 seconds at Châlons.

❧ 90 ❧

The Press Coverage of Wilbur's Flights

After the long delays in getting his machine ready, Wilbur made his first flight on Saturday August 8th 1908 at the small Hunaudières race-course, near Le Mans. The reception given, even to this first short trial flight—the first time Wilbur had ever taken this machine off the ground—can only be described as wildly enthusiastic. As Wilbur continued to make his flights, and especially after he was allowed to use the great military ground at Auvours, a few miles away, the reports continued in a paean of praise.

What chiefly struck everyone present was the mastery of manoeuvre which Wilbur displayed. No one, of course, had ever seen—or ever imagined—an aeroplane being handled like this; it will be remembered that it was only in January of this same year that Farman had made the first official circle in an aeroplane, a performance which itself was greeted with widespread enthusiasm. To see Wilbur Wright making circles as if they were part and parcel of his everyday flying—which they were—struck the spectators as astounding. There soon came to be one universally applied analogy—bird flight. Wilbur and Orville became known everywhere as the first 'bird-men' (les premiers hommes-oiseaux), and these words formed the title of a book on the brothers which was speedily produced by the great aviation journalist François Peyrey.

This is how readers of *The Times* (London) first heard of the event. The account appeared in the Monday (August 10th) issue and started:

"The Special Correspondent of *The Times* at Le Mans telegraphed late last night:

'Mr Wilbur Wright has made a remarkable flight this evening, lasting 1 minute 45 seconds, over a course of about 2,500 feet. He will resume his experiments on Monday. The average height maintained during to-day's flight was 30 feet.'

The news of this remarkable achievement, which took place in the presence of some of the leading members of the Aéro-Club, well-known aviators like M. Blériot, and aeronauts like M. Archdeacon, MM. Paul and Edmond Zens, and M. Peyrey, has been received with enthusiasm in the French Press. Such secrecy had been maintained with regard to the Wright aeroplane that a large number of Frenchmen were sceptical even as to Mr Wright's seriousness. All accounts, however, published in this morning's paper from the correspondents on the spot attest the complete triumph of the American inventor. All present affirm that, after yesterday's experiments, there can be no doubt that the Wrights possess a machine capable of remaining an hour in the air and almost as manageable as if it were a small toy held in the hand.

It was at half-past six that the flight took place. At the very first bound, obeying the handling of its pilot, the aeroplane rose 'stable, harmonius, and superb,' and, rising to some 30 feet or 40 feet twice without a hitch, glided round the Hippodrome, finally alighting gently, with the ease and grace of a wood-pigeon, to use the words of one of the correspondents, some 50 feet from its point of departure. 'Thereupon the enthusiasm was indescribable. The Frenchmen and the Americans present received Mr Wright, who had just won for his brother and himself the title of the real creator of aeroplanes, with the most extraordinary enthusiasm.' . . .

All accounts agree that the most admirable characteristic of yesterday's flight was the steady mastery displayed by Mr Wright over his machine. It is recalled that he and his brother are the sole constructors of this admirable apparatus, including the motor. . . ."

The Times, in the same issue, ran an Editorial, which commenced:

"Immense enthusiasm has been called forth in France by the successful flight achieved on Saturday by Mr Wilbur Wright at Le Mans. . . ."

In its issue of August 14th, *The Times* quoted the following passages by François Peyrey:

"These experiments were really remarkable. They proved over and over again that Wilbur and Orville Wright have long mastered the art of artificial flight. They are the public justification of the performances which the American aviators announced in 1904 and 1905, and they give them, conclusively, the first place in the history of flying machines, that rightly belongs to them. It was at nightfall on August 8 that I saw Wilbur Wright make his first flight. He had made no flights for some months, and yet his first experiment began with the most delicate of all manoeuvres in aviation—namely, circling. He rose forthwith to a height of about 30 feet, and the spectacle was marvellous and delightful. We beheld the great white bird soar above the race-course, pass over and beyond the trees from its shed to the winning-post of the course. We were able to follow easily each movement of the pilot, note his extraordinary proficiency in the flying business, perceive the curious warping of the wings in the process of circling and the shifting position of the rudders. When after 1 minute 45 seconds of flight Wright again touched the ground, descending with extraordinary buoyancy and precision, while cheers arose from the crowd in the tribune, I saw the man who is said to be so unemotional turn pale. He had long suffered in silence; he was conscious that the world no longer doubted his achievements. . . . The method of operation seems very simple. The most interesting feature consists in the 'working' of the extreme under part of the wings, whereby the flight of a bird is imitated and perfect lateral stability is secured. . . . Mr Wright has realised the most delicate problem of aviation—namely, the question of balance. To behold this flying machine turn sharp round at the end of the wood at a height of 60 feet, and continue on its course, is an enchanting spectacle. The wind does not seem to trouble him, Wright having flown in fairly stiff breezes. In a word, the Wright brothers are the first men who have succeeded in imitating birds. To deny it would be childish."

But it took the popular London *Daily Mirror* to voice the general view in less restrained, and more colourful, language, in its issue of August 13th:

"THE MOST WONDERFUL FLYING-MACHINE THAT HAS EVER BEEN MADE. . . . No such perfect control over a flying-machine has ever been known as that exercised by Mr Wilbur Wright, the young American aeronaut, over the aeroplane in which he has for the past few days been accomplishing some marvellous flights at Le Mans, in France. In these trials the aeroplane has travelled speedily and easily round a race-course at the command of its pilot, has executed figures of eight in the air, has sailed, soared, and swept round in loops and circles with the consummate ease and grace of a swallow. In the words of a spectator at Tuesday's trial flight, other aeronauts are 'babies by comparison; the American is their master'. "

But the French newspapers were also exuberant, as already seen in Peyrey's piece, and they amply made up for the campaign of criticism which had preceded August 8th. Here is Frantz Reichel in *Le Figaro* for August 11th:

"I've seen him; I've seen them! Yes! I have today seen Wilbur Wright and his great white bird, the beautiful mechanical bird which, for eight years, in the solitude of Virginia and Carolina, has accomplished so many prodigious exploits that, in the course of time, one has ended by believing that it was only a 'canard'.

But today, no longer; there is no doubt! Wilbur and Orville

Wright have well and truly flown.... Without wishing in any way to diminish the value of what our aviators—Blériot, Farman, Delagrange, Esnault-Pelterie, Zens, Mangin, Gastambide and others—have done in France, one is obliged to recognise that there is a whole world of difference between their machines and the Wrights' (il y a tout un monde entre leurs appareils et celui des Wright).

To speak only of the two best known, the cellular machines (les cellulaires) of Farman and Delagrange—whose official performances have revolutionised world opinion—are, in comparison with the Wright aeroplane, ingenious 'instruments', without doubt, but rudimentary, in spite of a complicated construction; and complicated because it was not known how to solve, simply and mechanically, the different problems that have arisen.

If we have the truth about monoplanes (si, dans les monoplanes, nous sommes dans la vérité), we are in error over multiplanes. Of that, there is every evidence ...

Moreover, it's enough simply to have seen the Wright machine in repose in order to appreciate it immediately; it is enough to have seen it fly, just once, to be convinced by it. The other day I gave the dimensions and particulars. But what I did not furnish was its robust appearance, its cleanness, the purity of its lines, the impressive simplicity of its controls—reduced to a minimum—the admirable mechanical conception, and the execution no less remarkable. I was impressed (frappé), surprised, abashed; and each of these impressions can be expressed in the cry—'How simple it is!' ... And, one adds—'How handsome it is!' ...

At seven-twenty ... the departure lithe and immediate. The aeroplane flies very steadily towards the end of the Hippodrome, banks, and with a soaring motion of incomparable beauty, turns to the right, a thrilling white spot which stands out against the black curtain of trees and the purple sunset; returning towards us diagonally, and passing us at 70 kilometres an hour, it heads towards the other end of the Hippodrome, turns again above the tops of the pine trees, and, having completed a figure of eight, comes in smoothly, and lands gently, with an unbelievable precision, between the pylon and a delapidated grand-stand which we thought it was going to hit.

Nothing can give an idea of the emotion experienced, and the impression felt, at this last flight, a flight of masterly assurance and incomparable elegance."

The issue of *L'Aérophile* for August 15th, stressed the manoeuverability of the machine:

"But the facility with which the machine flies, and the dexterity with which the aviator gave proof from the first, in his manoeuvering, have completely dissipated all doubts. Not one of the former detractors of the Wrights dare question, today, the previous experiments of the men who were truly the first to fly ..."

Then, throughout its reporting of Wilbur's and Orville's flights, we find a series of rapturous descriptions in *L'Aérophile*, such as:

§ "un merveilleux succès d'Orville Wright"
§ "son vol admirable"
§ "... une ovation frénétique..."
§ "ses merveilleuses performances"
§ "une sureté et une rapidité prodigiuses";
§ "des performances sensationnelles des Wright";
§ "les prouesses de Wilbur Wright";
§ "journées glorieuses".

⚛ 91 ⚛

French Recognition and Repentance : 1908

One of the most interesting results of Wilbur's triumphant flights in France, was the admission by the French—or rather by the unprejudiced French—that the Wrights had obviously done in the past what they said they had done; and were, in fact, the first in the world to fly. François Peyrey, the most respected aviation journalist in France—and highly praised by Archdeacon, the leader of the anti-Wright faction—wrote in *L'Auto* (August 9th 1908):

> "For a long time, for too long a time, the Wright brothers have been accused in Europe of bluff—even perhaps in the land of their birth. They are today hallowed by France, and I feel an intense pleasure in counting myself among the first to make amends for that flagrant injustice."

Then, in his book *Les Premiers Hommes-Oiseaux* (1908), he wrote:

> "It would also be quite as puerile to contest the first flight on December 17th 1903 in North Carolina . . . as it would be to deny the demonstrations in the Sarthe [*ie* near Le Mans]."

Here is *L'Aérophile*, in its issue of August 15th, 1908:

> "But the facility with which the machine flies, and the dexterity with which the aviator gave proof, at the first go off, in his manoeuvering, have completely dissipated all doubts. Not one of the former detractors of the Wrights dare question, today, the previous experiments of the men who were truly the first to fly . . . At last everybody renders justice to two admirable pioneers of the science, which has been debated for too long. It is for *L'Aérophile* a matter of profound pride to have been the first, in 1903, and later, to reveal the decisive importance of their work, and to have hastened the dawn of their glory."

Even the cautious Delagrange, who weighed his words very carefully after the first excitement, said in his article in *L'Illustration* (August 15th 1908):

"The man [Wilbur], one must honestly recognise, is indeed the father of aviation (L'homme, il faut le reconnaitre loyalement, est bien le père de l'aviation)."

Here is the Comte de La Vaulx (*op. cit.*) who wrote:

"The man who had just manoeuvred in the air with that facility and virtuosity, who had landed with that light grace, was not on his first test; he was a master; he had known how to fly for a long time; since the date he had announced; since 1903! . . . the Wrights keep the incomparable glory of an undeniable priority in the creation of a true flying machine, . . ."
"The moment has come to reveal, at least briefly, the characteristics of this machine which has just revolutionised the aviators' world."

And finally, here is Charles Dollfus:

"Wilbur Wright came to France, and gave at the Hunaudières racecourse, and then at the Camp d'Auvours—in the neighbourhood of Le Mans—a series of flights which were sensational, and constitute a decisive epoch of history. The first flight took place on August 8th 1908. Numerous flights followed, revolutionising aviation by the excellence of the pilotage, the manageableness, and the versatility of the machine."

92

First and second Thoughts of the French Pioneers

There was an excellent turn-out of French pioneers to see Wilbur's first flight on August 8th 1908, and among those present were: Archdeacon (president of the Aviation Committee), Hart O. Berg (representing the interests of the Wrights), Leon Bollée, Louis Blériot, François Peyrey, René Gasnier, Ernest Zens, Paul Zens, Pierre Gasnier, Robert Guérin, Dickins, Captain Zazerac de Forge, de Moy, and two Russian officers.

As may well be imagined, the anti-Wright faction, led by Archdeacon and the Voisin brothers, were glum about Wilbur's flights, and maintained that the French aircraft were just as good as the Wrights'. The *New York Herald* interviewed a number of well-known French pioneers, and this is what Archdeacon is reported as saying, in the issue of August 9th:

> "But I think that we can do as well with our machines in France. It must take a long time to learn to fly with the Wright machine. Also I consider our machines superior in the fact that they have wheels and that they can start wherever they may descend without the help of rails."

The Voisin brothers had a lot to say, and are noted separately in Section 93.

Meanwhile, three other well-known pioneers went on record in the *Herald* as follows:

Louis Blériot:

"I consider that for us in France, and everywhere, a new era in mechanical flight has commenced, I am not sufficiently calm after the event to thoroughly express my opinion. My view can be best conveyed in the words, 'It is marvellous!'"

René Gasnier:

"It is a revelation in aeroplane work. Who now can doubt that the Wrights have done all they claim? My enthusiasm is unbounded. The whole conception of the machine, its execution and its practical worth is wonderful. We are as children as compared with the Wrights."

Paul Zens:

"I would have waited ten times as long to see what I have seen to-day. Mr Wright has us all in his hands. What he does not know is not worth knowing. This machine proves that travel by aerial means is at hand. It would give me immense pleasure to go out with Mr Wright at any time."

In its issue of August 14th, the *New York Herald* carried another, and much longer, statement from one of the best known aeronautical figures of the day, the airship designer, M. Surcouf, who was reported as follows:

"It is marvellous! Wright is a titanic genius. The flights of this morning demonstrate his enormous lead over the rest of the world. I don't know when to admire him most—when in the air or when landing. The little accident is nothing.[1] It serves to show that the aeroplane is not a dangerous toy. Mr Wright's control is beyond all praise. Of course accidents will happen. They happen to automobiles and tramcars and every form of locomotion. If Mr Wright could escape the common lot in these things he would not be human. I say this all the more freely, inasmuch as I had always doubted the truth of the Wrights' claims. To-day I am so convinced of their genius that if I heard to-morrow, or at any time, that they had fulfilled their contract I should not even take the trouble to verify the statement. They can do the fifty kilometres when they like. Mr Wright has solved the problem of flight. . . . Mr Wright is as superb in his accidents as in his flights. He has shown that even when the motor stops there is very little danger, because he can glide down so smoothly. The broken wing is the punctured tire of the automobile. It is impossible to avoid such incidents. There is no danger in them. You see how easily he kept his seat through it all. The machine is beyond criticism—that is to say, I defy anyone for the moment to say how it can possibly be improved."

Léon Delagrange, although he was to cool down a little, and itemise various points—the usual ones—in which he found the Voisin machines

---

[1] The accident referred to took place on August 13th. Wilbur wrote: "I made a blunder in landing and broke three spars and all but one or two ribs in the left wings, and three spar ends of the central section, and one skid runner. It was a pretty bad smashup, but Kapferer, who was present, pronounced it as fine a demonstration of the practicability of flying as the flights themselves. It did not shake me up a particle".

superior, was one of the greatest admirers of the Wrights and their machines. He could not be present for Wilbur's first flight on August 8th, but he saw him fly on the 10th. Next morning, before he returned to Paris, a *Matin* reporter managed to have a word with him. The reporter wrote as follows:

> "Before his departure, he did not hide his admiration. 'We are beaten (nous sommes battus)' he declared this morning." (*Matin*, August 12th).

Although he virtually took back this particular remark later on, it was Delagrange who made some of the most flattering remarks on record about Wilbur and his aeroplane, as we shall see.

Delagrange was still very enthusiastic when interviewed by the newspaper *Siegl*, which published his comments. He is reported as saying to their representative:

> "I was truly amazed at what I have just seen at Le Mans, whence I arrived only a few hours ago, and I am trying to put my ideas in order for an article."

He went on to say that he had fully expected a very interesting machine to be produced by the Wrights,

> "However, what I have seen, I confess, has surpassed what I had expected to see. The Wright machine is the most astounding (stupéfiante) thing one can imagine from the point of view of simplicity. Everything is simple: . . ."

When asked why he thought the Wright machine was superior to the French, he replied:

> "In my opinion, what makes the machine superior is a better utilisation of the wings, the warping of the wings, . . . which gives the machine a stability which is a little lacking in ours."

Delagrange points out that the Wright engine is not the light motor that it is said to be, as it weighed just over three kilogrammes per horse power; but, he said,

> "their engine has one superiority, in that it possesses a reliability which is never known with ours. . . . The greatest difference which exists between the American aeroplane and our aeroplanes is in their propellers, of which there are two; but they have a much greater surface than those on our machines: they give twelve metres of effective pitch(?) (d'appui) in the air, as opposed to about three with ours."

Delagrange was not worried about the Wrights' rail and pylon launching; and when he came to describe the Wright machine and its flying, he was all but ecstatic:

> "If the machine is already amazing to look at in repose, especially when one knows it is an aeroplane, it is still more amazing (stupéfiante) to see it manoeuvre. I have no words to express my admiration. . . . it manoeuvres with a marvellous facility (souplesse). Its turns are particularly extraordinary. . . . At other times it accomplished repeated figures of eight with incomparable ease. It is marvellous, I assure you, marvellous. The Wright aeroplane lands like a bird, with the same gentleness, as it slides and comes to a stop with an unparalleled facility."

After making even more complimentary remarks about the Wright machine, Delagrange ended by saying:

> "Ah! If I had an aeroplane like that, I would long ago have beaten all the records, and carried off all the prizes offered . . ."

When he came to write his article in *L'Illustration* (issue of August 15th) he waited till near the end, and then asked: "Are we beaten? Must we, as has been said, acknowledge defeat?" And he answered himself with the words "Never! Quite the contrary! On many points we are well in advance." After all, he was a Frenchman!

When Henri Farman was interviewed by *Le Figaro*, and reported in their issue of August 26th, he said:

> "I believe that our machines are as good as his, . . ."

Then he went on to say that he thought that the Wright engine was superior—"he has solved the problem of the engine"—and he also conceded that the French machines did not "obey the hand" like the Wright, and that they had not sufficient "docility". Then he went on to compliment Wilbur, not on his lateral control, but on his rudder and elevator control, and on the sureness and authority of his flight-control in general. However, he said, "we possess, in compensation, and thanks to our 'cellule d'arrière', a greater stability".

But Farman was to be the one who, par excellence, was to read, mark, learn, and inwardly digest what he saw Wilbur perform in the air; and then to set out to rival him on his own ground, as we shall see later.

Blériot, having recovered from his spontaneous praise of the Wright machine, went on record in the paper *Siegl* for August 8th 1908, as expressing the view that the Wright machine possessed an undeniable superiority over the French, but only "a momentary superiority". He admired the large slow-turning Wright propellers, but criticised the propeller bearings, saying that he doubted whether the machine would be able to manage 50 kilometres without incident. He need not have worried, as Wilbur made eight flights of well over 50 kilometres, including two of over 90, and a third of 124 kilometres. Blériot also commented that the French machines were better finished (soignée) and more robust, a common observation at the time. He did not appear to have commented on the Wrights' flight-control; but he too, was to study the Wright *Flyer* very carefully, and take over its wing-warping technique in its entirety next year.

It was to be expected that the Voisin brothers Gabriel and Charles, would be sourness personified! They wrote a jointly signed letter to *Le Matin*, which published it in their issue of September 5th 1908. In effect, the letter was a somewhat pompous homily. The Editor headed it:

> "Where was aviation born? IN FRANCE. The Voisin Brothers defend the cause of our national genius."

The writers started by saying they had read in the *Matin* of August 31st, "the history of Wilbur Wright, narrated by himself". Then:

> "Without wishing to diminish at all the merit of the aviators of Dayton, we permit ourselves to make the observation that French aviation was not born uniquely by their experiments; and that if we have derived some information—moreover very little—from their tests, they have also profited from French genius in large measure.
>
> The history of mechanical flight started, in reality, from the researches of Mouillard and Pénaud. In that epoch, Tatin constructed, in model form, the first flying machine which, driven by a compressed-air motor, left the ground under its own power. . . .
>
> Research, since that epoch, has been going on ceaselessly."

They then go doggedly through the history of aviation, noting Langley's successful steam-driven model, Maxim's huge machine, and Lilienthal's "sensational experiments" which "he accomplished ten years before the Wrights, on a motorless biplane . . ." Lilienthal was "incontestably, the undisputed originator (promoteur) of gliding".

The Voisins then arrive at Ader, and with fulsome praise, relate first the undoubted free take-off he made in his *Éole* in 1890, before describing the Satory tests of *Avion III* in 1897 which, they rightly say, took place before a control committee. The Voisins then say that the *Avion III* flew for 200 metres; Ader himself claimed 300. Unfortunately, all this was proved to be a deliberate fabrication of Ader's when the official report of the military control committee was published for the first time in 1910, for it showed that the *Avion III* had not flown for a single yard! But the Voisins, like so many others, had swallowed Aders' claim to have flown. Then, giving the wrong date (it should be 1897) the Voisins continue:

> "The victory, the 'première' of aviation, which took place in 1898 [*ie* 1897], is a French victory, and it is as such that we wish to make it known to the public at large through the agency of the *Matin*.
>
> We have often made the pilgrimage to the Museum of Arts and Crafts, where one can see the glorious ancestor of the aeroplane of today, and we are certain that, even now, the *Avion* would fly like the Wright machine if it were to be flown by the experienced pilot of Le Mans (*ie* Wilbur).
>
> Aviation was born in France; it was in France that Chanute, the great 'precurseur' of the Wrights came to absorb the data about that admirable machine (*ie* the *Avion III*) from which the aviators of Dayton had learnt to fly.
>
> It was in France that Ader accomplished the first mechanical flight, and it is still France which leads the present scientific movement, with claims that cannot be disputed."

Suffice it to say that neither Chanute nor the Wrights did, or ever could have, learnt a thing from the *Avion III*, whose control system was so primitive that it was a mercy for Ader that the machine did not leave the ground. Incidentally, the *Avion III* had no roll-control at all. The whole sad story is discussed at length in my Science Museum monograph *Clément Ader, his Flight-Claims and his Place in History* (see Bibliography). The Voisin account goes on:

> "Wilbur Wright declared, in the course of the *Matin* article, that M. Archdeacon 'had asked two young mechanics, the brothers Voisin, to build him a machine of the Wright type.' There is here a slight twisting of history.
>
> M. Archdeacon, in fact, had had made by M. Dargent a 'modeleur', an aeroplane of the Wright type, of which drawings had appeared in *L'Aérophile*.
>
> At this time, we did not know M. Archdeacon; and it was at the moment when the experiments were started that we got to know him.
>
> We tested the machine in question, under the direction of Captain Ferber, while M. Esnault-Pelterie was making, on his part, analagous tests with a similar machine.
>
> The results were so little encouraging that the type of warping biplane was abandoned by both [experimenters] after two months.
>
> The first machine which we designed and built on M. Archdeacon's account was already fitted with a rear 'cellule', and it is the same [type of] machine which Farman at first—and then Delagrange— have piloted to victory, establishing official records which have never been beaten.

> We have followed a different road to that of the Wrights, and nothing in our machine can be related to that being tested at Le Mans. The principles and the forms differ in the most absolute sense."

They go on to say how they started in aviation, and how they first knew of the Wrights in 1904; then how they tested the Archdeacon float-glider on the Seine, which was already their definitive machine. Then back to the Wrights:

> "We have each created a school. The future will teach us which is the best.
> One is based on the equilibrium being kept by movements of the pilot; the other on the equilibrium obtained by the form of the machine itself. Take two gliders, a Wright and a Voisin. Release them from the height of the Eiffel Tower.
> The Wright, deprived of its pilot, will somersault in space, and smash itself on the ground.
> Ours will descend normally, retaining the attitude (position) for which it has been designed and built.
> Aviation is a French discovery, in both cases, from its origins to its final blossoming.
> The first mechanical bird which was able to leave the ground was French in conception and construction. The *Avion* of Ader, we repeat, had, as with the Wright machine, all the necessary controls for its equilibrium; it only lacked a brain, and the aviators of Dayton had taught us that eight years was needed to learn the necessary movements to manage such a machine.
> We were the only people to defend the Wrights up till the time of their experiments. But we also declare that French aviation still remains victorious, and it is to us, the French—to us alone—to whom belongs the glory of having created the first aeroplane in history which was able to fly."

Comment on this strange and rather pathetic document would be superfluous. The Voisins were living in a world of their own, and drawing their skirts ever tighter around themselves to protect them—and the French nation, so they thought—from the encroachment of foreigners.

93

Gabriel Voisin on Wilbur Wright in France: 1908

Gabriel Voisin's contempt for the Wrights knew no bounds when he came to the subject of Wilbur's visit to France in 1908: Here are some particularly choice quotations from his autobiography, to show to what lengths he was prepared to go (his italics):

"When this same machine, crowned for fifty years with the attributes of success, came to France on 8 August 1908 to do its first test there, at the Auvours camp near Le Mans, *it did not fly* although it was launched with a plethora of pylons and counterweights such as we had never used. Its first test ended inadvertently after one minute and forty seconds! On 6th July 1908, a month earlier, Delagrange, in a Voisin, had flown 18 kilometres (11 miles)! And we were on the point of making the first city-to-city flight which took place on 30th October 1908. . . . After the miserable demonstration in France on 8th August 1908, the buyers, who were not willing to fool about, had a French engine fitted to the Wright, and the Wright, artificially launched, flew at last. But its performance was immediately outdone by French performances and it went back to America." (page 241)

"At the moment when his boat left the shores of France—France which had welcomed him with an almost unbelievable enthusiasm—Wilbur Wright, had he been able to peer through the Channel fog, would have distinguished the shadow of the Blériot XII [*ie* XI], the aircraft which was to be carried to England on the wings of victory. Then he might have realised the uselessness of his efforts, the poverty of his devices and the futility of his secrets." (page 238)

"No technician of real standing can admit that the Wrights inspired anything at all, and that for two reasons. The first is important: the Wrights kept their secret so well that it remained impenetrable from 1903 to 8 August 1908, a date by which French aviation was definitely under way. The second reason also has its significance: *the Wright aircraft had no future.* When the Wrights came to France, financed by influential people and with resources we could never

The Rebirth of European Aviation 1902-1908

have hoped for, praised to the skies by a largely paid press, it taught us nothing. The efforts of the Wright consortium were ceaseless, but they brought no results. A few pupils, taught with difficulty, and a few imperfect machines, were to be the sole trace left by this mysterious aircraft from the other side of the Atlantic. Wilbur Wright had difficulty in beating our records. He had hardly completed his performances when we showed our clear superiority by improving on them. Wilbur Wright made his first flight in France on 8 August 1908. On 24 July 1909 he packed his bags and went back to America. The Wright aeroplane was dead and contributed not a tittle to future designs." (page 237)

"The Wright aircraft left no trace either in America or elsewhere." (page 242)

In the first quotation, Gabriel appears to think that the 1903 machine was brought to France; yet he seems to have grasped that there were later aircraft. It is impossible to know what he means by saying (in italics) *"it did not fly"*. The first test of the *Wright A (France)* did not end "inadvertently"; it was simply a trial run; but that trial run alone, with its two graceful circles, was such as nobody in Europe could conceive as possible, and the general excitement was universal. About the engine, suffice it to say that no French-designed engine was ever fitted; a Wright engine, built under licence by Bariquand et Marre, was fitted for convenience to Wilbur's machine for its flights from October 31st 1908 onwards, but after Wilbur had broken all the French records, and had made no less than four flights each of over an hour! The machine never went back to America; it continued to make spectacular flights until the end of the year at Auvours; then at Pau; and finally at Rome.

Gabriel might have paused before writing what appears in the second quotation, because the *Blériot XI* (not the *XII* as reported here) that conquered the Channel incorporated the most successful embodiment of the Wrights' wing-warping ever applied to the monoplane!

In the third quotation it is interesting to note that Gabriel says that by August 8th 1908, "French aviation was definitely under way"; and about time too, one might add! The jibe about finances was, as Gabriel must have known, totally untrue: the Wrights were never financed by anyone except themselves. The press, far from being paid, went nearly hysterical with praise when they saw Wilbur fly. And the remark that the machine "taught us nothing" is dealt with fully elsewhere in this book. As for Wilbur having difficulty in beating the French records, he beat them all very easily.

The last quotation is also fully answered in this book.

❧ 94 ☙

The Question of Assisted Take-off in History

One of the most curious 'accusations', so to say, against the Wright brothers is that from September 7th 1904 until 1909-10, they used an assisted take-off device: this comprised a weight—falling within a derrick—which was attached to a releasable rope running over three pulleys, that, by means of a tow-bar, drew the aircraft forward, the machine resting freely on a tandem-wheeled yoke running along a wooden rail, which was laid down into wind. In 1904, they had received permission to use a pasture of about ninety acres' area, some eight miles east of Dayton: it was known as the Huffman Prairie, after its owner Torrence Huffman. It was a very rough field, with many hummocks, some horses and cattle, and a barbwire fence surrounding it. It was also skirted on two sides by trees which not only shut off some of the wind, but also produced what Wilbur called a 'downtrend'. The brothers spent a long time laying down the track for take-off, and often found—by the time it was ready—that the wind had fallen or was blowing from another direction. In addition the track had to be a long one to enable the new machine to get off in calm weather.

Their new aircraft (*Flyer II*) with a new 15-16 hp engine, was a great improvement on the first *Flyer*, but with the track difficulties, they only managed to make some thirty-nine take-offs, with very brief flights, before they decided that they must be "independent of wind", and able to lay down a short rail, and take off quickly. Hence the take-off device described above, which they started to use on that day, September 7th: it was simple, effective and safe.

But the adoption by the Wrights of this method of assisted take-off has, over the years, provoked constant criticism and attack as not being the "proper" or "fair" method of effecting the take-off of a powered aeroplane; the so-called "proper" method of take-off being taken to mean running along the ground on wheels in order to take off.

Gabriel Voisin has given us his views in a typically egregious statement (*op. cit.* pp. 162, 163). He is referring here to Henri Farman's making the first kilometre circle in Europe on January 13th 1908:

> "The magic circle of the first closed circuit kilometre, officially timed and observed, was ours. The heroic era had ended. Aviation was about to enter the commercial era. For the first time in the world, under official observation, that is in a manner whose exactitude cannot be challenged, a machine, carrying on board the power needed for its flight and the man piloting it, had left the ground under its own power and, on a predetermined course, had accomplished a flight over a closed circuit with return to the point of departure. It had thus demonstrated the possibility of voyaging by a heavier-than-air craft and of manoeuvring in altitude and in azimuth, a thing which nobody, until that day, had been able to do.
>
> Some will say that the Wrights had achieved this at least two years earlier. The reply to that is simple:
>
> (a) the Wrights' performances were not officially observed;
>
> (b) their courses were not marked out;
>
> (c) the Wright aircraft—and this is of capital importance—did not leave the ground under its own power;
>
> (d) the Wrights were incapable of giving a demonstration on a day and at an hour fixed in advance because they were dependent upon the wind, their machine, until 1908, being nothing other than a motorised glider."

Here is Gabriel's autobiography again:

> "When Wilbur Wright finally came to France in 1909 [*ie* 1908], the 'trick' of artificial launching with exterior aids in the form of pylon, pulleys, rails and launching trolley was an eye-opener. The Wright aircraft were not leaving the ground under their own power and, after landing, they were immobilised on the ground. The Wrights, very shrewdly, had economised on landing gear and so dispensed with the additional power needed for take-off. In a word, they had not 'resolved' the problem in a definitive manner" (page 160).

Archdeacon was especially critical of the Wright machine, and he is here quoted from the *New York Herald* of August 9th 1908:

> "Also I consider our machines superior in the fact that they have wheels and that they can start wherever they may descend without the help of rails."

It could surely not have escaped him that there were precious few places where an aeroplane like the Voisin could land on perfectly smooth ground and then take off again. Archdeacon was just repeating the first excuse that came into the heads of most of the European pioneers.

Historically speaking, there can never be any proper objection raised about how a powered aeroplane was launched into the air, provided it could fly properly, and be fully controllable, once it was airborne: and indeed this is the only possible criterion. The business of an aeroplane is to sustain itself and fly through the air, and to be satisfactorily controlled in its flight. Provided the occupants remain safe and sound from start to finish, it does not matter in the least how the machine leaves the ground—whether under its own power or not—or how it lands.

When we read the remarks of those who feel that the Wrights gained some sort of 'unfair' advantage over the Europeans by their assisted take-off method, it becomes necessary to pose a vital question:—Why did not the Europeans themselves resort to similar assisted take-off methods during their early pioneering years? It was certainly not for any reason of "playing fair", by feeling that they had to run along the ground in order to take-off! Far from it: if they could have got their machines into the air by any conceivable means, and kept them there, they would not have been worried for a moment—nor should they have been—about how they achieved it.

The truth is, of course, that having abandoned the Wrights' philosophy of learning the proper flight-control of a glider before attempting powered flight, the Europeans simply did not dare to have themselves precipitated into the air at speed, since they would have had little or no idea of what to do with their aeroplanes if they had thus suddenly found themselves airborne. The casualties would probably have been disastrous.

In passing, it is interesting to note that, apart from gliders—which do not count in connection with assisted take-offs—there was, in those early days, one powered aeroplane which was given an assisted take-off, from an overhead cable; this was Ferber's powered glider of 1905, which was fitted with a small 12-hp Buchet engine: this engine could not, of course, sustain it in level flight, let alone permit of a powered take-off. But as soon as Ferber built a properly powered machine in 1908, he promptly eschewed all forms of assisted take-off, and made his tests by running along the ground.

In terms of the fruitless but perennial arguments about the ability of Langley's full-size piloted *Aerodrome* to fly before the Wrights in 1903, I feel certain that, had the *Aerodrome*—which was catapult-launched—succeeded in flying, we would have never heard another word about the non-acceptability of assisted take-offs!

"Assisted take-off", as a term, is highly equivocal; it generally applies to power-assisted take-offs where some form of catapult mechanism is applied to accelerate the aeroplane from a "standing start", so to say. But any means employed to augment the machine's ability to overcome inertia is really a form of assisted take-off, and this must include a take-off down an inclined ramp, or indeed any other means whereby friction is overcome as between the aircraft and the surface from which it takes off.

So the next point we should consider is this: if, for argument's sake, it was ruled by some international authority that flights following assisted take-offs were historically unacceptable, what then would be the criterion? Would an aeroplane—to qualify for its ability to fly—have to take off under its own power from rough, or from smooth ground? And how rough or smooth would the ground have to be? And how about ice? The Swiss would have had an unfair advantage over others with their ice-covered lakes, and could have taken off with only skids instead of wheels.[1]

Then what about the length of take-off ground to be permitted? Could the machine be allowed to run along for two miles before it took off? Or would it have to take off within five hundred yards? Or one hundred yards? Or one hundred feet? Would the ground have to be perfectly level all the way? And what about the wind? Would flights made after take-offs in anything but a dead calm be invalidated? If not, what wind velocity would be acceptable?

Next, what about landings, which are just as vital as take-offs? If it were to be held that wheels are *de rigeur* for an aeroplane, it might also be properly argued that—in the early days—wheels were by no means the landing gear of choice, especially when there were comparatively few places fit to accommodate flying-machines; for if the machine ventured by chance out of range of its airfield, or if it essayed a cross-country flight, a more or less severe accident was inevitable if an aircraft with a wheeled undercarriage had to land on rough ground; whereas the skid undercarriage of the Wrights would ensure a safe landing in almost any terrain. So, on balance, the skid-equipped aircraft was clearly superior, as it could be taken off from, and land on, almost any surface; whereas the wheeled type could only operate and land on, smooth surfaces. The first cross-country flight in history was made by Henri Farman on his much-modified Voisin—to which he had by now added large ailerons—when he flew from Bouy to Reims, on October 31st 1908 a distance of 27 kilometres (16.7 miles). I have driven along this route, and creditable as indeed the flight was, it was obvious that Farman had chosen a route where, at any point, he could have landed with ease on flat ground on either side of the road, or

[1] Curiously enough, the first machine to make a powered flight in Switzerland was Captain Engelhardt's German-built *Wright A*—with skids only—which landed on an ice-and-snow-covered lake at St Moritz in 1910.

even on the road itself: flat ground of some kind extended the whole distance to Reims.

In fact, the question of what landing-gear an aeroplane had, or how it left the ground, shows at every step that it makes no difference how an aeroplane is got into the air provided it stays there, and can land safely.

Another point which is often advanced by the 'anti-Wright faction' is that, by adopting a skid undercarriage, the Wrights 'unfairly' saved weight by not having any wheels and their necessary gear and accessories. This idea is, in fact, the purest nonsense; for, towards the end of 1909, some of the owners of the standard type A Wright biplanes, fretful at the traditional use of the Wrights take-off technique, themselves added four small wheels to the skid undercarriage, thereby taking off and landing with ease from the growing number of airfields being prepared.

The complaint about the weight saved, and the unfair advantage thus gained by the Wrights, was largely a case of sour grapes. If anyone in France could have been enabled to fly by any means whatsoever, they would have grasped it with both hands. By having abandoned the idea of mastering the flight-control of gliders, and thus being committed, as La Vaulx called it, to the 'direct' method, the French pioneers were forced to have robust undercarriages to withstand the constant heavy landings, after they hopped off the ground. The contemporary records are littered with descriptions of the early machines leaving the ground, and coming down with an "atterrissage brusque", "atterrissage très brusque", "atterrissage dur", "mauvais atterrissage", and so on; and sometimes, if the undercarriage did not stand up to the treatment, "machine destroyed by a 'brusque' landing (appareil detruit par atterrissage brusque)".

The notion of providing assisted take-offs for powered aeroplanes may be said to have started with Henson in 1843; and in some of the fanciful and charming views of what he envisaged in the future, he shows a down-ramp take-off technique being employed. We then find among others, Du Temple's full-size piloted machine running down a ramp to make the first full-size powered attempt to fly (*c.* 1874); Mozhaisky's piloted machine, which similarly descended a ramp in 1884; and Langley's *Aerodrome*, which was twice catapulted from the top of a house-boat on the River Potomac in 1903, only to immediately crash into the river on both occasions.

There can be no doubt whatsoever that the only rational and workable criterion for the flight of a powered aeroplane is that—once it is safely in the air—it be sustained, and controlled; that the machine should be able to sustain itself freely, and be satisfactorily controlled, in a horizontal or rising flight-path, without loss of airspeed, beyond a point where it could be influenced by any momentum (not velocity) which might have been built up before it left the ground;[1] and that it should be able to land safely at a point as high as that from which it took off.

In practice, these conditions could not have been satisfactorily fulfilled unless the machine had been airborne approximately for at least a quarter of a mile over the ground, in still air; or its equivalent in air distance: this distance is, of course, arbitrary; but unless the aircraft was well beyond this arbitrary minimum, it could not be said to have flown in any real sense.[2]

---

[1] The whole question of what constitutes a flight is dealt with in my Science Museum booklet *The World's First Aeroplane Flights*.

[2] It was to help clarify the historical perspective on claims to powered flying that I suggested in *The Times* that a minimum distance of about a quarter of a mile (1,320 feet; 402 metres) through the air should stand as a modest criterion of a simple powered and sustained flight. Dr G. W. H. (now Sir George) Gardner, the then Director of the Royal Aircraft Establishment at Farnborough, included these words in an answering letter to *The Times* (April 30th 1958): "Clearly it is necessary to distinguish between an undoubted sustained, powered, and controlled flight, and a 'powered leap'. . . We agree with others that nothing much less than a quarter of a mile would seem to remove all reasonable doubt that a flight was indeed sustained."

Could the Wright Machines only be flown by Acrobats?

One of the most widely held criticisms of the Wright machine was, of course, over its inherent instability, a quality which the French pioneers shunned as if it were the plague. Nothing can quite explain why the idea of inherent instability, with proper flight-control, took such a deep hold of horror on these early pilots in Europe. They did not even feel, until much later, that a compromise could be effected between stability and controlability; so, when they saw Wilbur manoeuvering his *Flyer* like a bird, and realised the machine was inherently unstable, the Continental pioneers attacked this aspect of Wilbur's aeroplane with vigour, saying that one had to be an acrobat to fly it; whereas on the stable Voisin, a man could learn to fly in a few minutes. What no one was too eager to admit was that in any weather but a dead calm, you just did not think of taking off at all.

And, naturally enough, they made this instability into a fine old stick to beat the Wrights with. Everyone would agree that such an aircraft, with such inherent instability, had only a limited future in that form; but its performances were so spectacular that the wiser heads among the Europeans—like Farman and Blériot—decided quietly to watch what Wilbur did in the air, and then to copy him, while at the same time being determined to produce a reasonably stable type of machine. It was put like this in *Automobilia and Flight* (May 1909):

> "Acrobats not needed for Wright Flyers. The idea that the Wright aeroplane could only be handled by the Wright brothers themselves, has been exploded by the very successful and long flights made by two of their pupils, Paul Tissandier and Comte de Lambert, each of whom had received less than a dozen lessons, have made flights alone lasting nearly half an hour, proving themselves to be complete masters of the aeroplane."

But, despite its instability, the Wright machine was not in fact difficult to fly, especially as it had dual control; and pupils had no difficulty in learning. Some of the Wright pilots compared flying the Wright with flying a Voisin, by reference to a bicycle and a tricycle; the bicycle, although it has only two wheels, being in the end by far the easier to ride, and by far the most useful.

Here is an amusing passage by M. Degoule in *L'Aérophile* for October 1st 1908:

> "There is attributed to the Wrights a virtuosity of manoeuvre bordering on acrobatics. . . . An exercise which can be prolonged for more than an hour and a half, without trace of muscular fatigue, or mental—such as Wright has given proof of—responds badly to the idea that it is due to an acrobatic feat. At least, it is acrobatics which seemingly ought to be within the reach of many people."

It is also interesting to read what one of the pupils himself says about learning to fly a Wright *Flyer*. Here is Paul Tissandier, quoted from *Le Figaro*, in *Automobilia and Flight* (June 1909):

> "It was at Pau, above the Pont Lond aerodome, that I was first initiated into the secrets of this artificial bird, while my flying companions Comte de Lambert and Captain Lucas-Gerardville had their first training at Auvours. I steered the aeroplane for the first time—Wilbur Wright remaining with his hands on his knees—on February 18, after seven lessons, making a total flight of 2 hours 20 minutes. The following day I again handled the aeroplane, the flight lasting 21 minutes, the performance being one that gave every satisfaction to my teacher.
>
> Wilbur Wright first allowed me to hold the elevation rudder, controlling the ascent and descent of the aeroplane. My first attempts at steering were easy, for they were all on a straight line, Wright keeping in his own hand the lever operating the vertical rudder and the flexing of the wing tips. When I was sufficiently advanced with the elevation rudder, Wilbur Wright allowed me to take the second rudder, the operation of which is more delicate, but with which I was soon familiar. After this my education was practically finished, for on February 18 and 19 I was in full charge, with my professor at my side, closely watching all my movements, and ready to interfere, if I made a mistake. But there was no need to interfere, for I made no mistake, and after this test I was allowed to take the mechanical bird aloft with my companion and friend the Comte de Lambert as passenger.
>
> It is worth repeating here that the handling of an aeroplane—a 'Wright' at any rate—is an easy matter. Personally I never had any serious difficulty. There is no doubt, however, that the practice of other sports, especially automobiling and ballooning, help in aeroplane apprenticeship. All exercises which train a man to have full

control of himself, to never allow his nerves to dominate when subjected to the sensations of speed and height, are an excellent preparation for aviation. There is no doubt about this; but in order to become an aviator in a very short space of time there is no need whatever to be an acrobat, or to have any acrobatic dispositions."

⚹ 96 ⚸

Flying in the Wind

As has already been pointed out, the few European pilots at the time would seldom venture out in anything but a dead calm, for the simple reason that they did not have full control over their machines. They could only control them properly in pitch and yaw, with stability in roll being left to the side-curtained wings, in the case of the Voisin. Blériot was already aware of the advantages of ailerons, but only as 'correctors'. In any case he knew nothing of the combined use of ailerons and rudder. None of the European pilots who were airborne in 1908 had any extensive experience of the air and its vagaries, since none of them had made more than a few glides in uncontrollable gliders, and only short powered flights.

This point was all too obvious to the men of the day, and not too much was said about it when it became evident that Wilbur was perfectly able and willing to tackle winds, as he and his brother had tackled them in their gliders. In this connection, I think a single quotation will suffice: it is from the well-known writer Frantz Reichel, appearing in *Le Figaro* for August 13th 1908:

> "Between the take-off and the landing, 6 minutes 56 seconds had elapsed. This time is official, and was taken with a chronometer of the Aéro-Club de Sarthe.... What is important to draw attention to, and underline, is that it [the flight] was made in a wind said to have been between 15 and 17 kilometres an hour.
>
> This evening, Wright resumed his experiments; the circumstances were less favourable, for the wind blew in violent gusts. The American aviator made two flights. The first lasted for 39 seconds, and covered 850 metres; the second was of one minute 45 seconds, and covered about 1700 metres. The latter was very successful, notably in the landing, which was effected with an astonishing precision and facility.
>
> Wright, who appeared very satisfied with this last endeavour—

the first he had attempted in a wind, and in a gusty wind—was very warmly complimented by M. Henri Kapferer, one of the construction engineers of the [airship] Ville de Paris, . . ."

Postscript. I feel it is also appropriate to repeat here some lines from Peyrey's account in *The Times* (see Section 90):

"The wind does not seem to trouble him, Wright having flown in fairly stiff breezes. In a word, the Wright brothers are the first men who have succeeded in imitating birds. To deny it would be childish"

97

Passenger-Carrying

One of the most miraculous things to the Europeans in 1908 was the way in which Wilbur took up passengers, and flew them about with as much ease and grace as he showed when flying solo. It is hard for us today to realise what an incredible impact this had on those who watched Wilbur at Auvours. What is more, the machine was built as a two-seater and yet the engine was only a 30-hp unit, as opposed to the 50 and 100 hp Antoinettes of the time in Europe, which, with the primitive propellers in vogue, could only fly with one man on board except for a few wavering efforts.

There are only a handful of passenger flights on record for the Europeans throughout the whole of 1908, the best being only 500 feet, by Delagrange, when he took up Thérèse Peltier at Turin on July 8th 1908. In January of 1909, an *Antoinette*, with Wright wing-warping, took up a passenger for 1,640 feet at Issy. And Farman then flew with a passenger in February, for 3.1 miles at Mourmelon. Only in mid-1909 did the Europeans, now reaping the advantages of what Wilbur had shown them of flight-control, start taking passengers for a few miles at a time. Whereas Wilbur had carried some 60 passengers before 1908 was out; he had carried one passenger for over 34 miles; another for 43.4 miles; and a fourth for about 46 miles, all by early October 1908.

Note. The *Antoinette* type of wing-warping differed in some detail of its operation from the Wright type, but was directly derived from it.

⚔ 98 ⚔

Gabriel Voisin and the Wright Engines: 1908

Among the more extraordinary statements made by Gabriel Voisin in his autobiography (see Section 1) are the following, in which I would ask the reader to note here only the remarks about the engines, as I have already dealt with the other aspects of these quotations (the italics are Gabriel's):

"With the passage of time and thanks to the publication of the 'Wright papers' we now know why the Wrights surrounded their work with so much mystery. From 1903 until 1908, the Wrights only used motorised gliders, quite incapable of leaving the ground under their own power and absolutely incapable of flying in 'still air'. When they came to France, they could only fly with the aid of French engines, made in France." (page 161)

"After the miserable demonstration in France on 8 August 1908, the buyers, who were not willing to fool about, had a French engine fitted to the Wright, and the Wright, artificially launched, flew at last. But its performance was immediately outdone by French performances and it went back to America." (page 242)

"The Wright aircraft was never able to leave the ground under its own power until the 1908 period, when it was fitted with a French engine, built in France by the firm of Bariquand et Marre of Paris. We have that categorical statement from an associate of the Wrights." (page 243)

"Orville and Wilbur Wright had never used in America (always without official observation) anything other than rather feebly motorised gliders. *They were only able to fly without the help of up-currents from the moment when they were able to use French Bariquand et Marre engines.*" (page 214)

Gabriel may well be believed by a number of readers, especially as this idea of French engines has now penetrated some other books; for

Gabriel is actually saying that it was French-*designed*, and French-built, engines which powered the Wright machines in France. He seeks pathetically to stiffen his statement by reference to a "categorical statement from an associate of the Wrights": one can only wonder what fuddled memory inspired that remark, since the truth about this matter can so easily be ascertained by anyone willing to consult the relevant documents. The facts are as follows:

(a) The firm of Bariquand and Marre in Paris were commissioned by the Wrights to build the standard Wright 30-hp engine under licence. So whenever there was a Wright machine powered by an engine built in France, *it was a Wright engine built under licence;*

(b) Wilbur Wright made some 80 flights at Hunaudières and Auvours during 1908 with his own Dayton-built engine, including four flights of over an hour each, before he broke this engine, and had to fit one of the Wright engines built by Bariquand and Marre; his first with a French-built Wright engine took place on October 31st.

(c) Orville Wright used a Dayton-built engine for all his flights at Fort Myer in September 1908, including four of over an hour each.

(d) All the flights made by the Wrights in 1903, 1904, and 1905, were also made with Wright home-built Dayton engines, *ie* what I call Dayton-built.

It was a great pity that Gabriel's passionate hatred of the Wrights should have led him to tell this absurd story, and others noted in this volume.

99

Gabriel Voisin
and the Wright Propellers

In his autobiography, Gabriel Voisin says, in so many words, that the Wrights received their data for propellers from Renard in France, by way of Ferber and Chanute. Suffice it to say that Ferber never wrote to Chanute, nor Chanute to the Wrights, giving them any details of Renard's work on propellers. Indeed, if there was any information worth having, why did not the French—and particularly Ferber himself—make better propellers than the primitive 'egg-beaters' which were in use right up to 1908? Here is Gabriel:

"Chanute was in constant touch with Ferber and, in consequence, the Wrights were kept informed of the work of Charles Renard on everything concerned with centres of pressure, propellers and empennages, three matters about which we know the importance, for it was this information which allowed us in 1905 to build machines capable of being controlled, while our competitors at the same period lost precious time discovering empirically that which Charles Renard had codified in 1890." (page 240)

"In the glider, they had mounted a four-cylinder engine inspired by the French Darracq and this engine drove two propellers by chains, the propellers having the characteristics which had been indicated to Chanute by Ferber." (page 241)

Here is Orville Wright, quoted from his patent deposition:

"We next proceeded with the construction of the parts to be used in this first power machine, and while we were doing this we began an investigation of screw propellers. At first we hoped to be able to procure a theory of the reactions on a screw propeller from works on marine engineering, but we soon found, after examining the few books we were able to secure in the Dayton Public Library pertaining to marine engineering, that water screw propellers at that time were not based upon theory but almost entirely upon empirical data.

> We had thought that we could adopt the theory from the marine engineers, and then by using our tables of air pressures, instead of the tables of water pressures used in their calculations, that we could estimate in advance the performance of the propellers we would use. When we found we could not do this, we began the study of the screw propeller from an entirely theoretical standpoint, since we saw that with the small capital we possessed we would not be able to develop an efficient air propeller on the 'cut and try' plan. As a result of this study we developed a theory from which we designed the propellers which we used in this 1903 power machine.
> These propellers had an efficiency of over 66 per cent, an efficiency, I believe, rarely exceeded by the marine engineers, and never approached by any of the aeronautical investigators up to that time..."

And here is Charles Dollfus on the Wrights' propellers:

> "At the same time, the Wrights came up against the unknown, where aerial propellers were concerned. From analogies with the aerodynamics of the wings, by experimentation, and by their dual quality of intuition and reason, they designed and achieved wooden propellers which perfectly met the requirements of the aeroplane of 1903."

Incidentally, it will be noted in the first passage by Gabriel quoted above, that he includes "centres of pressure" and "empennages (tail units)" in the items which he implies the Wrights learnt from Renard. Renard neither did, not could, contribute a single thing of value to the Wrights about either subject.

100

Wilbur Wright's Influence on Europe (Summary): 1908

The vital and far-reaching influence of Wilbur Wright on European aviation was due to:

(1) Full and proper three-axis flight-control; this resulted in perfect control and manoeuverability in the air, including the making of banked turns, circles, figures of eight, etc.

(2) The use of this flight-control not only to 'redress balance', but to initiate and execute aerial manoeuvres; in other words, the active use of flight-control as well as the passive;

(3) The use of geared-down propellers of large diameter, rotating slowly, and hence exerting far more thrust than the Europeans' air-screws, which were of far less efficient shape, and always rotated at engine speed;

(4) The use of less powerful engines to produce—*via* the geared down propellers—as much, or more, thrust than the more powerful European engines.

(5) The ability to fly in windy weather, as a result of full flight-control whereas the Europeans seldom ventured out in anything other than a dead calm.

(6) The ability to carry two men under all normal flying conditions.

❧ 101 ❧

The Influence of the Wright Machine

The 'influence-impact' of the Wright machine, as a result of its magnificent performances in France, from August to December 1908, may most simply and directly be gauged by what followed in European aviation as soon as the lessons were properly digested. The results will be noted later on. But it is significant that the greatest contemporary historian, the Count de La Vaulx, who was in the swim of every move and thought at the time, and the greatest aeronautical historian of today, Charles Dollfus—both of them, be it noted, are Frenchmen—agree that the Wright machine revolutionised aviation when it appeared in France. The main focus of interest was, of course, its manoeuverability, and it is true to say that all modern flight-control of aircraft derives from Wilbur's masterly display of full three-axis control.

Aviation could never be the same again after it had been 'impacted' by such flying as was seen at Hunaudières and Auvours in 1908. From regarding control in the air as consisting of inherent stability, along with elevator and rudder, and accompanied in the case of one or two men by the use of ailerons purely as corrective surfaces to 'redress the balance', they were turned by the Wright machine to the dynamic use of controls for manoeuvre, along with proper control in roll into the bargain.

Here is La Vaulx (*op. cit.*):

"Auvours became very quickly a place of pilgrimage for all those who were interested in the progress of aviation. . . . The moment has come to reveal, at least briefly, the characteristics of this machine which had just revolutionised the aviators' world (qui venait de révolutionner le monde des aviateurs)."

And here is Charles Dollfus:

"Wilbur Wright came to France, and gave at the Hunaudières race-course, and then at the Camp d'Auvours—in the neighbourhood of Le Mans—a series of flights which were sensational, and constitute a decisive epoch of history (et constitue une époque décisive de l'histoire). The first flight took place on August 8th 1908. Numerous flights followed, revolutionising aviation by the excellence of the pilotage, the manageableness, and the versatility of the machine (révolutionnant l'aviation par la qualité du pilotage, la maniabilité et la souplesse de l'appareil). The operation of the wing-warping was also a revelation."

Next, there is a passage from the journal *Flight*, in the course of their obituary of Wilbur Wright in the issue of June 1st 1912:

"Notwithstanding the veil that leaves his first conquest dim in the public eye, however, Wilbur Wright nevertheless began to fly in France at the psychological moment. By his achievements there, he gave to others working in his own field the greatest gift of all—confidence in their own ability to succeed. Of all stories of mundane progress, there is none perhaps quite so fascinating or romantic as that which tells of Wilbur Wright's first flying days in France. There you have an incredulous public watching with ever changing mien the steady but none the less graduated unfolding of his art by a past master. Wilbur Wright, with his characteristic caution flew first a little and then a little more; he cared nothing for the spectators, and thought neither of startling them nor of mystifying them. Seeing the flights of a beginner executed with the precision of an expert, the crowd knew not what to believe; they could only wait and watch. And so, in full view before the eyes of the world, did Wilbur Wright re-conquer the air—and who shall say that this belated, but still spontaneous recognition of his triumph did not, after all, accomplish most in the long run for the future of flying."

Finally, we have the succinct statement by that most respected of French aviation writers, François Peyrey, in his *Les Premiers Hommes-Oiseaux* (1908):

"The Wright brothers have sounded the reveille for French aviation."

As it was the ease of handling, and the perfect manoeuverability of the Wright machine, which made the greatest impact on the French pioneers, it was obvious that it would be in this direction that the brothers' last and major influence would be exerted. The first man to implement the Wrights' lesson on roll-control was, as the reader might guess, Henri Farman, who was at this time by far the most enterprising of the Europeans. In the authorised biography of Henri Farman by Jacques Sahel [1] the author, in Chapter X, describes Henri's determination to outdo the Wrights:

[1] Jacques Sahel, *Henry Farman et l'Aviation*. Paris, 1936.

"And now to work, to give 'la cellule' [*ie* Henri's Voisin] the qualities which would allow him to catch up, and then overtake, Wright. The first thing to seek is a means to give the turning the flexibility (souplesse) which the Americans obtained by the warping of their wings. . . . to turn, as with Farman, on a machine with rigid, non-deformable, wings, imposes great strains on the aeroplane, obliging it to turn in a large radius. The French pilots ease up the machine when banking,[1] or shift themselves slightly in their seats; but these solutions are wretched.

How, by what artifice, can the exertion be made supple, and help the turning wing to rise?

Farman ponders over the solution for a long time. He resolves the problem at the first attempt; amply; definitively; with a disconcerting simplicity of means; and in a manner so perfect, so absolute, that today (1936) there is not a constructor in the world who knows any other way.

He discovers the aileron. At the rear extremity of the wing. . . .

When it is lowered, and when the aeroplane is in flight, the air rushes into the angle which it (the aileron) forms with the fixed part of the wing, and raises it. The machine banks, and its wings help it to turn."

This vital modification, made in October 1908, was to his already much modified Voisin-Farman, which now became the *Voisin-Farman I-bis* (second modification) (Fig 53).

It is fascinating, to say the least, to read about these great pioneers, and to watch what they claim. Gabriel Voisin naturally claims the invention of the aeroplane in his autobiography! And Henri Farman is given almost as much credit by his biographer; but poor M. Sahel must have listened to Henri, quite enraptured by the latter's reminiscences, and then swallowed whole the delightful fantasy that he 'discovered' the aileron! As we have seen, the aileron came to France through the ingenuity of Esnault-Pelterie in 1904; then Blériot took it up, but still as a corrective device. Farman would have seen Blériot's ailerons in *L'Aerophile*—he simply could not help having done so; nor could he have failed to know that Santos had also used them in 1906. But, to do him justice, it seems certain that it was Farman who, as his biographer related, was first to think seriously about how to rival the Wrights; then to realise that these moveable wing-tip surfaces, which he had so often seen photographs of, were being used only as corrective surfaces. Thinking of the Wrights wing-warping, he suddenly saw the light, and conceived their active function, as he said, to raise the wings to aid a turn. He was the first to fit large and effective ailerons to an aeroplane.

---

[1] The banking being caused by the rudder being put over, and the 'outer' wing being thus made to move faster, and therefore given more lift. The pilot would ease off the bank by straightening the rudder, or even putting it over to the other side.

Next year, 1909, when he came to design and build his own famous *No III*, the ailerons were one of the main reasons for its world-wide success (Fig 54).

The Wrights, having inspired Farman to use his flight-controls actively as aids to manoeuvre, had an even more direct influence on Blériot, an influence for which we do not have to seek support from the literature of aviation. For with Blériot, he simply took over the actual mechanism of warping, not half-heartedly—like Esnault-Pelterie in his powered machines—but fully, with a negative angle of incidence on one side and a positive on the other. He fitted this fully-developed warping-system on his famous *No XI*, which crossed the Channel on July 25th 1909: this machine was first tested in January of 1909. There is no doubt that his warping was successful, as warping Blériots remained in service until World War I (Fig 58).

It is not, of course, suggested here that warping was as potentially efficient as ailerons; but it is the force of the Wright influence which is amply demonstrated, from which efficient ailerons were developed, once the Wrights had demonstrated the proper uses for roll-control.

The next strong influence which the Wrights exerted was on that great designer of aeroplanes and engines, Léon Levavasseur. For we find that after the *Antoinettes IV* and *V* which first flew in December 1908 (with ailerons), (Fig 56), all subsequent *Antoinettes* were fitted with wing-warping, *ie* from the *No VI* onwards, this machine making her first flights in April of 1909 (Fig 57). Latham, and other well-known *Antoinette* pilots, found the wing-warping very efficient, and their machines could fly successfully in quite strong winds.

Wing-warping was also used on numerous other machines, both biplanes and monoplanes.

Meanwhile, the standard Voisin biplane, which had become crystalised by the end of 1908, appeared throughout 1909 in its 'classic' form of box-kite wings and tail-unit, with a front elevator (Fig 55), showing clearly its Wright origins, *via* Ferber and Archdeacon. But the Voisin machines had no control in roll, despite the European lead given by Farman in 1908, when he fitted large ailerons to his much-modified Voisin-Farman as a result of having observed Wilbur Wright's mastery of flight-control during his spectacular performances at Auvours.

102

Conclusions

I hope the reader will agree, after reviewing the evidence I have offered, that the influence of the Wright Brothers on the Continental pioneers fully amounted to the initiation and precipitation of the rebirth of European aviation; or the birth of practical man-carrying powered aviation in Europe. And, as a result of Wilbur Wright's flying in France in 1908, that European aviation was finally put firmly on the right road; after which the French, then the British, proceeded to overhaul the Americans; and that after 1910 the Wrights' work was, historically speaking, completed. It is very seldom that the first successful exponents of a technological undertaking succeed in staying with the results of their work, and influencing it, for so long as the Wright brothers did.

I hope I have shown that the Wrights' gliders provided the initial and pivotal influence, by persuading Ferber, and then Archdeacon and others, to turn from Lilienthal-type gliders to gliders of the Wright type. Then the Wrights, by their powered flying in 1903, 1904 and 1905, exerted a tremendous influence on the French, although their machines were not physically in evidence: the Wrights became what the Germans call a "Gespenst im Hause" (skeleton in the closet) which lived with the Europeans, haunting them and plaguing them into action. "Must we", cried Tatin, "read one day in history that aviation, born in France, only became successful thanks to the Americans?" "There is still time", urged Ferber, "but we must not lose a minute."

We next come up against the unbelievable procrastination and stagnation of European aviation; an inexplicable and massive inertia, which in 1906 led the President of the Aéro-Club de France to appeal despairingly to his countrymen to shake themselves out of their "inexcusable torpor". This torpor, which was of paralysing power, continued to dog the French inventors for over five years, time enough—what with all the clues necessary to success which were strewn along their path—for France to have achieved a dozen successful powered aeroplanes, the

Conclusions

first of which should certainly have taken to the air, tentatively, at least as early as 1904; with a fully controllable powered machine by 1905.

It is, I feel, fair to surmise that, had it not been for the ghostly influence of the Wrights, erupting over and over again, French aviation would have remained in as deep a slough of despond as British aviation; the latter fell fast asleep after the death of Pilcher in 1899, and did not truly come awake until 1909. Both France and England had, during this decade, superlative scientists, designers, builders, and craftsmen in almost every department of technology except aviation.

The excuse of being pre-occupied by aerostation—lighter-than-air craft—will simply not do. Comparatively few men had their interest absorbed by balloons and airships. If anything, it might be said that automobilism had become the main obsessional interest among the adventurous: but there were more than enough men to spare who should have been embracing aviation.

We then have to consider the personal qualities of the European pioneers who, in retrospect, seldom seem to have measured up to the requirements for aviation pioneers until the arrival of Farman. On the surface, they were all that could be desired—an excellent mixture of types and backgrounds suitable for developing the flying-machine: there were engineers, automobile racers, army officers, artists, rich patrons, and so on. Yet most of them lacked what I have termed 'aero-talent'.

At this point it should be noted that the French, and other European pioneers, did not have anything like the problems to solve which the Wright brothers had. The Wrights designed and built—with one assistant—their own airframes, their engines and their propellers. They learnt successfully to pilot gliders with proper three-axis flight-control; then successfully to pilot powered aeroplanes, with the same fully-developed flight-control. Such an achievement was indeed highly remarkable.

Of the Europeans, only Esnault-Pelterie, Ellehammer and Levavasseur designed both their airframes and engines; their propellers were so primitive they don't merit the word "designed". Of the 'great' European pioneer pilots, Farman and Delagrange designed nothing, although the former should probably be credited with introducing excellent modifications. Blériot is still a puzzle, and it now seems doubtful whether he was fully responsible for the design of many of his machines: it is only necessary to note how he blundered about from one configuration to another before finally achieving fame with his *No XI*, which Dollfus tells me was designed by Saulnier.

Ferber's talents were certainly very limited; and Gabriel Voisin's talent as a designer was only of a modest order, and he did not become a pilot: he, too, had a technical friend—Colliex—who was very influential.

It was often said by the Europeans that Octave Chanute was of great technical assistance to the Wright brothers; but a study of the Wright-Chanute correspondence shows that the brothers did not owe any basic technical feature—except the Pratt-truss method—to their friend.

Then there is the curious note, running through these years 1902-08, of the alleged need of spurs, prizes, and competitions to egg on the flagging or stagnant pioneers. "To hasten this supreme triumph of the human intelligence, hoped-for from century to century, and now so near," wrote the President of the Aéro-Club, "the only difficulty for both private individuals and the Government itself, is to decide what rewards are to be offered to the pioneers." This came as quite a shock to the present writer, a shock which showed the continual procrastination in a new light. It would appear that the oft-repeated appeals to patriotism had themselves to be bolstered by what simply amounted to 'bribes'!

In criticising the French in such terms, one must always bear in mind that we British were far worse in every respect. Just as baffling—despite the brave efforts of Phillips—was our deeply and uncompromisingly negative attitude to aviation for the decade 1899-1909. With all the technical talent in this country, our monumental non-activity was a national disgrace, as indeed was pointed out by H. G. Wells.

The overall picture of this sorry trail of European stagnation, slapdashery, and procrastination must always remain baffling to the historian, leaving in its wake not one single successful glider; then the virtual abandonment of gliding altogether; and finally the creepy-crawling emergence of the powered aeroplane, which could not even remain airborne for sixty consecutive seconds until November of 1907, a month short of four years since the Wrights made their four first tentative flights on December 17th 1903.

Finally a brief excursion into historical speculation. If there had arrived a practical powered aeroplane in Europe about the year 1905—which there certainly should have been—what would have happened in 1914? The aeroplane, despite its great advance after the Reims meeting of 1909, was still in relative infancy when World War I broke out; and the military commanders on both sides regarded aircraft as nothing better than aerial observational platforms.

Germany, of course, took the bomber seriously, but had made the profound mistake of squandering a fortune on developing the airship as its bomb-dropper of choice, after having neglected the aeroplane almost as inexplicably as had the British.

If four years of development could have been added to aviation history before the War, one wonders what fears of retaliation might have been harboured by Germany and her allies, if she had been confronted with the sort of combined aerial battle-fleets as Britain and France should have been able to assemble by the year 1914.

❧ 103 ☙

Synopsis of Quotations on Aircraft, Influence and Procrastination: 1902-1908

There follow here brief extracts from the quotations previously set out in this book, in order to bring together some of the more significant words which were uttered during or about these seven years, starting with examples of the anti-Wright attitude outlined in Section 1. The numbers in brackets refer to the sections, not pages.

* * * * *

LILIENTHAL, IN 1896, ON THE NECESSITY OF PRACTICE

"One can get a proper insight into the practice of flying only by actual flying experiments.... The manner in which we have to meet the irregularities of the wind, when soaring in the air, can only be learnt by being in the air itself.... The only way which leads us to a quick development in human flight is a systematic and energetic practice in actual flying experiments." (1896)

GABRIEL VOISIN, ON THE WRIGHTS' INFLUENCE (1)

"Aviation was born in France, and not one of our great men, true pioneers of the air, borrowed anything at all from the men of Dayton."

"the totality of French constructors had no knowledge of the Wrights or of their work. In a word, the existence of the two Americans never influenced our researches in any way."

"No technician of real standing can admit that the Wrights inspired anything at all, ..."

"When the Wrights came to France ... it taught us nothing."

"The Wright aircraft left no trace either in America or elsewhere."

A CORRESPONDENT IN *Flight International* (1964) ON THE WRIGHTS' INFLUENCE (1)

"These facts emerge inexorably as a result of a dispassionate approach to the subject ... (a) the aeroplane as we know it today is wholly European (and primarily French) both in concept and development; (b) that if the Wright brothers had never lived, the aeroplane would still have been conceived in Europe by the same people at the same times and would have proceeded through the same stages of development, that, in fact, it did.

Any attempt to deny the foregoing is not only a distortion of history, but inflicts a grave injustice on the true creators of the modern aeroplanes..."

The year 1902

LA VAULX ON AVIATION IN FRANCE IN 1902-03 (2)

"While the American aviators, following the way opened by Lilienthal, arrived in a few years—by the 'detour' of gliding—at establishing the full-size powered aeroplane, what was happening elsewhere, and particularly in France? With us, Lilienthal did not create a school. Our aviators preferred to approach the problem direct, without passing through the school of gliding; ... Moreover, many inventors neglected the aeroplane altogether, ... In all this, there was no collective effort, no real feeling that there was a possibility of impending achievement. What went on was mostly theoretical studies, projects, and highly erudite discussions, but which did not appear to be conducive to a speedy solution. In a word, French aviation—despite the efforts of the early pioneers like Victor Tatin—seemed to be dozing. The present belonged to the balloonists, ... Into this conspicuous apathy cut the activity of a man who, perhaps alone in France, had properly understood the full range and interest of Lilienthal's experiments: this man was ... Ferber."

DOLLFUS ON FERBER (8)

"the great French precursor of practical aviation: he was the first in our country to build, and test methodically, full-size gliders".

FERBER ON WILBUR WRIGHT (8)

"Just think of it, that without this man (Wilbur Wright) I would be nothing; for I would not have dared, in 1902, to trust myself to a flimsy fabric if I had not known—from his accounts and from his photographs—that it would bear me. Think of it, that without him, my experiments would not have taken place, I would not have had Voisin as a pupil; the backers such as Archdeacon and Deutsch de la Merthe would not—in 1905—have offered the prizes you know about; the Press would not have sown the good seed everywhere; your journal [*L'Aérophile*] would not have quadrupled its circulation; and other specialised journals would not have been born."

CHAPTER-HEADING IN FERBER'S BOOK OF 1909 (8)

"FERBER IN PURSUIT OF THE WRIGHTS FROM 1902 TO 1906"

FERBER ON THE WRIGHT INFLUENCE (8)

"About the same time (February 1902), thanks to Mr O. Chanute, I received brochures and photographs on the work of the Wright brothers in 1900 and 1901; and they were so convincing and so remarkable that it was not difficult to gauge that they would arrive easily at a complete flying-machine. But at that moment, such was the state of discredit among the public of the idea of aviation, that the contents of these brochures only achieved understanding — following my campaign of popularisation—three years later! ... [the Wrights] had adopted a practical method of construction, and invented suitable controls, which constituted a great progress. To bring home the invention to France, it was therefore necessary to pursue them on the same lines, and overtake them. I constructed my aeroplane No 5 on their data, ..."

FERBER ON THE WRIGHT INFLUENCE (8)

"I believe, however, that I did not do wrong in trying first to realise that which already worked well elsewhere, for one must always take what exists in order to improve it afterwards."

PEYREY ON THE WRIGHTS (8)

"... the keen and long labour of the two brothers, of which the gliding at Kitty Hawk had sounded the reveille for French aviation."

FERBER ON WING-WARPING (9)

"As to warping, I did not wish to employ it in 1902, as I judged it useless to begin with; so my successors, having set off along my track, did not use it either ..."

WILBUR WRIGHT ON HIS AND ORVILLE'S GLIDES IN 1902 (10)

"The past five days have been the most satisfactory for gliding that we have had. In two days we made over 250 glides, or more than we had made all together up to the time Lorin left. We have gained considerable proficiency in the handling of the machine now, so that we are able to take it out in any kind of weather. Day before yesterday we had a wind of 16 metres per second, or about 30 miles per hour, and glided in it without any trouble."

The year 1903

FERBER ON LILIENTHAL'S DISCIPLES (12)

"There are actually five of Lilienthal's disciples in the world: Messrs. Chanute and Herring in Chicago, Messrs. Orville and Wilbur Wright in Dayton, USA, and Captain Ferber in Nice. The Englishman Pilcher died in 1899. One sees that Europe is insufficiently represented."

ARCHDEACON ON FERBER (12)

"well known in our particular sphere as the only man in France who is making gliding experiments in the manner of Chanute."

L'Aérophile ON FERBER (12)

"Captain Ferber, who was for a long time the only French disciple of Lilienthal, and who made the only gliding experiments which we could set against the American aviators."

ARCHDEACON ON THE WRIGHT NO 3 GLIDER (13)

"... their [*ie* the Wrights] last machine appears actually to be the model of its kind. I will therefore describe only this machine, for fear of unduly prolonging my article."

LA VAULX ON CHANUTE'S LECTURE (13)

"For most of his listeners, except Ferber and his friends, it was a revelation, and even one which was a little disagreeable; when we spoke at times in France, rather vaguely, about the flights of the Wright brothers, we were far from doubting their remarkable progress; but Chanute was now admirably explicit about them and indicated their full significance. The French aviators at last felt that, despite their own enlightened views, they had been resting on the laurels of their elders a little too long, and that it was time for them to get seriously to work if they did not wish to be left behind. These facts particularly impressed themselves on the mind of one man, who was to play a great role in the evolution of French aviation—Ernest Archdeacon. ... Anxious to keep for his country the glory of seeing born the first man-carrying aeroplane which would raise itself from the ground by its own power ... Ernest Archdeacon decided to shake our aviators out of their torpor and put a stop to the indifference of French opinion concerning flying-machines."

FERBER ON FRANCE AND AVIATION (14)

"... the aeroplane must not be allowed to reach successful achievement in America."

ARCHDEACON ON FRANCE AND AVIATION (15)

"The result of all this, my very dear 'colleagues in locomotion', is that the solution is approaching, and even approaching *very quickly*; also, that France, this great homeland of inventors, assuredly does not hold the lead *in the special science of AVIATION*, even when the majority of good minds are today convinced that this alone is the true way. Will the homeland of Montgolfier have the shame of allowing this ultimate discovery of aerial science—which is certainly imminent, and which will constitute the greatest scientific revolution that has been since the beginning of the world—to be realised abroad? You scholars, to your compasses! You, the Maecenases; and you, too, Gentlemen of the Government; put your hands in your pockets—or else we are beaten!"

L'Aérophile ON ARCHDEACON (16)

"Stirred by the first successes of the Wrights in America, he started an admirable and fruitful campaign on behalf of French aviation, by his words, by his pen, by his example, and by his liberality. He created the Aviation Committee of the Aéro-Club; stimulated both intellect and energy; and has become the promoter of our present-day successes."

CHANUTE ON FERBER (17)

"In Nice I met Captain Ferber, of whom I spoke to you, who has been trying experiments with a machine similar to yours. It is rudely made by a common carpenter, ... He says that he is not trying to invent a new system, but simply to experiment [with] the best that others have designed, and is much of the opinion that you are ahead of all others. He says that he is much inclined to go to America to take lessons from you, and that he wishes to purchase your 1902 machine, ..."

CHANUTE ON THE WRIGHTS (18)

"The two brothers glided alternately, and they soon attained almost complete mastery over the inconstancies of the wind. They met the wind gusts and steered as they willed. They did not venture to sweep much more than one quarter circle, so as not to lose the advantage of a head wind; but they constantly improved in the control of the machine, and in learning the art of the birds. Some 800 glides were made ... In point of fact the Messrs. Wright are now gliding very nearly as well as the vulture, ..."

CHANUTE ON POWERED AVIATION (18)

"the time is evidently approaching when, the problems of equilibrium and control having been solved, it will be safe to apply a motor and a propeller."

DOLLFUS ON WRIGHT INFLUENCE (18)

"On April 2nd 1903, Octave Chanute gave a lecture to the Aéro-Club de France, which was published in the journal *L'Aérophile*, along with drawings of the machines, that revealed the use and method of construction of the Chanute and Wright gliders. This communication had a decisive importance: it orientated French aviation towards an attempt at imitating the Wright gliders."

ARCHDEACON ON WHAT HE HOPED FROM HIS PROPOSED GLIDING COMPETITION (19)

"... in a word, to stimulate—from one end of France to the other—the great rivalry which is nearly always necessary to quicken progress."

THE WRIGHTS ON THEIR FIRST POWERED FLIGHT IN 1903 (23)

"The first flight lasted only twelve seconds, a flight very modest compared with that of birds, but it was, nevertheless, the first in the history of the world in which a machine carrying a man had raised itself by its own power into the air in free flight, had sailed forward on a level course without reduction of speed, and had finally landed without being wrecked."

DOLLFUS ON THE WRIGHTS' FIRST FLIGHTS (23)

"THE FIRST HUMAN FLIGHT—December 17th 1903 is the great day in the history of aviation. The Wright brothers made, in succession, four sustained flights with their powered aeroplane. This final 'consécration' took place at the Kill Devil sand-dune, ... It assures an imperishable glory for the two first aviators and for the United States."

CHANUTE ON THE WRIGHTS (23)

"Too much praise cannot be awarded to these gentlemen. Being accomplished mechanics, they designed and built the apparatus, applying thereto a new and effective mode of control of their own. They learned its use at considerable personal risk of accident. They planned and built the motor, having found none in the market deemed suitable. They evolved a novel and superior form of propeller; and all this was done with their own hands, without financial help from anybody.... In 1900, 1901, 1902, and 1903 they made thousands of glides without accidents and even succeeded in hovering in the air for a minute and more at a time. They had obtained almost complete mastery over their apparatus before they ventnred to add the motor and propeller. This, in the judgment of the present writer, is the only course of training by which others may hope to accomplish success. It is a mistake to undertake too much at once, and to design and build a full-sized flying machine *ab initio*, for the motor and propeller introduce complications which had best be avoided until in the vicissitudes of the winds bird-craft has been learned, with gravity as a motive power."

L'Aérophile WAITS FOR FURTHER INFORMATION (24)

"We shall wait before commenting, but may we recall to French aviators the cry of alarm uttered by Mr Ernest Archdeacon [*ie* in *La Locomotion* of April 11th 1903]: 'Will the homeland of the Montgolfiers have the shame of allowing that ultimate discovery of aerial science to be realized abroad?'; and this recommendation of Captain Ferber: 'The aeroplane must not be allowed to be perfected in America'. There is still time, but let us not lose a minute."

FERBER ON THE WRIGHTS (24)

"Although the results were less remarkable than was announced at first, the date December 17th 1903 nonetheless marks the day when a *piloted* flying machine *has really flown*, and the honour of this memorable experiment falls to the name of Wright." (Ferber's italics.)

The year 1904

L'*Aérophile* ON FRENCH EFFORTS (25)

"Gliding flight, so vigorously launched in France by Mr Archdeacon, will not be long in bearing its own fruit. What do we lack? A few specialists trained in the tricks of the trade."

FERBER ON THE WRIGHTS (25)

"When I learned on the 21st of December that Wright had succeeded with his motor, I was at first quite annoyed at not having been able to take this first step myself. But now, just think that this success of Wright is doing me lots of good, and is much to my advantage. I believe that people are now saying: 'Why, that Captain was not such a fool after all, as the other chap has met with success'. . . . Archdeacon is very active, and hence I believe that not fewer than 6 apparatuses of the 1902 Wright type are now being built in France. I believe we will see a great movement."

PHILIPPE REY ON THE WRIGHTS (25)

"It went for fifty-nine seconds against a wind of 10 meters a second. It travelled only 266 metres, four times in a row, on December 17. The cold drove away the experimenters, who will not begin again until next season. That gives us a respite of six months. The experiment is not as grand as we thought, but it nevertheless represents a new fact. For the first time, a heavier-than-air machine (338 kilograms!) has flown in horizontal flight for 266 metres."

ARCHDEACON COMMENTS ON HIS FIRST GLIDER (27)

"A machine which is, apart from subsequent modifications, exactly copied from that of the Wright brothers."

ARCHDEACON REPORTED ON THE OBJECT OF HIS PROPOSED GLIDING COMPETITION (28)

"The object of encouraging the efforts of those pursuing the path of aviation ... [he] concluded by appealing to his colleagues on the Committee, and praying them to be willing to promote it [the idea of gliding competitions] in order that France—which has always occupied the front rank in this field—shall not be outdistanced by the United States."

FROM A REPORT OF TATIN'S SPEECH AT THE AÉRO-CLUB (29)

"In closing, M. Tatin protested against the tendency we seem to have in France of slavishly copying the gliding machines of the Americans.... And then, where do such copies lead us? Does this not seem like a confession of our incapacity to make anything original ourselves? Nevertheless, it would appear that we still have in France some men of genius capable of successfully carrying out such work without putting ourselves in tow of foreigners who in any case do not seem to be on the best track."

TATIN SPEAKS (29)

"Must we one day read in history that aviation, born in France, only became successful thanks to the Americans; and that the French only obtained results by slavishly copying them? For us, that would indeed be glorious! Have we not already seen enough French inventions completed by foreigners, such as the steam-engine, gas-light, steam-ships, and many others. Alas, are we to add aviation to them? As we have engineers ready to go to work, I think it would be a disgrace to remain behind any longer, since there is certainly still time. It is in France that the first journey by a flying machine must be made. We need only the determination. So let us get to work."

L'Aérophile ON ARCHDEACON (30)

"M. Archdeacon has indeed chosen the best means to draw the attention of public opinion to aviation, in inaugurating in a practical manner his effective campaign by testing the machine [*ie* the Wright-type glider] which we already know about, the results of which are assured. But, in the spirit of their promoter, the first tests at Merlimont are, above all, intended to stimulate French aviators, to urge them to emulation; in brief, to promote individual study and research; to lead us, we hope—by the employment of new means and new apparatus, or by the better application of those we know about already—to the practical machine which everyone awaits, and which will allow man finally to annex to his domain the uninhabited vastness of the atmosphere."

ARCHDEACON ON AVIATION (30)

"For many years I have been possessed by an interest in the experiments undertaken by the late Otto Lilienthal, then by the Americans; and I have always dreamed of 'acclimatising' aviation studies in France. I have already said, and repeated everywhere, that we must make haste, on pain of being overtaken by the Americans, especially by Chanute and the Wrights, who have pushed much further ahead in this kind of study that we have. We are, however, exceptionally well placed to succeed, since we are the real fathers of the light motor. What I predicted has now come about. Despite various contradictions and a fair number of exaggerated reports published in the periodicals, it appears to have been established that the Wrights have recently succeeded in making a flight of 266 metres, with a machine of 350 kilograms, mounted on wheels (*sic*) and fitted with a motor *from which it derived the whole power of sustentation* [the italics are Archdeacon's]. It is certain, Gentlemen, that the results obtained are considerable, and—I do not cease to repeat—we must hurry if we wish to catch up with the enormous advance made over us by the Americans. . . . To obtain anything from inventors in France, it is necessary to stimulate rivalry; and, to stimulate rivalry one must—and the automobile has well proved this—establish races or competitions. . . .

So, as you see, we are going along as well as possible; and provided that we continue, we shall rapidly catch up with—and even overtake—the Americans. . . . But, to succeed in this difficult task, we must really have the help of all. I am therefore counting on you, Gentlemen, not only as possible participants in our competition, but as those to spread the word far and wide; to recruit 'proselytes'; and bring in subscribers. It is absolutely essential to ensure for France the glory of the ultimate conquest of the air, which the homeland of Montgolfier must not let fall into the hands of foreigners."

LA VAULX TO THE AÉRO-CLUB (32)

"A new branch of aeronautics, aviation—which had not yet received the official 'consecration' of our Society—has been able to win the 'freedom of the city' among our labours, thanks to the persevering studies of Captain Ferber and the impetuous generosity of M. Ernest Archdeacon."

L'Aérophile ON THE PROPOSED GRAND PRIX (34)

"The Grand Prix d'Aviation—still in process of organisation—will crown in a deserving manner the complete solution of the problem [*ie* of powered flight]. But this great discovery is still, without doubt, rather distant. It necessarily remains to encourage and assist the investigators, to reward the more modest results, it is true, but the immediately realisable and progressive stages toward complete success."

'PHILOS' IN *L'Aérophile* (36)

"M. Archdeacon also has the right to congratulate himself on the success of his crusade. If he has been minded to start with American machines, it is only the better to stimulate the activity and emulation of the French investigators by setting before their eyes immediate results. He has attained his object, seeing that on all sides, ideas and new machines arise which—and this is his dearest wish, and also ours—will retain, in the land which was the cradle of aviation, as it was of aerostation, the glory of finally conquering the air by means 'heavier than air'."

L'Aérophile ON FRENCH GLIDING (37)

"The aim of these first experiments was simple: that is, to study the qualities and defects of the machine in practice, and to become familiar with its difficult handling. To those who may be tempted to be surprised at the tardiness of these tests, one must answer that the best and quickest results are obtained, not by hurry, but (on the contrary) with method, and by resigning oneself to begin at the beginning.... To hope to surpass, at the first 'go'—with a new machine and insufficiently trained operators—results which were obtained by the masters of the subject, the Wright brothers, over long months, even years, of groping, would have been foolish. M. Archdeacon saw the temptation offered to his natural impatience, and he avoided it; he cannot be sufficiently congratulated.... As to the results themselves, without yet being comparable with those of the American aviators, they were none the less very encouraging, and showed good promise for the definitive tests."

ARCHDEACON ON HIS SECOND GLIDER (37)

"Our new machine, which will soon be ready, will allow us—I am certain—to do as well as the Wright brothers. One might therefore consider that the tentative phase, which is inevitable at the beginning of such an enterprise, is at an end. We are now entering the active period of tests, which will commence at Berck at the end of April. In a few weeks, French aviators will have nothing more to envy in their transatlantic rivals. They will even perhaps surpass them, to the greater profit of the 'aerial ideal'."

LA VAULX ON ARCHDEACON, FRENCH AVIATION AND THE WRIGHTS (37)

"Archdeacon thought that the most rapid means of catching up with the Americans—if they really were ahead—consisted of making use of their own work ... and reproducing their experiments in seeking to improve them. Considering the superiority of the French automobile industry, which was in a position to supply the lightest motors of that time, he did not doubt that when the moment came, the application of a motor to a perfected glider would be relatively easy, which would allow us to arrive well in front. The extraordinary 'tour de force' of the Wrights, in creating with their own hands the motor they lacked—and which the American industry was not in a position to supply—frustrated this calculation. This was not foreseeable; and Archdeacon's idea, which furthermore astonished an important group of French aviators—who intended to arrive at results by their own means and following the lines laid down by their national predecessors—was rational, if one recalled the necessity of succeeding quickly, indeed more quickly than others.'

L'Aérophile ON THE WRIGHTS (40)

"We, today, fall very much short of these results [*ie* of the Wrights' flying in 1904]. However that may be, it is certain that the progress of aviation is being keenly pursued in America. It is for the aviators of France to unite their efforts to those of M. Archdeacon—whose fine campaign to encourage similar experiments is well known—and not allow themselves to be overtaken."

THE WRIGHTS ON FLYING CIRCLES (40)

". . . a landing was made each time, without accident, merely to avoid passing beyond the boundaries of the field. On the fourth trial, made on the 20th of September [1904], a complete circle was made, and the machine was brought safely to rest after having passed the starting point. Thereafter we repeatedly made circles, and on the 9th of November made four circles of the field in a flight lasting a few seconds over five minutes."

AMOS ROOT DESCRIBES WHAT HE WITNESSED WHEN THE WRIGHT FLYER II FLEW ITS FIRST CIRCLE (40)

"When it first turned that circle, and came near the starting-point, I was right in front it; and I said then, and I believe still, it was one of the grandest sights, if not the grandest sight, of my life. . . . these two brothers have probably not even a faint glimpse of what their discovery is going to bring to the children of men. No one living can give a guess of what is coming along this line, . . ."

COLONEL CAPPER REPORTS TO THE WAR OFFICE IN LONDON AFTER VISITING THE WRIGHTS (45)

"Both these gentlemen impressed me most favourably; they have worked up step by step, they are in themselves both well educated men, and capable mechanics, and I do not think are likely to claim more than they can perform. . . . The work they are doing is of very great importance, as it means that if carried to a successful issue, we may shortly have as accessories of warfare, scouting machines which will go at great pace, and be independent of obstacles of ground, whilst offering from their elevated position unrivalled opportunities of ascertaining what is occurring in the heart of an enemy's country."

FROM THE PROCEEDINGS OF THE RE COMMITTEE OF THE WAR OFFICE IN LONDON (45)

"There appears little doubt but that the machine has done all that the Wright brothers claim for it, and that it is within their power to construct a similar machine capable of going much longer distances and carrying one or more passengers. . . . That the Wrights are ahead of the rest of the world in this matter seems to be absolutely certain, and by securing their services we should obtain a lead over other nations."

The year 1905

ESNAULT-PELTERIE ON HIS GLIDER (51)

"A great stride seemed to have been realised in this very arduous and delicate question of the conquest of the air. We confess that results so magnificent, especially coming from the other side of the Atlantic, have left us a little sceptical. But in scientific matters, scepticism has no value. When an experiment seems surprising, there is a very simple means of resolving the doubts, and that is to make a repeat-experiment. It was with this aim that we constructed an aeroplane scrupulously following the directions of the Wright brothers, directions and diagrams which, moreover, were published in *L'Aérophile*. Our machine was exactly like that of the American experimenters, as much as to the general dimensions as to the curvature of the ribs and the disposition of the controls. Only some questions of construction and detail were different."

ESNAULT-PELTERIE, AGAIN, ON HIS GLIDERS (51)

"It is nevertheless certain that all the difficulties involved can only be solved by long practice. It is therefore necessary to arm ourselves with patience. Methodical experiments will alone be able to produce the answers to such questions."

FERBER, ON LEARNING TO FLY (54)

"It is not possible to learn to walk, dance, skate, or ride a bicycle—let alone fly—in a minute; but that is what these [French] aviators wish to do. No, one must slowly create the necessary reflexes; one must, as M. Chanute puts it so well in our language, learn little by little the art of the bird; and that is why Lilienthal's method—which I like to call the method of 'step by step, leap by leap, flight by flight' —is so fruitful; for after every check, one starts afresh."

FERBER, ON THE WRIGHTS (54)

"It is thus that, taking solely to the Wright type in 1902 I am two years behind him [*ie* Wilbur Wright] and I have not yet been able to catch up, whilst I retain a similar advance over my pupils. The reason for this is that to achieve practicality is always a long, difficult, and costly business."

L'Aérophile ON FERBER'S POWERED GLIDER (54)

"has carried out the first glide with a piloted powered machine (May 27th 1905) which has been achieved in Europe, in the manner of the Wright brothers."

THE WRIGHTS, WRITING TO FERBER ON THEIR 1905 SEASON (56)

"But our experiments of the past month have shown that we can now build machines that are really practical and suitable for many purposes, such as military scouting, etc. On the 3rd of October we made a flight of 24,535 metres in 25 minutes and 5 seconds. . . . October 4th we made a distance of 33,456 metres in 33 minutes and 17 seconds. . . . On October 5th our flight had a duration of 38 minutes and 3 seconds, covering a distance of over 39 kilometres. . . . Witnesses to these flights have become so enthusiastic that they have been unable to hold their tongues, and, as a result, our experiments have become so public that we are compelled to discontinue them for the present, or at least until we find a less public place to carry them on. . . . We are prepared to furnish machines on contract, to be accepted only after trial trips of at least 40 kilometres, the machine to carry an operator and supplies of fuel etc. sufficient for a flight of 160 kilometres."

FERBER ON RECEIVING THE WRIGHTS' NEWS (56)

"I was the first in the entire world to know—long before the others— this sensational news. . . . I wished to let my country profit, first of all the army, naturally. . . . I made a report to my official chiefs, who did not believe a word of it, and treated me as a mild lunatic. . . . The two main reasons for doubt at this moment were (a) if men had really flown through the air, one would have known about it; and (b) how could a simple Captain of artillery know something of which the American journalists were ignorant, men with which it was a point of honour to be the best informed in the world!"

ARCHDEACON ON THE WRIGHTS (57)

"Whatever the respect I feel for the Wrights—whose first experiments without a motor are undeniable and of the greatest interest—it is impossible for me to accept as historical truth the report of their latest tests, which have not been witnessed, and about which they have voluntarily maintained the most complete obscurity. . . . It is only 'yesterday' that aviation has become physically possible, and not much time has been lost; but, on the other hand, it is now the moment to get going: we are at a turning point in the history of science; the goal is in view, and we must all prepare for the final rush to try and arrive first. . . . But if the question of the motor has today been resolved, the question of stability certainly has not. It is necessary, above all, to study this stability, and discover, at all events, a practical method of carrying out the training of the pilot without him breaking his bones . . ."

ARCHDEACON ON AVIATION (57)

"The problem [of aviation] has actually been so well propounded [by the late Colonel Renard] that it is more than half resolved. Let two or three men of real ability tackle it, and its final solution will only be a question of months, perhaps of days."

LA VAULX ON THE WRIGHTS AFTER THEY SENT A NEWS-LETTER TO FRANCE ABOUT THEIR 1905 SEASON (58)

"One may imagine the stupefaction of most of the French aviators—who did not want to believe in the flight of December 17th 1903—when they read the letter . . . When the first moment of stupor had passed, after the publication of this vital document, there was a general refusal in France to believe in the veracity of the Wright brothers."

ROBERT COQUELLE CABLES TO *L'Auto* IN PARIS (58)

"The Wright brothers refuse to show their machine; but I have interviewed the witnesses, and it is impossible to doubt the success of their experiments."

FERBER ON A PIRATED DRAWING OF A WRIGHT MACHINE (60)

"This drawing had great importance; it showed us the last details of which we were ignorant; and it was this drawing which caused the first aeroplanes of Delagrange and Farman—February and June of 1907—to have a forward cellular rudder [elevator]."

LA VAULX ON FRENCH NEGLECT OF AVIATION (61)

"In France on the other hand—after the trials over the Seine [by the Voisin float-gliders]—by the end of 1904, and during 1905, the ardour of the aviators seemed to slacken. One did not think any more about the Wrights, who showed no sign of life. One could only fear that, after a passing effervescence, aviation began again to fall into neglect. On the other hand, the airships were progressing, and attention was being concentrated on them. . . . This activity [*ie* various aviation work] had not passed beyond the threshold of the workshops, and no new experiments had been witnessed, when—at the end of 1905—the Wrights decided to break their silence."

THE SECRETARY-GENERAL OF THE AÉRO-CLUB, ON FRENCH STAGNATION (61)

"I ask myself if the attentive and detailed examination of the period elapsed since our last General Assembly has not revealed, despite appearances, some sinking, some stagnation—extremely natural and explainable, without doubt, but none the less disagreeable to record, when one has been in the habit of recording each year incontestable improvements and striking progress."

The year 1906

L'Aérophile REPORTS ACTIVITY AT THE AÉRO-CLUB (62)

"The great, the only, subject of conversation—l'affaire Wright; and naturally, also, the coming experiments of the Santos-Dumont helicopter. . . . But the question of the genuineness of the Wright experiments provokes—at the 'delectable moment of the cigar'—a debate between the sceptics and the convinced so passionate that it goes on till midnight."

L'Aérophile ON THE WRIGHTS' NEWS (63)

"From anyone else but the Wrights, the results announced would simply have been considered as 'bluff'. But their scientific past prevents us from holding this too summary opinion. Their previous experiments with gliders have never been seriously contested, and remain the finest to have been carried out . . ."

L'Aérophile ON THE WRIGHTS' NEWS (63)

"After this cable [from Coquelle to *L'Auto*, already quoted] it becomes more and more difficult, in our opinion, to challenge the magnificent powered aeroplane experiments announced by the

Wright brothers. If the results obtained by the Wrights are found to be definitely confirmed, let us not forget that aviation was born in France; ... If, by pure carelessness, it has come about that we have allowed ourselves to be forestalled, we shall catch up when we really wish to, and even surpass. ... If the news which we receive today from America is not true, it will be tomorrow. Let us remember the warnings we gave; and it is not long since we heard such authoritative voices as those of MM. Tatin, Archdeacon and Ferber; let us close with some words pronounced as early as 1903 by the distinguished President of the Aviation Committee of the Aéro-Club de France: 'You scholars, to your compasses! You, the Maecenases, and you too, Gentlemen of the Government, put your hands in your pockets—or else we are beaten!' "

L'Aérophile, AGAIN, ON THE WRIGHTS (64)

"After the detailed information gathered on the spot by our 'confrère', it became more and more difficult to doubt this cardinal point in the history of aerial locomotion—the complete success of the first powered aeroplane which, carrying a man, covered distances of several 'tens of miles'; was manoeuvred at the wish of the experimenter; and returned at will to its point of departure."

L'Aérophile IS FINALLY CONVINCED (64)

"Thus we have the final proof that aerial navigation by purely mechanical means has just made the decisive step. We can only regret that it was not to be 'chez nous'. But at least let us have the satisfaction of seeing, on this particular occasion, the inventors to whom humanity will be indebted for this progress (which has been hoped-for from century to century) derive from their labours—with a glory before which the whole world will bow—the material gains which they have a right to expect. Here is one occasion, at least, when the legend of the inventor as an onlooker—unrecognised and ruined—at the triumph of his ideas, will be absent; and it will be a profound satisfaction for *L'Aérophile* to have been the prime cause of this occasion."

LA VAULX ON THE WRIGHTS (65)

"The first moment of stupor having passed, after the publication of this cardinal document [the Wrights' first letter to *L'Aérophile*] there was a general refusal in France to believe in the veracity of the Wright brothers. ... The results obtained by the Wrights surpassed those achieved up to then in France to such a degree that—aided by amour-propre—it seemed quite simple to deny them. It was incomprehensible that the Wrights had succeeded so quickly where our aviators had failed, forgetting that the success of the aviators of Dayton was the logical result of a long series of efforts, starting with Lilienthal and continued by Chanute, which had cost the Wrights themselves five years of unremitting labour."

THE LONDON *Automotor Journal* ON THE WRIGHTS (65)

"The whole subject is of extreme interest, and it will ultimately, no doubt, be regarded as epoch-making in the highest degree. We are dealing with the first reports of absolutely the first successful attempts to accomplish mechanical flight, and it is not astonishing, therefore, that a large mass of material has already found its way into the Press. The discovery of exactly what the Wright Brothers have accomplished, and its publication to the world, under the circumstances, must certainly be regarded as a credit to the enterprise of French journalism."

TATIN NOW ADMITS DEFEAT AND EXHORTS THE FRENCH (67)

"The glory of having obtained the first results is therefore for ever lost to France, which was nevertheless the cradle of aviation, and was for such a long time at the head of other nations in the matter of research. Unfortunately, for some years, in spite of all that could be stimulating from the news reaching us of the partial success and the well-founded hopes of the Americans, we have remained in a regrettable state of expectancy, when we had here in France all that was necessary to resolve—better and more rapidly—the problem of which the solution abroad has today aroused us; but aroused us a little too late, alas! Nevertheless, we cannot accuse all the French engineers of apathy, and we render justice to the all-too-small number of those who—like MM Ferber, Archdeacon, and Henry Deutsch (de la Meurthe)—have made the most laudable efforts to shake off the indifference and inertia of their compatriots, either by their own work or by founding prizes to stimulate emulation. Since we now apparently wish to pull ourselves together, and make up for lost time, what still seems to me to be possible—despite our evident backwardness—is that we study a little the ways and means which are capable of leading us to the desired result. This study will be able to show us whether there are grounds simply for resorting to the same features as the Americans, and copying their machines, whilst at the same time seeking to perfect them—this, I believe, would be an admission of incapacity—or whether it would not be more worth while to rely on that originality of which French genius has so often given proof. ... It is in France that there must be made flying-machines both speedy and unbeatable, as with the automobiles we make already: it is only a question of our getting down to it."

L'Aérophile ON THE WRIGHTS (68)

"The Wright 'affaire', whatever may be the outcome, will at least have the advantage of shaking our aviators out of their torpor."

THE LONDON *Automotor Journal* ON SANTOS-DUMONT (68)

"*Santos-Dumont adopts the Aeroplane.* Among the secondary results of the revelations which have been extorted from the Wright brothers, may be counted the decision M. Santos-Dumont has embraced with some suddenness, of, if not abandoning the navigable balloon type of airship, betaking himself to experiments with machines of the aeroplane type . . ."

FERBER ON SANTOS-DUMONT (68)

"Santos Dumont at first was merely a balloonist, but after the considerable sensation caused in 1905 by the Wright brothers—and it may be remembered that this was solely due to the documents which I possessed and published at that time—he felt that the time had come for flying-machines."

THE PRESIDENT OF THE AÉRO-CLUB ON ARCHDEACON'S MANIFESTO (71)

"To shake us out of our inexcusable torpor, to attract in France that universal interest—and the active sympathy which its capital importance merits—in that ultimate problem of modern locomotion, that is to say, aerial navigation by purely mechanical means, M. Ernest Archdeacon. . . . has had the happy idea of appending to a manifesto he has written—and which is reproduced below—the authorised signatures of those savants, engineers, and pioneers best qualified by their world-wide reputation or their special work. . . . This affirmation of what will be—if we really desire it—the brilliant reality of tomorrow, is not just a profession of faith by one individual. It takes on a decisive importance, given to it by the leading scientists of our country. . . . Under such patronage, this manifesto on French aviation should make the great impact it deserves. It will stimulate the ardour and emulation of those inventors and 'savants' concerned with seeking the solution. . . . To hasten this supreme triumph of the human intelligence, hoped-for from century to century, and now so near, the only difficulty for both private individuals and the Government itself, is to decide what rewards are to be offered to the pioneers."

FROM ARCHDEACON'S MANIFESTO (71)

"The final discovery of heavier-than-air aerial navigation is imminent. If France wishes to do what is necessary, we can still arrive before the others, and present the first demonstration in public with a flying-machine. But we must hurry up. . . . Much has recently been talked about the experiments which the Wright brothers are *supposed* to have made in America. Although they have always worked in the greatest secrecy, there is certainly some truth in what has been reported. On the other hand, it seems more and more to have been confirmed that their great secret is in the handling of their machines, ... It is thus that they have arrived at learning, little by little, 'the difficult craft of the bird'. Nevertheless, everything leads us to believe that their machine is still only tentative, ... There is therefore still time for us to catch up on the slight advance which the Americans have been able to make.... We have, indeed, better and lighter engines than theirs, and a galaxy of learned specialists who are only waiting to be encouraged to get going. There must be found for aviation one or more Maecenases, more eager for glory than for pecuniary gain, ... Could there not be founded, to help aviation, a very large cash prize, a prize large enough to make certain that the pioneer who wins it will be generously indemnified for his expenditure? ... In conclusion, we can only repeat the following; *'the discovery of aerial navigation is imminent, but there is still just time for us to carry out in France the first public demonstration.'*"

FERBER ON SANTOS-DUMONT (73)

"The 23rd October 1906 will remain a memorable date. On this day, at 4.45 pm, Santos-Dumont, in his aeroplane No *14-bis*, and propelled by a 50-hp Antoinette engine, left the ground and traversed, in full flight, a distance greater than 50 metres, and under 100 metres. ... Henceforth there is a precise fact; leaving the ground, a man in a flying-machine has traversed more than 50 metres; this is not one of those apocryphal, or simply affirmed, results, like those of the Wrights. A propos of this, I have always believed in the success of the latter, because I realised, from my own experiments, what was actually possible; but today, I believe that if the Wright brothers will not make a public experiment, they will not only lose the profits they anticipate, but even the glory of being the first inventors. ... One salutes the triumphant one, and the new era which opens up for us. For there is no doubt about it, it is really a new world which opens before mankind. ... A new era commences ..."

ARCHDEACON ON SANTOS-DUMONT (74)

"Gentlemen, as I take the floor, I am experiencing today one of the purest joys of my existence. First, because I am celebrating the first truly decisive experiment in this science of aviation, of which I have made myself the apostle; ... If I had ever been capable of the sin of envy, I would envy my friend Santos-Dumont today, who has just assuredly gained one of the greatest glories to which a man can aspire in this world. He has just achieved, not in secret, or before hypothetical and obliging witnesses, but in broad daylight and before a thousand people, a superb flight of more than 60 metres at three metres above the ground, which constitutes a decisive step in the history of aviation... For me, after this experiment at Bagatelle on

October 23rd, having seen what I have seen, I predict that the question will now march with a giant's stride, faster even than I have hoped for in my most optimistic dreams."

The year 1907

FERBER, ON MEETING WILBUR WRIGHT (79)

"As to my impression, it was profound, and I grasped his hand and looked upon him with great emotion. Just think, that without this man I would be nothing, for I should not have dared, in 1902, to trust myself on a flimsy fabric if I had not known from his accounts and his photographs that it would carry me! Think, that without him, my experiments would not have taken place, and I should not have had Voisin as a pupil. Capitalists like Archdeacon and Deutsch de la Meurthe would not in 1904 have established the prizes you know of. The press would not have spread the good seed on all sides. Your magazine [*L'Aérophile*] would not have quadrupled its circulation, and other special journals would not have been born!"

THE PRESIDENT OF THE AÉRO-CLUB, REFERRING TO SANTOS-DUMONT, HENRI FARMAN AND OTHERS (80)

"It seems that to the genius of France is reserved the glorious mission of initiating the world into the conquest of the air. . . . In these researches [*ie* in aviation] our dear country can still claim the monopoly, and the Aéro-Club de France should be proud to count among its members the principal promoters of aviation."

ARCHDEACON ON FARMAN AND THE WRIGHTS (80)

"You can believe me, without difficulty, when I tell you that this evening is one of the most marvellous of my existence. . . . It is incontestible that the superb flight achieved by Henri Farman on October 26th marks an absolutely decisive stage in the history of aviation. . . . The famous Wright brothers may today claim all they wish. If it is true—and I doubt it more and more—that they were the first to fly through the air, they will not have that glory before History. They would only have had to eschew these incomprehensible affectations of mystery, and to carry out their experiments in broad daylight, like Santos-Dumont and Farman, and before official judges, surrounded by thousands of spectators. The first *authentic* experiments in powered aviation have taken place in France; they will progress in France; and the famous 50 kilometres announced by the Wrights will, I am certain, be beaten by us well before they will have decided to show their phantom machine."

The year 1908

THE PRESIDENT OF THE AÉRO-CLUB ON FARMAN (84)

"The date January 13th 1908, will henceforth be a historic date: it will recall to the generations to come the brilliant victory achieved this day by Henri Farman. It is he, indeed, who—the first among men—has succeeded in rising and steering in the air, thus resolving one of the most extraordinary problems posed since the beginning of the world.... The victory of Henri Farman ... shows us that men are capable of resolving the most difficult of questions—one is tempted to say the most impossible—when they are actuated by a burning conviction and by a never-failing energy."

FERBER ON FARMAN (84)

"On Monday January 13th, Henri Farman has astonished the entire world by victoriously demonstrating that a man can make a mechanical flight with a machine heavier than air; that he can depart from a fixed point, manoeuvre, and return to the point of departure. He has thus realised the dream which everybody has harboured for such a long time."

GABRIEL VOISIN ON FARMAN (84)

"It had thus demonstrated the possibility of voyaging by a heavier-than-air craft, and of manoeuvering in altitude and in azimuth, a thing which nobody, until that day, had been able to do."

ARCHDEACON ON FARMAN (84)

"One must indeed, for history, *fix the actual and indubitable advent, in France, of aerial locomotion by aviation.* Eh bien! I do not hesitate to say that date ought to read January 13th, 1908.... It is *Farman who is the first, incontestibly, to win the mastery of the air by aeroplane* [the italics are Archdeacon's]."

THE LONDON *Flight* ON FARMAN (84)

"When Henry Farman won the first Grand Prix... by flying a circular kilometre, the world perforce hailed it as the greatest thing that had yet been done, because so far as the world at large was concerned, that was the first circular flight that mankind had 'officially' observed. Three years previously, Wilbur Wright had already flown distances exceeding *twenty miles* in length, and when in 1908, he also went to France and showed what he could do with the machine that had for so long been idle in its crate, the world at last no longer doubted his attainments in the past."

Synopsis of Quotations on Aircraft, Influence and Procrastination: 1902-1908

THE LONDON *Times* ON WILBUR WRIGHT (90)

"Mr Wilbur Wright has made a remarkable flight this evening, lasting 1 minutes 45 seconds, over a course of about 2,500 feet.... All accounts, however, published in this morning's papers from the correspondents on the spot, attest the complete triumph of the American inventor. . . . the enthusiasm was indescribable. The Frenchmen and the Americans present received Mr Wright, who had just won for his brother and himself the title of the real creator of aeroplanes, with the most extraordinary enthusiasm."

FRANCOIS PEYREY ON THE WRIGHTS (90)

"These experiments were really remarkable. They proved over and over again that Wilbur and Orville Wright have long mastered the art of artificial flight. They are the public justification of the performances which the American aviators announced in 1904 and 1905, and they give them, conclusively, the first place in the history of flying machines that rightly belongs to them. . . . To behold this flying machine turn sharp round at the end of the wood at a height of 60 feet, and continue on its course, is an enchanting spectacle. The wind does not seem to trouble him, Wright having flown in fairly stiff breezes. In a word, the Wright brothers are the first men who have succeeded in imitating birds. To deny it would be childish."

THE LONDON *Daily Mirror* ON WILBUR WRIGHT (90)

"THE MOST WONDERFUL FLYING-MACHINE THAT HAS EVER BEEN MADE. . . . No such perfect control over a flying-machine has ever been known as that exercised by Mr Wilbur Wright."

FRANTZ REICHEL IN *Le Figaro* ON THE WRIGHTS (90)

"I've seen him! I've seen them! Yes! I have today seen Wilbur Wright and his great white bird, the beautiful mechanical bird which, for eight years, in the solitude of Virginia and Carolina, has accomplished so many prodigious exploits that, in the course of time, one has ended by believing that it was only a 'canard'. But today, no longer; there is no doubt! Wilbur and Orville Wright have well and truly flown. . . . Without wishing in any way to diminish the value of what our aviators—Blériot, Farman, Delagrange, Esnault-Pelterie, Zens, Mengin, Gastambide and others—have done in France, one is obliged to recognise that there is a whole world of difference between their machines and the Wrights'. . . . Nothing can give an idea of the emotion experienced, and the impression felt, at this last flight, a flight of masterly assurance and incomparable elegance."

L'Aérophile ON THE WRIGHTS (91)

"But the facility with which the machine flies, and the dexterity with which the aviator gave proof, at the first go off, in his manoeuvering, have completely dissipated all doubts. Not one of the former detractors of the Wrights dare question, today, the previous experiments of the men who were truly the first to fly. . . . At last everybody renders justice to two admirable pioneers of the science, which has been debated for too long. It is, for *L'Aérophile*, a matter of profound pride to have been the first, in 1903, and later, to reveal the decisive importance of their work, and to have hastened the dawn of their glory."

MORE PRAISE FROM *L'Aérophile* (90)

"un merveilleux succès d'Orville Wright; . . . une ovation frénétique; . . . ses merveilleuses performances; . . . une sureté et une rapidité prodigieuses"; . . . des performances sensationelles des Wright; . . . journées glorieuses".

FRANÇOIS PEYREY, AGAIN, ON THE WRIGHTS (91)

"For a long time, for too long a time, the Wright brothers have been accused in Europe of bluff—even perhaps in the land of their birth. They are today hallowed by France, and I feel an intense pleasure in counting myself among the first to make amends for that flagrant injustice."

"It would also be quite as puerile to contest the first flight on December 17th 1903 in North Carolina . . . as it would be to deny the experiments in the Sarthe [*ie* near Le Mans]."

DELAGRANGE ON WILBUR WRIGHT (91)

"The man [Wilbur], one must honestly recognise, is indeed the father of aviation."
"Before his departure, he [Delagrange] did not hide his admiration. 'We are beaten (nous sommes battus)' he declared this morning."
(*Matin*, August 12).

LA VAULX ON THE WRIGHTS (91; 101)

"The man who had just manoeuvred in the air with that facility and virtuosity, who had landed with that light grace, was not on his first test; he was a master; he had known how to fly for a long time; since the date he had announced; since 1903! . . . the Wrights keep the incomparable glory of an undeniable priority in the creation of a true flying-machine, . . ."
"The moment has come to reveal, at least briefly, the characteristics of this machine which had just revolutionised the aviators' world."

Synopsis of Quotations on Aircraft, Influence and Procrastination: 1902-1908

BLÉRIOT ON WILBUR WRIGHT (92)

"I consider that for us in France, and everywhere, a new era in mechanical flight has commenced, I am not sufficiently calm after the event to thoroughly express my opinion. My view can be best conveyed in the words, 'It is marvellous!'"

RENÉ GASNIER ON THE WRIGHTS (92)

"It is a revelation in aeroplane work. Who now can doubt that the Wrights have done all they claim? My enthusiasm is unbounded. The whole conception of the machine, its execution and its practical worth is wonderful. We are as children as compared with the Wrights."

CHARLES DOLLFUS ON WILBUR WRIGHT (101)

"Wilbur Wright came to France, and gave at the Hunaudières racecourse, and then at the Camp d'Auvours—in the neighbourhood of Le Mans—a series of flights which were sensational, and constitute a decisive epoch of history. The first flight took place on August 8th 1908. Numerous flights followed, revolutionising aviation by the excellence of the pilotage, the manageableness, and the versatility of the machine."

Appendices

Appendix A

A brief Synopsis of Aviation prior to the Death of Pilcher in 1899

Apart from the famous Renaissance figures of the Italian Leonardo da Vinci (1452-1519)—who designed ornithopters and was the first to design a parachute—and the Englishman Robert Hooke (1635-1703), both of whose work was not to be published until it was too late to promote progress, there was little talented, sustained, or constructive thinking about aviation (heavier-than-air flight) until the end of the eighteenth century. In the field of aerostation (lighter-than-air flight) there was one outstanding, though mistaken, design for a "flying ship" by the Portuguese Jesuit Father Francisco de Lana (1670), who believed that it would be possible to float large copper spheres in the atmosphere, once they had been evacuated of air: he also wrote vivid and accurate prophecies of aerial warfare—of bombing, troop-carrying, and invasion.

But one should not forget the continuous but faltering line of brave or foolhardy men who, since time immemorial, have fitted themselves with artificial wings, and risked their lives in jumping from towers and other points of vantage, and floundering down to earth, thus keeping fresh among mankind the image of adventure and airmindedness, an image also sustained by countless works of fiction in which the heroes—and often heroines too—successfully navigated the air.

Then, in 1783, the balloon was born; most unexpectedly born, so to say, out of a clear sky; and born without any prophecy or preamble. Men could suddenly "enclose a cloud in a bag", and float off through the air with ease and safety, sustained by hot air or by hydrogen; a little later, coal gas was introduced (1800). The first human balloon voyage was made at Paris by Piltâre de Rozier and the Marquis d'Arlandes in a Montgolfier hot-air balloon on November 21st 1783; and the second, also at Paris, in a hydrogen balloon, invented and flown by J. A. C. Charles, on December 1st of the same year. This astonishing experience of actual aerial locomotion revolutionised intelligent thinking about human flight; it not only encouraged a favourable climate of opinion towards all aspects of aeronautics, but initiated a century of aerial experimentation which was to prove an invaluable prelude to the arrival of the aeroplane. Hard on the heels of the balloon came the first of A. J. Garnerin's courageous and successful descents from balloons by parachute in 1797 over Paris.

The helicopter model was brought into a new prominence with the making, in 1784, of a twin-rotor model—operated by a bow-drill mechanism—by the French savants Launoy and Bienvenu. It was this model that Sir George Cayley saw or heard of, and of which he made a variant in 1796, with feathers stuck in corks for the rotor blades. Cayley's model was published in 1809, and given wide and continued publicity; it is the true ancestor of every helicopter flying today.

But a quiet revolution was now taking place in the still neglected field of heavier-than-air flight. For between 1799 and 1809, Cayley had made himself the master-pioneer of a subject till then virtually unknown, that is to say the aerodynamics, construction, and flight-testing of fixed-wing aeroplanes; he published details of these in 1809-10, from which the whole of modern aviation derives. Cayley, in studying the bird, decided that the ornithopter (flapping-wing machine) must be replaced by an aircraft with fixed wings, in which the principle of lift was separated from the propulsion system, and in which inherent stability, as well as tail-unit control-surfaces, must be incorporated. He built and flew both models and full-size machines incorporating these ideas, the latter mostly in ballast (1804-53).

The lack of suitable engines all but completely thwarted the would-be aviators, for the steam engine was too heavy and the muscles of men too weak. But, in one way, this was to prove a disguised blessing, as the attention of some of the greatest pioneers became concentrated upon gliding as the best preparation for powered flight, which they were confident would ultimately follow. Also—and it should not be underestimated—there was the steady popularity of ballooning and parachute-jumping throughout the century, which maintained air-mindedness among the public, and created a body of men who braved the air and learnt its ways. It was from these adventurous aeronauts that aviation was later to draw many of its pilots.

The nineteenth century also saw a prodigious amount of wasted effort among aviation pioneers, owing to duplication of effort, ignorance of what others had done in the past, and the continued presence of much misdirection, and even more fantasy.

An interesting development occurred in rocketry early in the century. Artillery rockets had been invented in China about 1100 AD, and had later reached Europe: here their military application declined, but their pyrotechnic use prospered. As a result of the effective use of war rockets by Tipu Sultan in India against the British army at the end of the eighteenth century, Sir William Congreve in Britain effectively re-invented the artillery rocket for Europe about the year 1805. This weapon—which had marked success when correctly used—then its successor, the finned Hale rocket, were in action sporadically right through into the present century, and were to lead ultimately to the conquest of space.

In the first half of the nineteenth century there was little activity in aviation except by Cayley, until W. S. Henson published his brilliantly prophetic design for a monoplane *Aerial Steam Carriage* (1843), with double-surface wings, which was widely and repeatedly republished throughout the century, and did much to condition the thinking of later pioneers. In 1847 Henson completed a steam-powered model of his design, but it could not sustain itself in the air. Next year (1848) his friend John Stringfellow—following Henson's lead—built and tested an improved steam-driven model monoplane; but although this machine was well designed, it could not quite sustain itself, but it was a great advance on his friend's aircraft.

There also appeared one important oddity; this was the first published design, which was never built, for a tandem-wing monoplane by the Englishman Thomas Walker, published in 1831, which probably influenced D. S. Brown, and through him, S. P. Langley, in later years.

Probably inspired by Henson, the 1850s saw the first Continental designs for powered aeroplanes; Michel Loup's in 1853, and a highly sophisticated patent by Félix Du Temple in 1857, whose clockwork-powered model was the first model to fly, either in that year or the next. In 1858-59, F. H. Wenham in England carried out his influential tests with multiple high aspect-ratio wings (published 1866-7), which showed that a cambered wing, at a small angle of incidence, derived most of its lift from the front portion, and hence a long narrow wing was best for lifting.

The descendants of the brave tower-jumpers of old now entered the field just as bravely, but better informed; the Frenchman Louis Letur was killed in 1854, after some safe descents in his curious parachute-type glider; and in 1857 another Frenchman, the sea-captain J. M. Le Bris, made the first of his bold attempts to fly in gliders based on the albatross. This decade also saw the first tentative airship flight (1852) by the great French engineer Henri Giffard, where the craft was propelled by a steam-engine; but it could only make some 5 mph.

The 1860s saw an important change in the type of men taking up aviation; the subject now attracted many more professional, and mechanically-minded, men; and as if to mark this change came the foundation in 1863 of the Société d'Aviation, to encourage heavier-than-air flight, followed by the Aeronautical Society (now Royal) of Great Britain, in 1866. At the first meeting of the latter, in June 1866, Wenham read his paper on "Aerial Locomotion" in which his advocacy of high aspect-ratio wings was to have a far-reaching influence. There was also much activity in France, where the first properly thought-out design for a jet-propelled aeroplane was made by Charles de Louvrié (1865), whose fuel was to be "vaporised petroleum oil". This was followed in England in 1867 by the first delta-wing jet designs by J. W. Butler and E. Edwards: none of these machines were built. Also in France there arose a widespread addiction to making helicopter models, many of which were successful and helped promote the idea of aviation in the public mind; but they did not advance aviation technically. Back in Britain, the Aeronautical Society staged the world's first aeronautical exhibition at the Crystal Palace in 1868, in which Stringfellow exhibited his steam-powered model triplane, derived from a suggestion by Cayley. Although not itself successful, this model was—like Henson's design—widely and repeatedly published, and led directly to the growth of the superposed wing concept (biplanes, triplanes, etc) in aviation. It is also important to note that the aeroplane power-plant of the future had its birth in 1860, with the invention of the gas engine by Lenoir in France.

The 1870s saw a significant advance in aeroplane design, as well as the arrival of the popular elastic-powered model (1870), a power-unit which is still, and probably always will be, in use; it was first publicly seen (1871) in Alphonse Pénaud's model monoplane at Paris—called a 'Planophore'—in which he also first displayed Cayley's principles of inherent stability, and with which he gave to the would-be pioneers one of the most influential demonstrations in history; longitudinal stability was achieved by having the tailplane set at a negative angle of incidence, and lateral stability by the wing-tips being turned up to form a dihedral angle. In 1876 Pénaud patented designs for a remarkable amphibious monoplane, with double-surface wings, elevators, fin and rudder, glass-domed cockpit, retractable undercarriage, and a single control-column for the combined operation of elevators and rudder. This decade also saw a number of other important events; the first wind-tunnel for research by Wenham and Browning (1871); experiments by D. S. Brown with tandem-wing model gliders (1873-4), which probably influenced Langley, and certainly influenced Hargrave in his invention of the box-kite (see later); the death of the Belgian Vincent de Groof in his unproductive ornithoptering parachute (1874); the first powered take-off—but accelerated by a down-ramp run—by a full-size manned aeroplane, designed and built in *c.* 1874 by Du Temple; a partially successful—but not influential—demonstration of a large powered tandem-wing model by Thomas Moy (1875); and a successful compressed-air-driven model monoplane by Victor Tatin (1879), which powerfully reinforced the influence of Henson on later inventors. The model helicopterists were also very active, but to no very great effect. The aircraft power-plant of the future—without the aviation pioneers being aware of it—also saw a notable advance in this decade when the German engineer N. A. Otto achieved a practical four-stroke petrol engine (1876).

The 1880s provided a strange lull before the widespread inventive explosion of the next decade. But one vital illustrated and published item was issued in 1884—Horatio Phillips' patent for his double-surface aerofoils: he was the first to demonstrate that, on a double-surface wing —with a curvature of the upper surface being greater than that of the lower—by far the major amount of lift is created by the low pressure (partial 'suction') produced by the pronounced curvature of the upper surface, and the lesser amount by the pressure of the air on the flatter under-surface; this was an epoch-making discovery—which Cayley had long before surmised without being able to prove it—which was to vitally influence all mature designers thereafter. In the same year (1884) Russia's claim to the first powered flight was based on a down-ramp take-off by Alexander Mozhaiski's monoplane, powered by an English steam-engine; but it was only airborne for a second or two, due entirely to its assisted take-off, and could not sustain itself in the air.

Of the utmost future significance to aviation was the appearance in 1885 of the first practical petrol motor car, built by Carl Benz; and even more important, the first Daimler car next year, powered by Daimler's excellent high-speed petrol engine, derived from Otto's four-stoke cycle motor of 1876. For aviation was to call heavily on the crafts of the motor car constructor, and its driver, in the epoch shortly to open. In 1888 Karl Woelfert was the first to fit a petrol motor to an airship, which drove it tentatively, the motor being a Daimler.

As the 1890s dawned, the well-informed pioneers realised that the conquest of the air by powered aeroplanes was now in sight, and would be realised during the first or second decades of the new century. The inventors in aviation could now be said to divide into two separate streams; (a) the 'chauffeurs' who put their faith in building powered machines straight away, and believed that—once in the air—it was mainly a question of their steering the machines around the sky as if they were winged automobiles; and (b) the true airmen, who realised that they must become an integral and living part of the aircraft they piloted; and that full flight-control in gliders must be mastered before committing themselves to the air in powered machines. Needless to say, the chauffeurs—to a man—failed to fly; and it was two of the true airmen who first learned properly to control gliders in the air, and then found it comparatively easy to fly the powered machines to which they soon progressed.

We will take the 'chauffeurs' first. In 1890, the French telephone engineer Clément Ader made the world's first take-off in a powered aeroplane, entirely under its own power, but the airborne distance was only 50 metres (say 164 feet): the machine, called the *Éole*, was very elaborately built, but very primitively conceived, and was incapable of being sustained or controlled in the air; although very creditable, this performance was not a flight in any sense of the word. Next came the American expatriate Hiram (later Sir Hiram) Maxim, with a giant biplane test-rig—his research and construction cost him £30,000—which made its best trial in 1894, and just raised itself from the rails on which it ran: it, too, did not fly.

In the United States, the first great American pioneer was the astronomer S. P. Langley: after some years of effort, he succeeded, during 1896, in making two of his steam-powered tandem-wing model aeroplanes fly well. He had intended to abandon experiments after this, but was persuaded to undertake a full-size man-carrying *Aerodrome* (as he mistakenly called his machines): this machine—also a tandem monoplane—was completed, and twice launched by catapult from a houseboat on the River Potomac in 1903. On both occasions it fouled the catapult mechanism and crashed into the river. The machine was then abandoned, amongst much ill-deserved criticism and ridicule. Some years later (in 1914) Glen Curtiss, having secretly and drastically altered this machine, attempted to fly it, but could not stay in the air for more than a few seconds at a time; he did this in a forlorn attempt to spite the Wrights, and show that the Langley *Aerodrome* had been "capable of flight" when it was originally tested in 1903. In fact, it had never been capable of sustained and controlled flight.

Now, back in 1897, when Ader completed a new machine, the *Avion III*—again elaborately built, but again primitively conceived—which was twice tested in October, but did not once leave the ground on either occasion. Nine years later (in 1906), as a result of Santos-Dumont's first 'hops', Ader claimed that in 1897 he had made an uninterrupted flight of 300 metres (say 984 feet): but this claim is now known to have been a deliberate falsehood, put out by Ader to bolster his reputation.

The giant of the second half of the century was a true airman, who, by his brilliant pioneering work with gliders, inspired and precipitated singlehanded the final phase in the conquest of the air. He was a German engineer, Otto Lilienthal, who from 1891 to 1896 (when he was killed flying) designed and constructed a number of hang-gliders, including three biplanes; in these machines he was suspended by his arms, and swung his torso and legs in any desired direction to shift the centre of gravity, and thus exercise a limited amount of flight-control. Lilienthal was the first man in history to get up into the air and fly properly in gliders; his influence—greatly aided by the publication of many excellent photographs of him gliding—was widespread and profound; he summed up the aspirations of the century behind him, and pointed the immediate way to triumph ahead.

A follower of Lilienthal's, and the only man who might have beaten the Wright brothers in the race to fly, was the Scottish engineer Percy Pilcher; he had progressed rapidly in the flight-control of hang-gliders by 1896; in 1899 he had already had built the engine for his proposed powered machine, when he, too, was killed gliding on his *Hawk*.

Far away in Australia was an inventor of prime quality, Lawrence Hargrave, who would undoubtedly have exerted much greater influence in the history of aviation had he lived nearer the main stream of endeavour. But his contribution was still of vital importance; for in 1893 he invented the box-kite—as a result, as he said, of combining the well-known concepts of the tandem-wing idea and the biplane idea—which he first published to the world in the same year, and examples of which he brought to Europe in 1899; this device was to provide the Europeans with a new and definite conception of inherent stability, which—from 1905 onwards—formed the basis of the standard Voisin aeroplanes of 1908-10, and many others which followed.

Appendix B

Mr Chanute in Paris (1903)*
by Ernest Archdeacon

*This is a translation of the article by Ernest Archdeacon which appeared in *La Locomotion*, April 11th 1903, pages 225-227. This translation appears in *The Papers of Wilbur and Orville Wright* (see Bibliography) and is published here by kind permission of The Library of Congress.

A great flurry in the 'aerial' world was provoked in recent days, by the arrival in Paris, and reception at the Aéro-Club, of Mr Chanute, the American inventor who, if not extremely well known among ordinary mortals, is a veritable celebrity among 'aviators.'

For several years, Mr Chanute has been working with indefatigable earnestness on the difficult problem of aerial navigation by heavier-than-air means.

A great German scientist, Lilienthal, was the first in this field to make lastingly memorable experiments. He built machines with stationary wings which were attached to the body and permitted him, by launching himself from the top of a small hill, to glide for a certain time in space, exactly like soaring birds. Lilienthal had dedicated several years of his life to this study and had attained very remarkable results. He had made over two thousand gliding experiments, succeeding at times in rising higher than the starting point, and covering by this means distances up to 300 metres.

Success seemed more and more to encourage his efforts; so that he constructed another machine, much more important than his preceding ones, which, it was his hope, would produce excellent results. Unfortunately at the first trials of his new apparatus, having risen to a good height, he, for some unexplained reason, crashed in full flight. The inventor, hurled violently to earth, died a few hours later.

Immediately after the death of Lilienthal, which occurred in 1896, Mr Chanute resumed his own experiments with tireless patience and precision of method, never allowing himself to swerve from the unalterable path he had laid out.

I shall not speak here of the innumerable apparatus built successively by Mr Chanute, each new one marking a slight improvement over the former.

Moreover, Mr Chanute, unlike most inventors who wish to keep for themselves alone the whole glory of their ideas, understood very well the necessity of helping one another in this tremendous task.

Admitting that he was no longer very young, he took pains to train young, intelligent, and daring pupils, capable of carrying on his researches by multiplying his gliding experiment to infinity.

Principal among them, certainly, is Mr Wilbur Wright, of Dayton (Ohio), who for his part has made some extremely remarkable experiments, and whose most recently built machine now seems to be the model for its type. I shall therefore only describe this last one, lest my article become excessively long.

In Lilienthal's machines, the man was upright and obtained equilibrium by shifting the lower part of his body in the appropriate direction; in those of Mr Wright, on the other hand, the man is placed in horizontal position on the machine.

In order to regulate the equilibrium in the transverse direction, he operates two cords which act, by means of warping, upon the right and left sides of the wing and, simultaneously, by the movement of the vertical rear rudder.

In order to steer the machine in the longitudinal direction, he can, in the same manner, using the cords, operate the horizontal forward rudder, which must be vigorously pointed vertically downward when the speed is to be killed for landing.

We know that, in order to benefit from the full supporting action of the wind, the 'glide' is always made against the wind; it will be understood, then, that in pointing the horizontal rudder vertically downward, the machine turns up suddenly, presents an enormous surface to the opposing wind and finds itself stopped dead.

Mr Wright's present machine weighs about 53 kilos.

The total supporting surface is about 28 square metres, divided into two surfaces, each about 10 by 1.40 metres each.

The distance between the two surfaces is 1.50 metres.

The curvature of the wing is about $\frac{1}{20}$ of its chord, the point of deepest curvature being $\frac{1}{3}$ of the chord back from the front.

Here, in summary, are the results obtained at the present time:

The experimenters are launched from the top of a hill of about 30 metres, the slope of which is very gentle, not exceeding 10°.

Their longest 'glides,' to use Mr Chanute's picturesque expression, have reached 200 metres, and they have several times risen higher than their starting point. Much longer flights could be made *with equal ease* by starting from a higher point.

In order to launch the machine, which the experimenter cannot himself do by running, as Lilienthal did, two assistants are sufficient: carrying the machine together, they take several steps against the wind and then let go when they feel the lifting action of the wind take effect.

With good conditions, a mean angle of descent of about 6° can be achieved, and this, according to Mr Chanute, will diminish considerably with the incessant progress of the machines.

What is especially remarkable in these experiments is that they have been conducted with such prudence and so methodically that, from the beginning, Mr Chanute has not had to report a single accident—except a pair of torn trousers! And this although a very large number of people, even but slightly initiated, have already attempted and accomplished 'glides.'

It is also true that the experimenters, with their extraordinary care, have marvellously chosen their ground, which is all of soft sand and like that of our dunes. A terribly violent shock would be necessary on such soil to lead to serious bruises.

Mr Chanute is justly proud of this absence of accidents in his trials. Moreover, it is the only proud thing I know about him, for he is the most modest man in the world, always willing to attribute his own merits to others.

Finally, here is what Mr Chanute said in closing:

"Our experiments, methodically conducted, will permit us, little by little, to learn completely 'the art of the bird'—an art which, without seeming so, is extremely difficult.

"In a year, then, or perhaps not for two, we will know it thoroughly.

"Until then, it is useless, and even dangerous, to burden oneself with a motor, and I much prefer to use such an untroublesome and simple motor as gravity.

"Besides," added Mr Chanute, "these glides provide the most original and most enticing of sports; as proof of which, several of my friends, great sportsmen and hunters, have put aside their favourite sports to devote themselves with enthusiasm to aerial glides!"

Mr Chanute, on his return to America, is to send us complete plans of his latest machines, so as to allow us to execute similar ones in France.

Moreover, I do not doubt that, before long, some of our dauntless automobile sportsmen will happen to make themselves some similar machines and seek out somewhere a favourable spot for competing in these exciting glides.

The machine fully equipped comes to about 350 francs. A little cheaper than an automobile!

If there are any orders, I will undertake to handle them. Let's go, gentlemen, take out your subscriptions. Just write me at *La Locomotion*.

Mr Chanute in Paris (1903)

On leaving Mr Chanute, as I was speaking to him of the experiments of Professor Langley, he informed me of some very interesting news. Here I start a parenthesis, to inform those who did not know it, that Professor Langley is, along with MM. Chanute and Wright, the most eminent aviator in America and probably the whole world.

Even in 1896, an aeroplane of his construction, weighting 11 kilogrammes, and equipped with a motor of 1 horsepower, was launched over the Potomac River, It covered, *in free flight*, a distance of 1,600 metres in 1 minute 45 seconds, that is to say, at an average speed of 13.33 metres a second.

That is certainly a most remarkable result, the like of which no one else has succeeded in approaching.

Well! That is not the end of it, for Mr Langley last November announced to Mr Chanute that he at last had the definitive machine; that he had entirely *solved the very difficult question of equilibrium*, and that he was counting on showing the universe, in the course of 1903, the *first self-propelled, man-carrying aeroplane!!!* . . .

I add that all the labours of Mr Langley *are carried out with money of the American government;* and it is known that, in that country, when necessary they are not niggardly!

That, then, is *an indispensable condition of success*, for definitive flying machines assuredly cannot be built without very long and very costly trials.

From all this it results, my very dear colleagues in locomotion, that the solution is approaching, and approaching even very quickly; but also that France, this great homeland of inventors, assuredly does not hold the lead *in the special science of* AVIATION, even when the majority of good minds are today convinced that that alone is the true way.

Will the homeland of Montgolfier have the shame of allowing that ultimate discovery of aerial science, which is assuredly imminent, and which will constitute the greatest scientific revolution that has been seen since the beginning of the world, to be realized abroad?

Gentlemen scholars, to your compasses! You, you Maecenases, and you too, gentlemen of the government, your hands in your pocket . . . or else we are beaten!

At the moment of sending the present article to press, I have just received from Captain Ferber (well known in certain circles as the only man in France who is engaged in gliding experiments in the Chanute manner—see *Aérophile* of February 1903) a very curious letter.

This letter is extraordinarily curious in the sense that it shows once again how certain ideas are in the air (it is certainly a case of saying so). Written by Captain Ferber, *whom I have never had the pleasure of seeing*, it expresses ideas such as one would swear had been agreed upon in advance with me . . . .

I extract the interesting passage:

"As I know that you are more an aviator than a balloonist, I permit myself to write and ask you to use your influence in the Aéro-Club to have it establish as soon as possible a prize for distance for gliding machines.

"From experience with the automobile, we know that it is races which improve the machines, *and the aeroplane must not be allowed to be perfected in America ..."*

Really, it is impossible to say it better, and I am immediately going to lay before the Committee and the Technical Commission of the Aéro-Club a project toward this end.

It is not a question of working out a serious rule and finding a site.

In view of the great interest of this project, here is a very fine opportunity for the Maecenases of locomotion to display their generosity.

Here it is a matter of machines costing 300 francs to build, and even relatively modest prizes will be broadly sufficient to produce colossal emulation.

As for myself, I am quite ready to take the lead with a subscription for a sum of 3,000 francs, which will be definitely transmitted the day the competent committee will have completed its organization plans and judged the total subscriptions sufficient to assure success.

Appendix C

Aerial Navigation in the United States*

by Octave Chanute

* This is a translation of the article by Octave Chanute which appeared in French in *L'Aérophile*, August 1903, pages 171-183. There is a copy of Chanute's original unedited draft in English in The Library of Congress, and this draft has been consulted in producing this translation, which appears here by kind permission of The Library of Congress. This translation was first published in *The Papers of Wilbur and Orville Wright* (see Bibliography). The Figure references have been adapted to refer to those in the main part of this work.

Americans have done but little in aerial navigation. They deem it of small practical application for transportation. They believe that balloons will furnish one of the solutions, but that they will always be too slow and frail to render important service. Here and there projects for dirigible balloons have been advanced by inventors with more imagination than judgment, but such experiments as have been made have proved unfruitful.

Americans have rather favoured aviation. As elsewhere, many visionary projects have been advanced, but nevertheless the beginning of a practical solution seems to dawn on the horizon. It is well known that Mr Maxim, an American engineer residing in Great Britain, produced in 1894 a huge steam-driven aerial apparatus which might have flown had its equilibrium been adequate; that Professor Langley, Secretary of the Smithsonian Institution, constructed several working models driven by steam, of which two, in 1896, performed aerial flights of over one kilometre free in air. There remains to apply the same principles and the same methods of construction to an apparatus carrying a man and it is understood that Professor Langley has been engaged in doing this during the last three years. It is said that his apparatus is finished, and will shortly be tested.

A few other American inventors are also experimenting with dynamic flying machines, but they have not proceeded very far.

There is a small group, however, which believes that it is not best to undertake too much at once; that the motor and the propeller introduce complications, and that it is best first to solve the questions of equilibrium, to make it automatic if possible, before passing to the construction of a complete apparatus.

Lilienthal was the master of this school. He utilized gravity as a motive power before venturing out with an artificial motor. In fact, when he tried the latter, he soon gave over. During several years he carried on remarkable experiments, making upwards of 2,000 glides or flights, but he fell and lost his life in 1896, probably in consequence of having neglected to keep his apparatus in good order.

He was followed by Mr Pilcher, an English marine engineer, who made some improvements upon Lilienthal's practice, but who also fell and lost his life in 1897, in consequence again of an apparatus in bad repair. [P.S. It was 1899—CHGS].

The strains arise from the irregularities and turmoils of the wind. The current does not come as an evenly flowing stream, but as a series of swirling waves, as shown in the smoke issuing from a chimney. These waves strike the apparatus either on one side or on the other, from below or from above, and constantly tend to upset it. The velocity and force with which the machine is struck by the wave depend both upon the distance from the centre of rotation of the latter and upon the speed of the current, and this is probably the reason why anemometers show such varying pressures. A flying machine must meet and overcome all these vicissitudes and this must be done instantly.

In 1896 and 1897, being impressed that the equilibrium of the bird was partly automatic, and that the problem of stability in the wind was the first which must be solved, I undertook some experiments near Chicago, Illinois, with full-sized gliding machines carrying a man, in order to study equilibrium, and that alone. I caused to be built five machines of four different types. The first was a Lilienthal apparatus, in order to start from the known before passing to the unknown, and three of the other machines were based upon the reverse of the Lilienthal type, that is to say that instead of re-establishing the equilibrium (when compromised by the variations of the wind) by displacing the body of the operator, and therefore that of the centre of gravity, as did Lilienthal and Pilcher, the new machines were based upon the theory that it was possible so to arrange the carrying surfaces that they should move automatically under the action of the wind and bring back the centre of pressure vertically over the centre of gravity—a condition absolutely necessary in order to maintain equilibrium.

One of these machines, called the "multiple-wing," is shown in Fig 2. The supporting wings are at the front and are superposed and braced together. They turn upon ball bearings marked *B*, and are restrained in front by rubber springs which allows a certain amount of horizontal movement. If the *relative wind* increases, thus requiring a lesser angle of incidence to sustain the weight, the wings are blown backward, the apparatus oscillates towards the front, and the angle becomes smaller. Once the squall passed, the springs bring the wings back to their normal position. The rear wings are flexible and merely aid in balancing, and an aeroplane, which might be replaced by an additional pair of wings, surmounts the whole. The whole apparatus weighed 15.2 kilograms, the supporting area of the front wings being 13.33 square metres and that of the rear wings 2.74 square metres. This machine proved very nearly automatic, the operator having to move but 25 millimetres, in ordinary glides, as against 127 millimetres with the Lilienthal and 63 millimetres, with the "two-surface" machine, which is next to be described.

The "two-surface" machine gave the best results and the longest glides. It is shown in Fig 4. The supporting surfaces are of varnished silk and are affixed to a framework similar to a bridge truss, and the operator is below in an upright position, being sustained under the armpits by two horizontal bars. Equilibrium is obtained through a horizontal tail, similar to that invented by Pénaud, but with elastic fastenings devised by Mr Herring. This tail generally makes an angle of 7 to 8 degrees with the carrying surfaces, and by receiving the relative wind below or above automatically changes the angle of attack according to the exigencies of the moment. The frame was of wood and the weight was 11 kilogrammes and the surface 12.45 square metres, which easily sustained an operator weighing 71 killogrames. More than one thousand glides were made with these machines in 1896 and 1897 without accident, the following tables giving an idea of the results with each of the machines:

SOME GLIDES OF MULTIPLE-WING MACHINE (WEIGHT 86 KILOGRAMMES) MOUNTED

Operator	Length (metres)	Time (seconds)	Speed Metres per sec.	Remarks
Herring	45.11	7	6.45	
Avery	53.04	7.6	7.00	
Herring	50.60	7.5	6.72	
Avery	55.77	7.9	7.00	Angle of descent not measured but about 10° to 11°
Herring	52.43	7.8	6.70	

SOME GLIDES OF TWO-SURFACE (WEIGHT 82 KILOGRAMMES) MOUNTED

Operator	Length (metres)	Time (seconds)	Angle descent	Height fallen (metres)	Speed Metres per second	Rate of descent	Kg	Remarks
Avery	60.64	8	10°	10.5	7.62	1 in 5.75	106	
Herring	71.32	8.7	7½°	9.3	8.23	1 in 7.69	87	
Avery	77.11	....	10½°	14	....	1 in 5.50		Time not taken
Herring	72.84	....	11°	14.1	....	1 in 5.24		Time not taken
,,	67.06	9	....	....	7.42	......		Angle not taken
,,	71.62	10.3	....	....	7	......		Angle not taken
Avery	78.03	10.2	8°	10.8	7.6	1 in 7.18	86	
Herring	109.42	14	10°	18.9	7.8	1 in 5.75	110	

The slope of the hill being steep and the wind ascending, it was estimated that the resistance required 2 horsepower, instead of the 1⅓ hp shown by the work done as calculated. The drawings of these various machines were published in the *Aeronautical Annual* for 1897, edited by Jas. Means, Boston, USA, and amateurs were invited to repeat and to improve upon the performances.

In 1902 I caused to be built an apparatus to obtain automatic stability by a third method. This consists in pivoting the sustaining surfaces about $\frac{4}{10}$ of their width from the front, and restraining them by springs. If the relative wind increases, the centre of pressure moves backward and tends to give the surfaces a smaller angle of attack. Only a few tests were made in 1902, but more are to be carried on in 1903. The apparatus is shown in Fig 3 for the benefit of amateurs and a photograph of one of the glides is herewith reproduced.

All these experiments were made without accident. It is true that in the beginning all possible precautions were taken. The practice ground selected was on the soft sand hills which border the southern edge of Lake Michigan, some 50 kilometres from Chicago, where there are but few trees or bushes. No glides were made in very gusty winds, or those exceeding 50 kilometres to the hour, and Messrs Herring and Avery, my assistants, were alone allowed to experiment. Later on, more confidence was gained and visitors were allowed to make short glides under instruction from the experts. All succeeded well, and even the cook became almost an expert in a short time. It must, however, be recognised that this sport is dangerous and that all who would engage in it must take all possible precautions before venturing upon it.

The invitation to amateurs to repeat these experiments remained unacted upon till 1900, when Messrs Wilbur & Orville Wright of Dayton (Ohio), took up the question. They have accomplished such advance upon all previous practice that the rest of this paper will be devoted to giving an account thereof. The improvements which they have introduced are the following:

1st. Placing the horizontal rudder or tail at the front, a position which proves more efficient in acting upon the air.

2nd. Placing the operator prone on the machine, thus diminishing by $\frac{4}{5}$ the resistance due to his body.

3rd. Warping the wings to steer to right or left.

The practice ground chosen by the Wright brothers, after an inquiry made of the National Signal Bureau of the United States, is far superior to the hills near Chicago. It is situated at Kitty Hawk near the ocean in North Carolina, and consists of a soft sand hill about 30 metres high, on a tongue of land, without a single tree, bush or grass. At its foot a sand beach one kilometre wide extends to the sea, and regular sea breezes blow daily from the ocean. Moreover the spot is a desert and quite inaccessible to the curious, who are always in the way when one is seeking for unknown phenomena.

The Wright apparatus of 1900 measured 5.64 metres across by a width of 1.52 metres, there being two surfaces aggregating 15.6 square metres, and weighing 21.8 kilogrammes. It chiefly served to study the result of placing the horizontal rudder at the front, the apparatus being loaded with sandbags or weights and flown as a kite. Having ascertained that the machine worked well, the Wrights ventured to make a few glides with good success, and postponed to the next season learning the difficult art of the birds.

The method of conducting the experiments necessarily differed from those adopted by Lilienthal and his imitators. In those machines the operator, being on his feet, carries the apparatus, poises it in the wind, and runs forward until he leaves the ground. With the two-surface machine, for instance, which weighed but 11 kilogrammes (for 12.45 metres surface) the aviator could walk about with his wings on his shoulders. The Wrights, desiring to build a larger and heavier machine and to place the aviator horizontally, had to devise a new method of getting under way. This consisted in employing two assistants each grasping one extremity of the machine, and running forward against the wind. The aviator, placed in the centre, first remains on his feet, runs a few steps and, as speed and supporting power increase, tips forward prone on the framing, hooking his feet on the rear transverse bar, and as soon as he feels himself well carried by the air, he cries, "Let go," From this on gravity serves as a motor; he glides along down the slope of the hill, keeping near the ground, meeting the irregularities of the wind by the action of the front rudder and steering to the right or left, by the torsion of the wings, which are framed loosely, and by the vertical rudder behind. When the foot of the hill is reached, a vigorous action of the front rudder causes the machine to shoot upward a little, thus increasing the resistance and diminishing the speed. The machine then alights upon the sand and stops after a short slide upon its shoes.

The 1901 machine was 6.7 metres across by a width of 2.13 metres, the two surfaces being spaced 1.42 metres vertically and giving a surface of 27.1 square metres, with a weight of 45.4 kilogrammes. This apparatus performed several hundred glides without accident, but although the head resistance was less, by reason of the horizontal position of the aviator, the angles of descent were nearly the same (8° to 10°) as with the machines in which the aviator stood upright. Now the important thing for a gliding apparatus is that the fall through the air shall be the least possible. We can always lengthen the glides by starting from a higher point, but the angle of descent shows at once the ratio between the propelling force (the weight) and the horizontal resistance, and the latter should be reduced to a minimum before proceeding to other details.

The horizontal resistance consists of three factors:

1st. The *drift*, depending upon the angle of incidence.
2nd. The *head resistance* of the various thicknesses.
3rd. The *tangential* force, recognised by Lilienthal.

The *drift* is the horizontal resultant of the normal pressure and has been thoroughly written upon heretofore, so that it need not be here discussed. Recent experiments seem to show, however, that the friction of the air upon the surfaces is not negligible, and should be separately added.

The *head resistance* is the most important, and is the one in which improvements may be achieved, by shaping the framing so as to give the lowest possible coefficients, by diminishing the number of parts, and by reducing all guy wires and similar vibrating parts to a minimum.

The *tangential force* only applies to arched surfaces, and presents the curious property that at certain angles of incidence it acts as a propulsive force. Those who may wish to learn more about this are referred to Lilienthal's book, *Der Vogelflug als Grundlage der Fliegekunst,* and to his chapter in Moedebeck's *Taschenbuch für Flugtechniker und Luftschiffer.*

We arrive at the aggregate horizontal resistance by three different processes, as follows:

1st. The speed required for support and the angle of incidence with the relative wind, being first ascertained as accurately as possible, the normal pressure is computed by the aid of Lilienthal's table for arched surfaces, which will be found in Moedebeck's *Taschenbuch*, and this normal is multiplied by the sine of the angle of application to obtain the *drift*. To this is added the *head resistance* due to the framing, the aviator and adjuncts, which is obtained by measuring accurately all the thicknesses and applying to each a coefficient due to its form. It is interesting in this connection to know that beams fishlike in cross section offer but one sixth the resistance on a plane of the same area as their "master section" while vibrating wires offer about twice the resistance due to their "master section". All this is best obtained by drawing up a tabular schedule of all the parts and deducing therefrom the equivalent or fictive area of resistance of the apparatus, which, when multiplied by the air pressure due to the speed, gives the aggregate *head resistance*; to the sum of the latter with the *drift* is added or deducted the *tangential* force, as obtained from Lilienthal's table, according to whether it is positive or negative. In other words:

Resistance = drift + head resistance ± tangential. This process is tedious and slow, but it is well worth the trouble, for it indicates those parts of the apparatus in which resistance can be reduced.

2nd. The second process furnishes a check upon the first and consists in floating the mounted machine in a wind of sufficient intensity or towing it by a cord at sufficient speed to obtain support and measuring the actual pull with a hand scale. This cannot be done very often as it involves considerable preparation, but it is a check not to be disputed.

3rd. The third process consists in simply multiplying the weight by the sine of the angle of descent. In fact, as the speed of the glides is almost always uniform it is evident that the forces balance, and the resistance along the path is to the weight as the sine of the angle to the radius. It is found as the result of a great many such computations that the three processes agree very closely.

The Messrs. Wright concluded from their experiments in 1901 that there were a number of defects in the apparatus used that year. That the shape of the surfaces was not the best possible and that they became deformed under the pressure of the relative wind. During the succeeding winter they made a whole series of laboratory experiments upon various surfaces, and built a third machine for experiment in 1902.

This last apparatus was 9.75 metres across, by a width of 1.52 metres, the two surfaces being spaced 1.42 metres apart, and giving 28.4 square metres of supporting surface. The front rudder was 1.4 square metres in area, and the weight was 53 kilogrammes. Figure 9 shows this apparatus. The main pieces are shown full size. They were all of wood, the front spar and the main arm were of an American wood for which pine would be a substitute; these parts were either imbedded in the cloth, or in a sheath or strip glued on afterwards. The ribs were of ash, steamed and bent, with a curvature of $\frac{1}{20}$ one third of the chord from the front. All fastenings were made by lashing the sticks together, and it is well to dampen the twine with thin glue when wrapping it around. The surfaces consisted of tightly woven cotton cloth such as that used for balloons. It was not varnished but would have been the better for it. The braces were of piano wire and it was generally aimed to have a factor of safety of 10 times the breaking weight. A light pair of shoes or skis was attached to the framing and spar at the front.

The machine was tested on the practice ground already described, in September and October 1902, and it gave very superior results to those obtained with preceding machines. The angles of descent were flatter and the weight sustained per horsepower was greater. The two brothers glided alternately and they soon attained almost complete mastery over the inconstancies of the wind. They met the wind gusts and steered as they willed. They did not venture to sweep much more than one quarter circle, so as not to lose the advantage of a head wind, but they constantly improved in the control of the machine and in learning the art of the birds. Some 800 glides were made of which the following are representative:

SOME GLIDES AT KITTY HAWK: OCTOBER 8TH 1902

No	Distance (metres)	Seconds	Speed over ground	Wind	Total speed	Angle of descent	Sine of angle	Resistance	Total speed	Work done kg	hp	Kg per hp
Wilbur Wright; Wt 61.4 kg + machine 53 kg = Total 114.4 kg.												
6	54.7	12	4.56	5.68	10.14	7° 20′	0.12764	14.60	10.14	148	1.97	58.1
7	54.7	10	5.47	5.68	11.15	7° 10′	0.12480	14.28	11.15	159	2.12	53.9
8	41.1	10¼	4	5.68	9.68	7° 10′	0.12480	14.28	9.68	138	1.84	62.1
9	68.6	11⅔	6	6.68	11.68	6° 10′	0.10740	12.28	11.68	143	1.90	60.2
10	71.6	13	5.5	5.68	11.18	6° 20′	0.11031	12.62	11.18	141	1.88	60.8
14	49.4	9½	5.2	3.73	8.93	6° 20′	0.11031	12.62	8.93	113	1.50	76.0
Orville Wright; Wt 65.2 kg + machine 53 kg = Total 118.2 kg.												
15	45.1	8¾	5.15	4.12	9.27	6° 50′	0.11900	14.06	9.27	130	1.73	68.3
16	46	8¾	5.25	4.72	9.97	7° 20′	0.12760	15.08	9.97	150	2.00	59.1
17	51.8	9⅛	5.63	3.90	9.53	6° 30′	0.11320	13.38	9.53	127	1.70	69.5
18	50.3	9	5.48	4.47	9.95	6° 20′	0.11030	13.14	9.95	131	1.74	62.1
19	48.7	7	6.95	4.47	11.42	6° 40′	0.11610	13.72	11.42	156	2.08	56.8
20	47.2	6	7.87	4.47	11.54	6° 40′	0.11610	13.72	11.54	158	2.10	56.3

The variations mainly result from more or less skill in each glide and consequent slight swerving.

The last column in the table is that which best indicates the advance made over previous practice. The machine of Mr Maxim supported

12.7 kilogrammes, per horsepower, that of Professor Langley 16.6 kilogrammes, while previous glides from Lilienthal down showed 45 kilograms per horsepower. It is quite true that when a motor is added the weight sustained per horsepower is diminished about one half by reason of the inevitable losses due to the motor, the transmission and the propeller, but Messrs. Wright have so far diminished the resistances that they now sustain 62 kilogrammes per horsepower, and the time is evidently approaching when, the problems of equilibrium and control having been solved, it will be safe to apply a motor and a propeller.

In point of fact the Messrs. Wright are now gliding very nearly as well as the vulture, which generally descends one metre in ten (5° 45′) in calm air. It is therefore not impossible that man shall eventually master sailing flight and learn, when once well in the air and at speed, to imitate the soaring birds which circle and rise through the force of the wind alone when circumstances are favourable; which conditions chiefly consist in an ascending trend of the wind, such as is frequently found in hot countries, those being the regions which are frequented by sailing birds.

This does not mean that the future flying machine, if such is developed, can entirely do without a motor. This adjunct will be indispensable, not only to fly in those regions where the wind seldom has an ascending trend, but also to restore the speed or the equilibrium very quickly under many circumstances, such as a very sudden wind gust; or a descent too near to the ground. It is easily perceived, however, that when the motor need only be used occasionally the length of possible journeys of flying apparatus will be greatly increased in the countries where circumstances favour sailing flight.

It is thus seen that there has been a gradual evolution from Lilienthal onward. That gliding machines have been made much safer and are now fairly under control. There is doubtless a chance for further improvement, and this can only be accomplished by experiment. The present indications are that such experiments will be extensively undertaken, and a word of warning may well be given in concluding this article.

Let all exert the utmost possible prudence in conducting gliding experiments. First select suitable ground, a soft sand hill is the best, so isolated from other hills as to have no cross currents or whirling winds. Begin practice very gradually, starting from a small elevation and learning by degrees the control of the machine and the best way of alighting safely. Follow the slope of the hill closely, so as to have but a short distance to fall if a false manoeuvre is made. Examine the apparatus after each glide, and if anything has gotten out of order, repair it at once before making another glide. Try no feats which involve being high in air except over an appropriate sheet of water, but do not resort to that method of experimenting upon ordinary occasions. It involves too much loss of time in drying oneself and the machine, and the latter is likely to become distorted.

More important than all, do not try to beat previous records. This leads to taking risks and to producing accidents. It is well to have competitions at which several amateurs practice together, because they learn from each other and because they are then more disposed to have a surgeon present, a precaution which I have always taken, but we must always remember that the important things to be secured are control of the machine and safe landings, without regard to the distance glided over, the latter mainly depending upon the elevation started from.

Progress and safety will be greatly promoted by beginning with small machines just as the Wrights have done. An apparatus 6 metres by 1.3 metres, giving with two surfaces arched to $\frac{1}{25}$ of their width some 15 square metres of supporting area, is ample to carry a man and will be found far more manageable than the last machine of Messrs. Wright of which the drawings are here given. As the cost is not great, being from 500 to 1,500 francs, according to the amount of care bestowed on the apparatus, it is feasible to build new machines from time to time in order to introduce the improvements which constantly suggest themselves. It is hoped that when many experimenters get at work such progress shall be made as materially to advance the time when aviation shall become practical.

Bibliography

Bibliography

Note. The works are arranged chronologically by date of publication. Those in English, French and German should be understood as having appeared in London, Paris or Berlin unless otherwise stated.

CHICAGO: International Conference on Aerial Navigation, 1893. *Proceedings.* New York, 1894.
CHANUTE, O. *Progress in Flying Machines.* (First published serially in *The American Railroad and Engineering Journal*, October 1891 to December 1893) New York, 1894.
LILIENTHAL, O. [A number of his papers are available in English, and may be found in the three volumes of James Means' *Aeronautical Annual* (see below under 1895-1897)].
MOEDEBECK, H. W. L. *Taschenbuch zum praktischen Gebrauch für Flugtechniker und Luftschiffer.* Berlin, 1895 (2nd edition 1904). (Translated from the 2nd edition as *Pocket-Book of Aeronautics*, 1907).
AERONAUTICAL ANNUAL. Edited by J. Means. 3 volumes, New York, 1895-97.
WRIGHT, W. *Some Aeronautical Experiments.* Reprint from *Journal of the Western Society of Engineers.* [Chicago, Western Society of Engineers, 1901.]
LECORNU, J. *Les Cerfs Volants* (Kites). 1902.
LECORNU, J. *La Navigation Aérienne: Histoire documentaire et anecdotique*, 1903.
WRIGHT, W. *Experiments and Observations in Soaring Flight.* (Reprint from *Journal of the Western Society of Engineers.*) Chicago: Western Society of Engineers 1903.
FULLERTON, J. D. *Report on Aerial Navigation.* Chatham, 1904.
WRIGHT, O. and W. *MM Orville Wright et Wilbur Wright résidant aux États-Unis d'Amérique.* Brevet d'invention No 342.188. Perfectionnements aux machines aéronautiques. Demandé le 22 mars 1904. Délivré le 1er juillet 1904. Publié le 1er septembre 1904. [Paris] Office National de la Propriété Industrielle, 1904.
ROOT, A. I. "Our Homes—What hath God wrought?" [his eyewitness account of the Wrights' circular flight of September 20th 1904] in *Gleanings in Bee Culture*, January 1st 1905.
WRIGHT, O. and W. *Orville Wright and Wilbur Wright, of Dayton, Ohio. Flying-Machine.* No 821,393. Specification of Letters Patent. Application Filed March 23rd 1903. Patented May 22nd 1906. Washington: United States Patent Office, 1906.
AERO CLUB OF AMERICA. *Navigating the Air: a scientific Statement of the Progress of Aeronautical Science up to the Present Time.* New York, 1907.
HILDEBRANDT, A. *Airships Past and Present*, etc. (translated from *Die Luftschiffahrt*, etc., 1907). 1908.
MAXIM, Sir H. S. *Artificial and Natural Flight.* 1908.
PEYREY, F. *Les premiers 'Hommes-oiseaux': Wilbur et Orville Wright.* 1908 (New edition 1909).
COMMITTEE OF IMPERIAL DEFENCE. *Aerial Navigation: Report of a Sub-Committee of Imperial Defence, appointed* [in 1908] *by the Prime Minister.* 1909.
DUMAS, A. *Stud Book de l'Aviation: Ceux qui ont volé et leurs Appareils.* 1909.
FERBER, F. *L'Aviation: ses Débuts, son Développement.* 1909. (2nd edition 1910).

Rumpler, E. *Die Flugmaschine.* 1909.
Peyrey, F. *L'Idée aérienne: Aviation: Les Oiseaux artificiels.* 1909.
Berget, A. *La Route de l'Air.* 1909. (Translated as *The Conquest of the Air*, 1909).
Lougheed, V. *Vehicles of the Air.* 1909 [1910 given on title-page].
Jane, F. T. *All the World's Air-Ships: Aeroplanes and Dirigibles.* 1909, 1911.
Ferris, R. *How to Fly; or the Conquest of the Air.* 1910.
Grand-Carteret, J., and Delteil, L. *La Conquéte de l'Air vue par l'Image (1495-1909).* 1910.
O'Gorman, M. *Problems relating to Aircraft.* (Institution of Automobile Engineers publication.) 1911.
Frankfurt a.M: Exhibition, 1909. *Katalog der historischen Abteilung der ersten Internationalen Luftschiffahrts-Ausstellung. . . .* 1909 [including bibliog.]. Frankfurt, 1910.
La Vaulx, H. de (Comte). *Le Triomphe de la Navigation Aérienne.* 1912.
Dollfus, E. H. *Petits Modèles d'Aéroplanes.* 1912.
Maxim, H., and Hammer, W. J. *Chronology of Aviation.* New York, 1912.
Dumas, A. *Les Accidents d'Aviation.* 1913.
Wright, O. and A. [Special combined number of the *Aeronautical Journal*, July-September, 1916, on the Wright brothers.]
Vivian, E. C., and Marsh, W. Lockwood. *A History of Aeronautics.* 1921.
Magoun, F. A., and Hodgins, E. *A History of Aircraft.* New York, 1931.
Dollfus, C., and Bouché, H. *Historie de l'Aéronautique.* 1932. (Revised edition 1942).
Supf, P. *Das Buch der deutschen Fluggeschichte.* 2 volumes. Stuttgart, 1935. (2nd edition 1956).
Sahel, J. *Henry Farman et l'Aviation.* 1936.
Kelly, F. C. *The Wright Brothers.* 1944.
Wright, W. and O. *Miracle at Kitty Hawk: the Letters of Wilbur and Orville Wright.* Edited by F. C. Kelly. New York, 1951.
Wright, Orville. *How we invented the Aeroplane.* Edited by F. C. Kelly [Orville Wright's deposition in the Montgomery trial]. New York, 1953.
Wright, W. and O. *The Papers of Wilbur and Orville Wright.* Edited by Marvin W. McFarland. 2 volumes. New York, 1953.
Pritchard, J. L. "The Wright Brothers and the Royal Aeronautical Society." In *Journal of the Roy. Aeron. Soc.*, December 1953.
Kármán, T. von. *Aerodynamics: selected Topics in the Light of their Historical Development.* Cornell (USA), 1954.
Gibbs-Smith, C. H. "The Origins of the Aircraft Propeller." In *Quarterly Review (Rotol and British Messier Journal)*, 1959.
Voisin (G), *Mes dix mille Cerfs-Volants.* 1961. Translated (by O. Stewart) as *Men, Women and 10,000 Kites.* 1963.
Dollfus, C. *Les Avions.* 1962.
Gibbs-Smith, C. H. *The Wright Brothers: a Brief Account of their Work* (Science Museum publication). 1963.
Means, J. H. *James Means and the Problem of Manflight during the Period 1882-1920.* Washington, D.C. (Smithsonian Institution), 1964.
Parkin, J. H. *Bell and Baldwin: their Development of Aerodromes and Hydrodromes at Baddeck, Nova Scotia.* Toronto, 1964.

Young, Pearl I. (ed). *The Chanute-Wenham Correspondence, September 13, 1892 to January 23, 1908*. (Privately duplicated) Lancaster, Pa (USA), 1964.
Dollfus, C., Beaubois, H., and Rougeron, C. *L'Homme, l'Air, et l'Espace: Aéronautique, Astronautique.* 1965.
Gibbs-Smith, C. H. *The Invention of the Aeroplane, 1809-1909.* 1966.
Gibbs-Smith, C. H. *A Directory and Nomenclature of the First Aeroplanes, 1809 to 1909* (Science Museum publication). 1966.
Hart, C. *Kites: an historical Survey.* 1967.
Penrose, H. J. *British Aviation: the Pioneer Years.* 1967.
Wright, O. and W. *Wilbur and Orville Wright: a Bibliography commemorating the Hundredth Anniversary of the Birth of Wilbur Wright.* Washington DC [Library of Congress]. 1968.
Gibbs-Smith, C. H. *Clément Ader: his Flight-Claims and his Place in History* (Science Museum publication). 1968.
Gibbs-Smith, C. H. *Aviation: an historical Survey from its Origins to the End of World War II* (Science Museum publication). 1969.
Walker, P. B. *Early Aviation at Farnborough: the [official] History of the Royal Aircraft Establishment.* 2 volumes. 1971-73.

NEWSPAPERS AND PERIODICALS QUOTED

The Aeronautical Journal
L'Aérophile
The American Magazine of Aeronautics
L'Auto
L'Automobile
Automobilia and Flight
The Automotor Journal
Cassier's Magazine
Century Magazine (USA)
Flight (and *Flight International*)
Gleanings in Bee Culture (USA)
The Illustrated London News
L'Illustration
Illustrierte Aeronautische Mitteilungen
Interavia
Journal and Record of Transactions of The Junior Institution of Engineers
Journal of the Western Society of Engineers (USA)
Le Figaro
La Locomotion
Le Matin
McClure's Magazine (USA)
New York Herald
Revue d'Artillerie
Revue Scientifique (*Revue Rose*)
Revue Générale des Sciences
Siegl
Les Sports
The Times

Conversion Tables

Conversion Tables

METRES TO YARDS AND FEET
(taken to the nearest yard and foot for lengths over 20 metres)

metres	yards	feet	metres	yards	feet
1	1·09	3·28	70	77	230
2	2·19	6 56	80	87	262
3	3·28	9·84	90	98	295
4	4·37	13·12	91	100	299
5	5·47	16·40	100	109	328
6	6·56	19·69	150	164	492
7	7·66	22·97	200	219	656
8	8·75	26·25	250	273	819
9	9·84	29·53	300	328	984
10	10·94	32·80	400	437	1311
15	16·40	49·21	402	440 (¼ mile)	1320
20	21·87	65·62	500	547	1641
25	27	82	600	656	1968
30	33	98	700	766	2298
35	38	115	800	875	2625
40	44	131	805	880 (½ mile)	2640
45	49	147	900	984	2952
50	55	164	1000 (1 km)	1094	3282
60	66	197	1609	1760 (1 mile)	5280

KILOMETRES TO MILES

kilometres	miles	kilometres	miles
1	⅝th (3280 ft)	20	12·4
1·6	1	25	15·5
2	1·2	30	18·6
3	1·9	35	21·7
4	2·5	40	24·9
5	3·1	45	28·0
6	3·7	50	31·1
7	4·4	75	46·6
8	5·0	100	62·1
9	5·6	125	78·0
10	6·2	150	93·2
15	9·3		

SQUARE METRES TO SQUARE FEET
(including common wing areas in early aviation)

square metres	square feet	square metres	square feet	square metres	square feet	square metres	square feet
1	10·8	10	107·6	22	236·8	43	462·8
2	21·5	12	129·2	24	258·3	48	516·7
3	32·3	13	139·9	25	269·0	50	538·2
4	43·0	15	161·5	26	279·9	52	559·7
5	53·8	17	183·0	30	322·9	55	592·0
6	64·6	18	193·8	35	376·7	60	645·8
9	96·9	20	215·3	40	430·6		

LENGTHS AND DISTANCES OF SOME FAMILIAR ITEMS AND PLACES

	km	metres	miles	yards	feet
Length of cricket pitch		20·1		22	66
Length of lawn tennis court		23·8		26	78
Length of American football field		91·4		100	300
Length of international soccer field		109·7		120	360
Length of British soccer field		118·8		130	390
In London					
Breadth (east-west) of the Round Pond in Kensington Gardens		201·2		220	660
Length of Westminster Bridge		246·9		270	810
From Hyde Park Corner (Apsley House)					
to Dorchester Hotel, Park Lane		402	¼	440	1320
to (just beyond) Grosvenor House		805	½	880	2640
to Marble Arch	1·2	1207	¾	1320	3960
to Exhibition Road	1·6	1609	1	1760	5280
to (a few yards short of) Leicester Square	1·6	1609	1	1760	5280
In Edinburgh					
Length of Princes Street	1·6	1609	1	1760	5280
In Paris					
Length of the Champs Élysées, from the Place de la Concorde to the Arc de Triomphe	2·0	2012	1¼	2200	6600
In New York					
Length of the Brooklyn Bridge		486·2	⅓	531⅔	1595
The English Channel (at its narrowest)	33·8		21		

Index

Index

Note. The reader is asked to refer to the list of Section headings on pages vii to xii for the General Subjects dealt with in this book.

Accelerated take-off. See Launching.
Accidents and Fatalities. See Fatalities
Acrobats, 301
Ader, Clément, 7, 224, 291, 351, 352
Aerial Experiment Association, 134
Aero Club of Great Britain. See Royal Aero Club
Aéro-Club de France, 8, 137, 180, 189, 190, 209
    Aviation Committee formed, 67*ff*
    Chanute's lecture, 56*ff*
    tribute to Ferber and Archdeacon, 112*ff*
    manifesto, 213*ff*
    See also Prizes for aviation
Aero-engines. See the names of the makers
Aéronaute(L'), 211
Aeronautical Annual, 30
Aeronautical Journal, 10, 11, 38, 139
Aeronautical Societies, first, 349
Aeronautical Society of Great Britain. See Royal Aeronautical Society
Aérophile(L'), 8
    and Ferber, 43, 53*ff*, 136, 164, 165
    and Chanute, 56*ff*, 71*ff*, 114*ff*
    and Esnault-Pelterie, 152*ff*
    and the Wrights, 41, 60, 91, 98, 99, 106*ff*, 121, 130, 145, 181, 190*ff*, 193*ff*, 199*ff*, 201*ff*, 241
    and box-kites, 49
    and Archdeacon, 66, 74*ff*, 103*ff*, 105, 108*ff*, 116*ff*, 123, 147, 157, 181
    and model competition, 146
    and Santos-Dumont, 204, 205, 219, 220*ff*, 225*ff*
Ailerons (and elevons), 128, 133*ff*, 154*ff*, 208, 227*ff*, 237
    See also Flight-control
Air-brakes (steering). See Flight-control
Airmen and 'Chauffeurs', attitude of, 12*ff*, 81
Airscrews, See Propellers
Airships, 9
Alexander, Patrick, 209
American Magazine of Aeronautics, 134
Anti-Wright faction, 1*ff*   [Levavasseur
Antoinette aircraft and engines. See

Anzani engines, 238
Archdeacon, Ernest (1863–1957)
    hears from Ferber, 64*ff*
    calls to action, 65*ff*
    forms Aviation Committee, 67*ff*
    first Wright-type glider, 103*ff*, 122*ff*
    efforts to promote gliding, 64*ff*, 105*ff*, 116*ff*, 137
    on the Wrights, 108*ff*, 149*ff*, 173*ff*, 180*ff*, 245
    hears again from Ferber, 111*ff*
    Aéro-Club tribute to, 112*ff*
    second Wright-type glider, 147*ff*
    float-glider, 157*ff*, 174
    attacks the Wrights, 149*ff*, 171, 173*ff*, 180*ff*, 345
    his manifesto, 213*ff*
    his prizes for aviation, 137, 191, 212, 220*ff*
    on Farman, 244*ff*
    on Santos-Dumont, 222*ff*
    on Chanute in Paris, 354*ff*
Arms (whirling). See Whirling arms
Assisted take-off. See Launching
Associated Press, 97*ff*
Automobilia and Flight, 301
Automobilism, 8, 351
Auto (L'), 40, 172, 177, 207
Automobile(L'), 111
Automotor Journal, 10, 11, 139, 196, 197, 205, 227, 228
Auvours, Camp de, 272*ff*
Avery, W., 15
Avion III. See Ader, C.
Avro aircraft. See Roe, Sir A.V.

Baden-Powell, Colonel B. F. S., 139
Bagatelle, 218, 220, 225
Balloons, 8
Balsan, 109
Bariquand and Marre, 308
Benz, Carl, 351
Berck-sur-Mer, 122
Berg, Hart O., 242
Berget, A., 211
Besançon, George, 8, 112*ff*, 115, 116, 172, 177, 180, 190, 241
Bienvenu. See Launoy and Bienvenu

381

Biplanes, first, 29*ff*
Blair Atholl, 239
Blanchet, G., 114, 147
Blériot, Louis (1872–1936), 134, 135, 208, 235, 236, 248, 258, 267, 268, 286, 294
Bollée, Léon, 272
Bonnecasse, 120
Boulton, M. P. W., 133
Box-kite gliders, 159*ff*
Box-kite powered aircraft, 235*ff*, 253*ff*
Box-kites, 49*ff*
Breguet aircraft, 240
Bris, J. M. Le, 349
Bristol 'Box-kite', 183
Britain, aeronautical stagnation in, 139, 209
'British Army Aeroplane No 1.' See Cody, S. F.
Brown, D. S., 349
Browning, J., 350
Bryan, G. H., 10, 37
Burdin, M., 136
Butler and Edwards, 350
Buzzards observed by the Wrights, 18

Cailletet, L. P., 209, 213*ff*, 222, 244, 262
Capper, Colonel, J. E., 134, 135, 141, 256, 257
 on the Wrights, 141, 142
Carillon Park (Dayton), 167
Cassier's Magazine, 10
Cayley, Sir George, Bt, 210, 211, 348
Century Magazine, 85
Chanute, Octave (1832–1910), 196
 his gliders, 14*ff*
 pursues automatic stability, 14
 Progress in Flying Machines, 14
 and Ferber, 37*ff*, 69
 did not have movable control surfaces, 14
 treats Wrights as pupils, 24, 25, 62, 57*ff*, 71*ff*
 and the Wrights, 23*ff*, 56*ff*, 71*ff*, 76*ff*, 86*ff*, 92
 gives vital Paris lecture, 56*ff*
 articles in France, 71*ff*, 76*ff*, 359*ff*
 on powered aviation, 72, 80
 on need for caution, 73, 162
 does not understand Wrights' control system, 76*ff*
 how to fly a Chanute glider, 114*ff*

 on Langley, 210
 Ferber on, 269
Circles flown, first
 by Wrights, 129, 264
 by Farman, 261*ff*
Cléry, A., 225
Cody, S. F., 134, 135, 238, 239, 255*ff*
Congreve, Sir William, 349
Control (flight). See Flight-control
Coquelle, R., 179, 192, 193
Cormon, F. A. P., 262
Cornu, Paul, 240
Crystal Palace, 350
Curtiss, Glenn Hammond, 352

Daily Mail, 209
Daily Mirror, 281
Daimler, G., 351
Dargent, 103
Da Vinci. See Vinci
Deaths (aviation). See Fatalities
De Caters. See Caters
Degoule, 302
De Groof. See Groof
Delagrange, Léon (1873–1910), 237, 254*ff*, 306
 on the Wrights, 285, 288
Demoiselle. See Santos-Dumont
De Pischoff. See Pischoff
Descazes, Vicomte, 9
Deutsch-Archdeacon prize, 137
Deutsch de la Meurthe, H., 91, 116, 137, 191, 202
Dirigibles. See Airships
Dollfus, Charles
 on Cayley, 211
 on Chanute, 73
 on the Wrights, 3, 85, 129, 167, 310, 312
 on ailerons in Europe, 135
 on Ferber, 35, 40
Drzewiecki, Stefan, 92, 93, 109, 120, 180
Dunne, John William, 239
Du Temple. See Temple
Dutheil-Chalmers engines, 237

Edwards, E., 350
Elevators and elevons. See Ailerons; Flight-control
Ellehammer, J. C. H., 209, 232, 240, 256
Engelhardt, Captain, 278, 298
Enghien, Lake d', 208

382

Index

Engines. See the names of the makers
England. See Britain
Éole. See Ader, C.
Esnault-Pelterie, Robert (1881–1957).
    128*ff*, 152*ff*, 258
  REP No 1, 236
  REP No 2, 258
  wing-warping and ailerons, 133, 154*ff*
Etrich-Wels glider, 121, 240

FAI. See Fédération Aéronautique
    Internationale
Farman, Henri (1874–1958), 306
  first flights, 238, 244*ff*, 248
  his nationality, 238
  flies first circle in Europe, 253, 261*ff*
  Voisin-Farman aircraft, 238, 244*ff*,
    248, 253*ff*
  Flying Fish, 260
  use of ailerons, 135
  spelling of his name, 240
  his former career, 262
Farnborough. See Cody, S. F.
Fatalities (aviation), 257, 275
Fédération Aéronautique Internationale
  9, 173
Ferber, Captain Ferdinand (1862–1909)
  8, 10, 146
  Lilienthal-type glider (1901), 35
  first channel of Wright influence
    (1902), 35*ff*
  Wright-type gliders (1902–05), 35*ff*,
    41*ff*, 69*ff*, 136*ff*, 145*ff*
  did not copy Chanute gliders, 40
  powered Wright-type (1903), 70
  article in *L'Aérophile* (1903), 53*ff*
  letter to Archdeacon (1903), 64*ff*
  letter to Chanute, 100
  gliding in 1903, 69*ff*
  letter to Archdeacon about Wrights
    (1904), 111*ff*
  French Aéro-Club tribute to (1904),
    112*ff*
  first article in *Revue d'Artillerie* (1904),
    117*ff*
  tailed glider (1904), 136*ff*, 145
  work in 1905, incl. second article in
    Revue d'Artillerie (1904), 161*ff*
  Wrights' letter to (1905), 170*ff*
  on Santos-Dumont, 220*ff*
  tributes to Wilbur Wright (1907),
    118, 163, 164, 241*ff*
  lectures in London (1908), 265*ff*
  work in 1908, 43, 255
  death, 257
  Dollfus on, 40
  Chanute on, 69
Figaro (Le), 281, 304
Flight (and *Flight International*), 5, 264,
  313
Flight-control, 14*ff*, 17*ff*, 21, 41*ff*, 76*ff*,
  133*ff*, 154*ff*, 197, 227*ff*, 248, 312*ff*
Float-gliders, 159*ff*
Flying Fish. See Farman, H.
Fordyce, M. A., 278
France, Aviation in. See the list of
  Section headings on pages vii to xii
French Aero Club. See Aéro-Club de
  France
Fritsche, Lieutenant, 260
Furnas, C. W., 278

Galerie des Machines (Paris), 146
Garnerin, A. J., 348
Gasnier, Réné, 256
  on the Wrights, 287
Gastambide-Mengin aircraft. See
  Levavasseur
Geneva, Lake of, 158, 174
Giffard, Henri, 349
Girardot, 109
Gleanings in Bee Culture, 130, 131
Gliders, See the names of the designers
Gliding competitions. See Archdeacon
Grand Prix. See Prizes for aviation
Groof, Vincent de, 350
Gust-dampers, 16

Hang-gliders. See Lilienthal; Pilcher;
  Ferber
Hargrave, Lawrence (1850–1915),
  49*ff*, 159*ff*, 353
Helicopters. See Breguet; Cornu,
  Santos-Dumont
Henri Farman I. See Voisin, G.
Henry Farman III. See Farman, H.
Henson, William Samuel, 299, 349
Herbster, 120
Herring, Augustus Moore, 15, 16
Hooke, Robert, 347
Huffman Pasture (Prairie), 129*ff*, 166*ff*

Illustrated London News, 263
Illustration(L'), 206, 264
Illustrierte Aeronautische Mitteilungen,
  38, 102, 151

383

Instability. See Flight-control
Interavia, 50, 126
Issy-les-Moulineaux, 147, 238*ff*, 244*ff*
  256

JAP engines, 255
Jarrett, Philip, 50
Jatho, Karl, 7
Jet propulsion, first design for, 350
Junior Institution of Engineers, 265

Kapferer, H., 180, 237, 305
Kelly, Fred C., 44
Kill Devil Hills. See Kitty Hawk
Kites, Box. See Box-kites
Kitty Hawk and the Kill Devil Hills,
  17*ff*, 44*ff*, 82*ff*
Kress, Wilhelm, 7

Lahm, Frank, 171, 194
Lambert, Comte Charles de, 302
Lana de Terzi, F.de, 347
Lanchester, F. W., 239
Langley, Samuel Pierpoint, 210, 297,
  299, 349, 352
Lateral control. See Flight-control
Latham, Hubert, 146
Launching of aircraft, 16, 17, 147, 173,
  295*ff*
Launoy and Bienvenu, 348
La Vaulx, Comte Henri de (1870–1930)
  210, 240, 267
  on French aviation, 9, 63, 112, 124,
    125, 179, 195
  founds FAI, 9, 173
  on Archdeacon, 124
  on the Wrights, 285, 312
  on Ferber, 9, 10
  on Chanute, 63, 122, 124, 125
Lavezzari, 120
Lebaudy airship, 8
Le Bris, See Bris
Lecornu, J., 49, 159
Legagneux, G., 256
Lenoir gas engine, 350
Leonardo da Vinci. See Vinci
Lesh, L. J., 135
Letur, L., 349
Levavasseur, Leon (1863–1922), 120,
  210
  Gastambide-Mengin aircraft, 259
  Antoinette aircraft, 255, 259
  Antoinette engines, 210, 212, 218,
    235*ff*, 253*ff*, 259

Libellule. See Blériot
Lilienthal, Otto (1848–1896), 7, 53,
  108, 269, 352
Locomotion (La), 24, 57, 58, 64, 66, 354
Loup, M., 349
Louvrié, C. de., 350
Lucas-Gerardville, Captain, 302

Mallet, 109, 120, 240
McClure's Magazine, 17
Manifesto, the great, 213*ff*
Masfrand, Albert de, 123, 219
Matin (Le), 288
Maxim, Sir Hiram S., 10, 352
Means, James. See *Aeronautical Annual*
Mensier, General, 224
Merlimont, 122
Meurthe. See Deutsch de la Meurthe
Molson, K., 135
Model aircraft competition, 146, 229
Montgolfier balloon, 347
Montgomery, J. J., 44
Motorcars. See Automobiles
Motors (engines). See the names of the
  makers
Moy, T., 350
Mozhaiski, A. F., 351
Multiplanes (Chanute's). See Chanute

Neuilly, 212
New York Herald, 102, 287
Newspapers. See individual names

Ohio, buzzards in, 18
Otto, N. A., 351

Painlevé, P., 278
Parachute, first in use, 348
Passenger-carrying, first, 278, 306
Patent action (Wrights) against
  French, 18
Paulhan, L., 146
Peltier, Mme Thérèse, 306
Pénaud, Alphonse, 270, 350
Peugeot engines, 161
Peyrey, François, 40, 245, 280, 284,
  313
Phillips, Horatio F., 238, 351
'Philos', 121
Pilcher, Percy Sinclair (1866–1899),
  7, 50, 353
Pilots: airmen *versus* chauffeurs. See
  Airmen

Index

Pischoff, A. de, 238
Prandtl, L., 239
Pratt truss, 15, 29
Press. See the names of the individual papers, etc.
Prizes for aviation, 137, 191, 212, 220*ff*, 225*ff*, 244, 253, 254
Propellers, 247, 309*ff*

Reichel, F., 281, 304
Renard, Colonel, 103, 127, 174, 310
Renault engines, 254
REP aircraft and engines, See Esnault-Pelterie
Revue d'Artillerie. See Ferber
Revue Scientifique (Revue Rose), 36
Revue Générale des Sciénces, 76*ff*
Rey, P., 100, 101, 111
Richet, C., 37
Robart, 120
Roe, Sir A. V., and Avro aircraft, 238, 255
Root, Amos I., 129, 130, 131
Royal Aero Club, 8
Royal Aeronautical Society, 10
  See also *Aeronautical Journal*
Royal Aircraft Establishment. See Cody
Rudders. See Flight-control
Rue(de), pseud for Ferber

Sagan, F., 12
Saint Louis Exposition(1904), 16, 141
Santos-Dumont, Alberto (1873–1932), 7, 134, 236, 237
  takes up aviation, 204*ff*
  his *14-bis*, 134, 212*ff*, 218*ff*, 220*ff*, 225*ff*
  his *No 19*, 237
  banquet for, 222*ff*
Selfridge, Lieutenant T. E., 134, 275
Ski-jump ramp. See Launching
Societies (aeronautical) first, 349
Sports(Les), 172, 191, 224
Stability. See Flight-control
Steering air-brakes. See Flight-control
Stewart, Oliver, 4
Stringfellow, John, 16, 349
Surcouf, 287
Switzerland, landing on ice in, 298

Tatin, Victor (1843–1913), 229, 230, 240
  on the Wrights, 106*ff*, 210
  on copying the Americans, 106*ff*
  admits defeat by the Wrights, 201*ff*
  Ferber on, 267
Tatin-de La Vaulx monoplane, 240
Taube aircraft, 240
Temple, Félix du, 349
Templer, Colonel J. L. B., 141
Terzi. See Lana de Terzi
Tethered aircraft, 209
Times (The), 300
  on the Wrights, 279, 280
  on Farman, 263
Tippoo Sahib. See Tipu Sultan
Tipu Sultan, 348
Tissandier, P., 277, 302
Trussing (Wing), 29*ff*
Turgan, 148

Undercarriages, 298
United States, dealings with the Wrights, 167

Van den Born, 278
Vaulx, de La. See La Vaulx
Vinci, Leonardo da, 347
Voisin, Gabriel (b.1880), 147, 148
  anti-Wright statements, 4, 148, 293*ff*, 168, 169
  enters aviation, 126*ff*
  float-gliders, 157*ff*, 174
  and Farman, 261*ff*
  and the Wrights, 148, 293*ff*, 168, 169, 206
  and the Wright engines, 307*ff*
  and the Wright propellers, 309*ff*
  in *Interavia*, 50, 159
  his Chanute-type glider, 49, 126, 160, 237
  his powered aircraft (*Voisin-Farman*, etc.), 235*ff*, 248, 253, 255
Vuia, Trajan, 207, 227, 232, 236

Walker, P. B., 141, 142, 257
Walker, T., 349
Warp-drag, 21
Warping. See Flight-control
Weaver, H., 194
Weiss, J., 146
Welferinger, R., 259, 260
Wells, H. G., 267
Wels-Etrich glider, 121, 240
Wenham, F. H., 16, 99, 349

Western Society of Engineers, 11, 38
Whirling arms, 69
Wind, flying in, 304
Wind-tunnels, first, 350
Woelfert, K., 351
Wright, Katharine, 48, 86
Wright, Orville (1871–1948) and
 Wilbur (1867–1912)
 the anti-Wright faction, 1*ff*
 spheres of influence, 1*ff*
 'airmen' and 'chauffeurs', 12*ff*
 first ideas on warping, 18
 the word 'warp', 18
 warping kite (1899), 19
 relations with Chanute, 23*ff*, 56*ff*,
  71*ff*, 86*ff*, 92
 debts to Chanute, 17
 wing-trussing, 29
 three gliders (1900–02), 17, 44
 invention of warp-rudder linkage, 21
 influence of Nos 1 and 2 gliders, 35*ff*
 Wilbur's Chicago lectures (1901–2),
  11, 38
 influence of No 3 glider, 44
 influence in Europe starts with
  Ferber, 35
 work described by Chanute in
  France (1903), 56, 76*ff*
 said to be Chanute's pupils, 56*ff*, 71*ff*
 further mention in France (1903),
  71*ff*, 74*ff*
 confusion with Chanute gliders, 60
 powered *Flyer I* (1903), 82
 world's first powered flights (1903),
  84, 86*ff*, 97*ff*
 article in *Century Magazine*, 85
 relations with Chanute (1903), 76*ff*,
  86*ff*
 impact on Europe of first flights,
  91*ff*, 97*ff*
 pirated sketch of *Flyer I*, 102
 influence on Archdeacon (1904),
  103*ff*, 122*ff*
 further impact on Europe (1904),
  108*ff*
 influence on Esnault-Pelterie (1904),
  128*ff*
 Flyer II (1904), 129*ff*
 first assisted take-off (1904), 129
 first circle flown (1904), 129*ff*, 264
 first eyewitness account (1904), 130,
  131

 visited by Colonel Capper (1904), 141
 impact of 1904 flights, 129
 influence on Santos-Dumont
  (1904–06), 204
 influence on flight-control, 133*ff*
 impact on Britain (1904–05), 141*ff*
 further influence on Archdeacon
  (1905), 147*ff*
 challenged by Archdeacon (1905), 149
 further impact on Europe (1905),
  177*ff*, 189*ff*, 190*ff*, 193*ff*
 Flyer III (1905), 166*ff*
 first half-hour flight (1905), 165
 send news of *Flyer III* to Ferber
  (1905), 170*ff*
 attacked by Archdeacon (1905), 173*ff*
 another pirated sketch (1905), 155
 new impact on Europe (1905), 177*ff*
 pirated sketches of *Flyer III* (1906),
  182, 183, 207
 discussion of, at French Aéro-Club,
  (1906), 189*ff*
 first article on, in *L'Aérophile* (1906),
  190*ff*
 second article on, in *L'Aérophile*
  (1906), 193*ff*
 scepticism and belief in Europe
  (1906), 195*ff*
 re-publication of Wright patent in
  Europe (1906), 199*ff*
 impact on Europe (1906), 189*ff*
 Ferber's tribute to (1907), 241
 influence on Cody, 257
 work in 1908 (survey), 271
 make world's first passenger flights
  (1908), 306
 Wilbur in France (1908), 273
 revolutionises European aviation
  (1908), 273*ff*, 279*ff*, 284*ff*
 Orville in USA (1908), 275
 death of Selfridge (1908), 276
 the question of records, 277*ff*
 European press coverage of (1908),
  279*ff*
 French recognition and repentance
  (1908), 284*ff*
 pioneers' opinions of (1908), 284*ff*
 Gabriel Voisin on (1908), 290, 293,
  298
 question of assisted take-off, 295
 could only acrobats fly Wright
  machines? 301

Wright, Orville and Wilbur *cont.*
  flying in wind, 304
  passenger-carrying, 306
  Voisin and Wright engines, 307*ff*
  Voisin and Wright propellers, 309*ff*
  influence in 1908 and 1909, 311*ff*
  Dollfus on, 3, 85, 129, 167
  Wright-type gliders, 35*ff*, 103*ff*, 122*ff*, 128*ff*, 136*ff*, 147*ff*, 152*ff*, 157*ff*
  La Vaulx on, 285
  Blériot on, 286
  Gasnier on, 287
  Zens on, 287
  Delagrange on, 288
  Voisin brothers on, 290*ff*

Young, Pearl, 37

Zanonia plant, 239
Zens, E. and P., 256, 287
Zeppelins, 8

Printed in England for Her Majesty's Stationery Office by Eyre & Spottiswoode Ltd, Thanet Press, Margate
Dd 505769 K28